T0304673

Thermal Properties of Nanofluids

Thermal Properties of Nanofluids presents emerging prospects for understanding and controlling thermophysical properties at the nanoscale. It covers a comprehensive study of recent progress concerning these properties from the solid state to colloids and, above all, a different look at the effect of temperature on nanofluids' thermal conducting.

Introducing various techniques for measuring solid-state properties, including thermal conductivity, thermal diffusivity, and specific heat capacity, this book presents modeling approaches developed for predicting these properties by molecular dynamic (MD) simulations. It discusses the main factors that affect solid-state properties, such as grain size, grain boundaries, surface interactions, doping, and temperature, and the effects of all these factors.

This book will interest industry professionals and academic researchers studying the thermophysical behavior of nanomaterials and heat transfer applications of nanofluids. It will serve graduate engineering students studying advanced fluid mechanics, heat transfer, and nanomaterials.

Thermal Properties of Nanofluids

Taher Armaghani and Ramin Ghasemiasl

CRC Press is an imprint of the
Taylor & Francis Group, an **informa** business

Designed cover image: Shutterstock

First edition published 2025
by CRC Press
2385 NW Executive Center Drive, Suite 320, Boca Raton FL 33431

and by CRC Press
4 Park Square, Milton Park, Abingdon, Oxon, OX14 4RN

CRC Press is an imprint of Taylor & Francis Group, LLC

ISBN: 978-1-032-66406-4 (hbk)
ISBN: 978-1-032-66409-5 (pbk)
ISBN: 978-1-032-66411-8 (ebk)

DOI: 10.1201/9781032664118

Typeset in Times
by codeMantra

To my wife, Razieh

And my son, AmirAli

Taher Armaghani, Associated Prof of Mechanical Engineering, WTIAU, Iran

To my patient wife, Ziba

Ramin Ghasemiasl, Assistant Prof of Mechanical Engineering, WTIAU, Iran

Contents

Preface....................xii
About the Authors....................xiv

Chapter 1 Introduction....................1

1.1 An Overview of Nanoscience....................1
1.2 Defining Solid and Colloidal Nanomaterials....................2
1.3 The Benefits and Drawbacks of Nanostructured Materials....................3
1.4 Nanostructured Materials' Application....................4
References....................5

Chapter 2 Solid State....................8

2.1 Thermal Conductivity....................9
2.1.1 Theory....................10
2.1.2 Experimental Measurement Techniques....................17
2.1.2.1 T-Type Probe Method (for Nanotubes, Nanowires)....................17
2.1.2.2 3ω Method (for Nanotubes, Nanowires, Nanofilms)....................18
2.1.2.3 3ω-T Type Method (for Nanotubes, Nanowires)....................20
2.1.2.4 H-Type Method (for Nanofilms)....................21
2.1.2.5 Raman Spectroscopy Method (for Nanofilms)....................21
2.1.2.6 Time-Domain Thermoreflectance....................22
2.1.3 Experimental Studies....................24
2.1.3.1 Effect of Grain Size....................24
2.1.3.2 Effect of Grain Boundary....................25
2.1.3.3 Effect of Surface Interactions....................25
2.1.3.4 Effect of Doping....................27
2.1.3.5 Effect of Defects....................28
2.1.3.6 Effect of Temperature....................29
2.1.4 MD Simulations....................31
2.2 Thermal Diffusivity....................33
2.2.1 Theory....................33
2.2.2 Experimental Measurement Techniques....................35
2.2.2.1 Laser Flash Technique....................35
2.2.2.2 Laser Flash Raman Spectroscopy Method....................36
2.2.2.3 Transient Electrothermal (TET) Technique....................36

2.2.2.4 Photothermal Resistance Technique 38
2.2.2.5 Infrared Thermography 38
2.2.3 Experimental Studies 39
2.2.3.1 Effect of Porosity 39
2.2.3.2 Effect of Temperature 40
2.3 Isobaric Specific Heat Capacity 41
2.3.1 Theory 41
2.3.2 Experimental Measurement Techniques 44
2.3.2.1 3ω Method 44
2.3.2.2 AC-Calorimetric Technique 45
2.3.3 Experimental Studies 46
2.3.3.1 Effect of Size 46
2.3.3.2 Effect of Grain Boundary 47
2.3.3.3 Effect of Interfacial Component 48
2.3.3.4 Effect of Temperature 49
References 50

Chapter 3 Colloidal Nanofluids Thermal Conductivity 65

3.1 Theoretical Models 65
3.1.1 Role of Nanolayer Thickness 68
3.1.2 Nanoparticle Size Effect 70
3.1.3 Temperature Influence 71
3.1.4 Investigation of Nanoparticle Layer Thickness Change and Particle Diameter According to Yu and Choi Study 73
3.1.5 The Results of Examining Different Calculation Formulas for the Same Conditions 73
3.1.6 Aggregation Effect 74
3.1.7 Evolution of Advanced Models 76
3.2 Experimental Measurement Techniques 79
3.2.1 Transient Hot-Wire Approach 79
3.2.2 3ω Method 81
3.2.3 Laser Flash Method 82
3.2.4 The Steady-State Parallel Plate Method 82
3.2.5 Transient Plane-Source Method 83
3.3 Experimental Studies 84
3.3.1 Nanoparticle Concentration Effect 84
3.3.2 Temperature Effect 89
3.3.3 Nanoparticle Size Effect 90
3.3.4 The Impact of Nanoparticle's Shape 91
3.3.5 Nanoparticles Aggregation Effect 92
3.3.6 Effect of pH 93
3.3.7 Sonication Time Effect 94

3.4 Hybrid Nanofluids 95
3.4.1 Theory 95
3.4.2 Experimental Studies 96
3.5 Thermal Conductivity of Suspensions (Complementary) 98
3.5.1 Equations Based on the Theory of Effective Media.... 98
3.5.2 Exchanges Based on the Brownian Movement 100
3.5.3 Formulas Based on Interface.................................. 102
3.5.4 Two Constraining Items Related to the Structure of Nanoparticles 103
3.6 MD Simulations.. 104
Referenccs 108
Appendix 122

Chapter 4 New Look at Nanofluid Thermal Conductivity Correlations........... 129

Maysam Molana

4.1 Introduction 129
4.2 Methodology.. 131
4.3 Results and Discussion 135
4.4 Model by Aberoumand et al. 135
4.5 Model by Fakoor Pakdaman et al. 140
4.6 Model by Ahammed et al. 142
4.7 Model by Patel et al. 144
4.8 Models by Li and Peterson 147
4.9 Model by Hemmat Esfe et al. 151
4.10 Models by Hemmat Esfe et al. 153
4.11 Model by Hemmat Esfe et al. 156
4.12 Model by Hemmat Esfe et al. 159
4.13 Model by Harandi et al. 161
4.14 Model by Zadkhast et al. 163
4.15 Model by Kakavandi and Akbari 164
4.16 Model by Karimi et al. 167
4.17 Model by Karimipour et al. 170
4.18 Model by Ranjbarzadeh et al. 173
4.19 Model by Keyvani et al. 174
4.20 Model by Afrand et al. 177
4.21 A Comprehensive Correlation 183
4.22 Conclusion 184
References 197
Appendix 200
A.1 Nonlinear Regression 200
A.2 Nonlinear Regression Formula..201
A.3 Different Types of Nonlinear Regression.............................202
A.4 An Example of Nonlinear Regression.................................202

Chapter 5 Viscosity 204

5.1 Classical Models 204
5.1.1 Evolution of Classical Models 204
5.2 Advanced Recent Models 206
5.3 Experimental Measurement Techniques 208
5.3.1 Capillary Viscometer 208
5.3.2 Concentric Cylinders 209
5.3.3 Cone and Plate Approach 210
5.3.4 Other Measurement Methods 211
5.4 Experimental Studies 211
5.4.1 Effect of Temperature 212
5.4.2 Effect of Nanoparticle Concentration 213
5.4.3 Effect of Nanoparticle Size and Shape 214
5.4.4 Effect of pH 215
5.4.5 Sonication Time Effect 216
5.5 Hybrid Nanofluids 217
5.5.1 Theory 217
5.5.2 Experimental Studies 220
5.6 MD Simulation 221
References 223

Chapter 6 Statistical Study and Overview of Nanofluid Viscosity Correlations 230

A. Barkhordar

6.1 Introduction 230
6.2 Strategy 233
6.2.1 Analysis of Variance 233
6.2.2 Physical Analysis of Correlations 255
6.3 Sensitivity Analysis with Monte Carlo Test 259
6.4 Overall Analysis of Empirical Correlations 264
6.5 Proposing a Viscosity Model and Its Validation 266
6.5.1 Preliminary Analysis 266
6.5.2 Proposed Correlation 267
6.5.3 Evaluation of BAG Viscosity Model 268
6.6 Conclusion 277
References 279
Nomenclature 284

Chapter 7 Other Thermal Properties of Nanofluids 285

7.1 Isobaric-Specific Heat Capacity 285
7.1.1 Theory 285
7.1.2 Experimental Measurement Techniques 286
7.1.3 Experiments vs. Theory 287
7.2 Density 289

7.2.1 Theory 289
7.2.2 Experimental Measurement Techniques 290
7.2.3 Experiment vs. Theory 290
References 291

Chapter 8 Selecting a Thermophysical Model for Numerical Modeling of Nanofluids 296

8.1 Introduction 296
8.2 An Overview of Numerical Modeling Methods of Nanofluids 296
8.2.1 Single-Phase Approach 297
8.2.2 Two-Phase Approach 297
8.2.3 Boungiorno Semi-Two-Phase Model 298
8.2.3.1 Brownian Diffusion 299
8.2.3.2 Thermophoresis 299
8.3 The Most Widely Used Thermophysical Equations of Nanofluids 300
8.4 Evaluation of Thermophysical Relationships Based on Numerical Modeling 304
8.4.1 TiO_2–Water Nanofluid 304
8.4.2 Al_2O_3–Water Nanofluid 305
8.4.3 SiO_2–Water Nanofluid 307
8.5 Conclusion 308
References 308

Index 313

Preface

Nanomaterials have excellent optical, electrical, magnetic, mechanical, and thermal properties and are considered by many as some of the most promising materials for a multitude of applications. To understand the current research status and prospects of understanding and controlling thermophysical properties at the nanoscale, a comprehensive study of recent progress is performed in this book concerning these properties from the solid state to colloids and, above all, a different look at the effect of temperature on nanofluids' thermal conducting. First, various techniques for measuring solid-state properties, including thermal conductivity, thermal diffusivity, and specific heat capacity, are introduced. A presentation of modeling approaches follows this developed for predicting these properties by MD simulations. The main factors that affect the solid-state properties are grain size, grain boundaries, surface interactions, doping, and temperature, and the effects of all these factors are discussed in detail. Thereafter, methods for measuring and modeling (via MD simulations) colloid properties, including thermal conductivity, dynamic viscosity, specific heat capacity, and density, are presented. The main parameters affecting these properties, such as size, shape, concentration of nanoparticles, aggregation, and sonication time, are considered in our discussion.

Furthermore, the properties of not only simple nanofluids (advanced colloids, which are a mixture of nanoparticles and conventional liquids) but also hybrid nanofluids (composed of more than one type of nanoparticle) are also considered. Finally, research gaps, challenges, and prospects in this space are introduced. The novelty and main contributions of this book arise from (a) the comprehensive presentation and detailed discussion of both solid-state and colloid thermophysical properties, which goes beyond all previous studies available in the literature; (b) the consideration of both experimental approaches for the measurement of the properties of interest, complemented by studies aimed at modeling these properties, performed using MD simulation approaches; (c) a different look at the effect of temperature on the nanofluids thermal conductivity; (d) a different look at the effect of temperature on the viscosity of the nanofluids; (e) a detailed presentation of research in this space, recent developments in the field, and a future outlook; and (f) challenges of thermophysical property utilization.

Many of the relations obtained from the experimental values are not accurate enough for the thermophysical properties of the nanofluids. These topics have been fully explored in separate chapters. This book is mainly presented in eight chapters. The first chapter is an introduction to the nanotechnology background and a brief description of nanostructured features. The second chapter mainly introduces the latest research progress on the thermophysical properties of solid nanostructured materials. This chapter focuses on experimental measurement techniques, theory vs. experiments, and the main factors affecting various thermophysical properties.

The third chapter mainly introduces the latest research progress on the thermal conductivity of nanofluids, including various mathematical models (from classical to refined look), experimental measurement techniques, theory vs. experiments, and the main factors affecting nanofluids' k and heat transfer rate. In addition, MD simulations have also been introduced to research the thermophysical properties of solid and colloidal nanomaterials.

The fourth chapter mainly introduces a different look at the effect of temperature on the nanofluid thermal conductivity and its correlations, which is novel progress in Nanoscience.

Furthermore, one of the most critical applications of the thermophysical properties of nanofluids is their use in modeling the nanofluid heat transfer. Moreover, the fourth chapter will examine precisely which combination of thermophysical equations predicts accurate behavior and which combination of relationships shows significant deviations from experimental results.

The fifth chapter mainly introduces the classical and latest research progress and proposed models on the viscosity of nanofluids, experimental measurement techniques, theory vs. experiments, and the main factors affecting nanofluids' viscosity and heat transfer rate. The sixth chapter offers a fundamentally different look at nanofluid viscosity, which is novel progress in Nanoscience. Then, the seventh chapter mainly introduces other thermal properties of nanofluids. Finally, the eighth chapter explains which thermophysical model is more suitable for the numerical modeling of nanofluids especially in natural convection heat transfer.

This book provides an essential source for understanding future directions and hotspots for nanoscale solid-to-colloid state study. At the end of the chapters, an attempt is made to provide a comprehensive and complete relationship in which the available experimental values are included.

We would like to express our gratitude to Professor Omid Mahian, Professor of Mechanical Engineering at Ningbo University, China, for his invaluable efforts and insightful advice in the formation of this book.

Lastly, we express our gratitude to Mahtab Nazarahari, our MSC student, for her valuable efforts.

Taher Armaghani
Associated Prof of Mechanical Engineering,
WTIAU, Iran

Ramin Ghasemiasl
Assistant Prof of Mechanical Engineering,
WTIAU, Iran

About the Authors

Taher Armaghani is an Associate Professor of Mechanical Engineering at the West Tehran Branch of Islamic Azad University. He earned his BSc, MSc, and Ph.D. in Mechanical Engineering from Shahrood University of Technology, Shahrood, Iran. He teaches advanced heat transfer in the nanoscience field in the Ph.D. grade. He has published more than 60 ISI in the nanoscience field. He has published two books in the nanoscience field about applications of nanofluids and nanofluid applications in microelectronics. He is also an editor of the *Journal of Nanofluid* and a guest editor on special topics of nanofluids in some journals. Additionally, he serves as a reviewer of more than 100 scientific journals in the nanoscience field. In current studies, the modern experimental heat transfer of nanofluids will be explored in the advanced heat transfer laboratory of the West Tehran branch of Islamic Azad University instituted by Dr. Taher Armaghani and Dr. Ramin Ghasemiasl.

Ramin Ghasemiasl has been an Assistant Professor of Mechanical Engineering at the West Tehran Branch of Islamic Azad University since 2010. He earned his Ph.D. in Mechanical Engineering from Tarbiat Modares University (TMU), Tehran, Iran. He has published more than 40 WOS papers, more than 50 papers in conferences, and 12 books. His research interests include renewable energy, the multi-generation system for product electricity, fresh water and hydrogen, and nanofluids applications in modern technologies. In current studies, the modern experimental heat transfer of nanofluids will be explored in the advanced heat transfer laboratory of the West Tehran branch of Islamic Azad University instituted by Dr. Ramin Ghasemiasl and Dr. Taher Armaghani.

1 Introduction

1.1 AN OVERVIEW OF NANOSCIENCE

The history of nanotechnology is provided for familiarization with nanomaterials' distinct thermophysical characteristics. Richard Feynman, a Nobel laureate physicist, introduced a novel nanotechnology idea in 1959. Using scanning tunneling microscopy, an atomic cluster was noticed in 1981, heralding the beginning of nanotechnology. Iijima [1] discovered carbon nanotubes for the first time in 1991. Carbon nanotubes' (CNTs') outstanding thermal and electrical features and mechanical characteristics have since made them a popular topic in nanotechnology study disciplines. Scientists are presently discussing nanotechnology's future perspectives. Multidisciplinary work has resulted in the twenty-first century's likely nanotechnologies. Multidisciplinary fields include nanoelectronics, nanomechanics, engineering technology, medical science, nanochemistry, physics, nanobiology, and nanofabrication [2–4]. As a result, the primary task is to develop more innovative nanomaterials and technologies with diverse applications, such as nanogenerators, nanomedicine, and nanoelectronics. The advent of nanotechnology's developing concept has sparked much interest. Compared to classic engineered materials, they have outstanding electrical and optical features and mechanical, thermal, and magnetic characteristics.

Other investigations into the effects of particle migration, magnetic field inclination, internal heat sources, and porous media on nanofluid heat transfer and entropy generation are discussed in different studies. Through extensive reviews of experimental, numerical, and theoretical studies, valuable insights into the complex interplay between nanofluid properties, geometric configurations, and heat transfer mechanisms are provided. Additionally, emerging research trends and future directions in the field of nanofluid heat transfer are identified and discussed, offering guidance for researchers and practitioners engaged in the exploration of this rapidly evolving interdisciplinary domain [5–16].

Nanomaterials are described as having at least one nanoscale dimension. Preparing nanomaterial and describing its performance is now the focus of nanotechnology studies. Nanotechnology applications are in microelectronics, computer science, medical and healthcare science, the aerospace industry, environmental preservation, biotechnology, and agriculture. The main objective is to improve the poor thermophysical features by developing specific substances that are lightweight, last longer, and are stronger in hardness. Moreover, they have an unexampled higher electric and thermal conductivity and possess biomaterials and biomimetic substances that are not present in nature.

DOI: 10.1201/9781032664118-1

1.2 DEFINING SOLID AND COLLOIDAL NANOMATERIALS

The most prevalent nanomaterials are solid nanomaterials, broadly divided into four kinds based on their appearance: film, fiber, powder, and bulk. Nanomaterials are classified into four types based on their dimensions: (a) nanostructures with zero dimension (0-D) with three dimensions on the nanometer length scale, like nanoparticles and fullerenes; (b) one-dimensional nanostructures (1-D) with just a single dimension at the nanoscale like nanotubes, nanowires, and nanofibers; (c) nanostructures that are two-dimensional (2-D), with two dimensions at the nanoscale, like graphene and nanofilms; and (d) nanomaterials that are three-dimensional (3-D) and are also popular as bulk nanomaterials, possessing the properties of nanomaterials, like nanocrystalline materials and composite nanomaterials. In Figure 1.1, the solid nanostructured materials are depicted [17–19].

The working fluid serves as the heat transfer fluid in thermal devices like heat exchangers, utilized in different applications like automobile radiators or electronic equipment cooling. The thermal equipment may be badly damaged if the working fluid is not capable of heating or cooling. Adding solid nanoparticles to ordinary liquids is a method for altering their potential and improving the efficient thermal conductivity of the liquid. In 1993, scientists published research in which the k and dynamic viscosity of some suspensions were considered based on water. They had SiO_2, Al_2O_3, and TiO_2 nanoparticles whose sizes ranged between 1 and 100 nm [20]. Nanofluids refer to the remaining state-colloids nanostructure produced by physical or chemical processes. In 1995, Choi [21] introduced nanofluids, which are produced through the mixing of nanoparticles and ordinary liquids. A simple nanofluid is a colloidal suspension of one kind of nanoparticle plus a regular fluid.

On the other hand, hybrid nanofluids are made up of more than one kind of nanoparticle plus a base fluid. One of the basic nanofluids created by the Argonne National Laboratory group is ceramic nanofluids. Shortly after, the advent of metal-based nanofluids represented a significant advancement in the nanofluids field. For the first time, Xuan and Li [22] combined copper and transformer oil nanoparticles to create nanofluids. Nanofluids have since been researched extensively in the

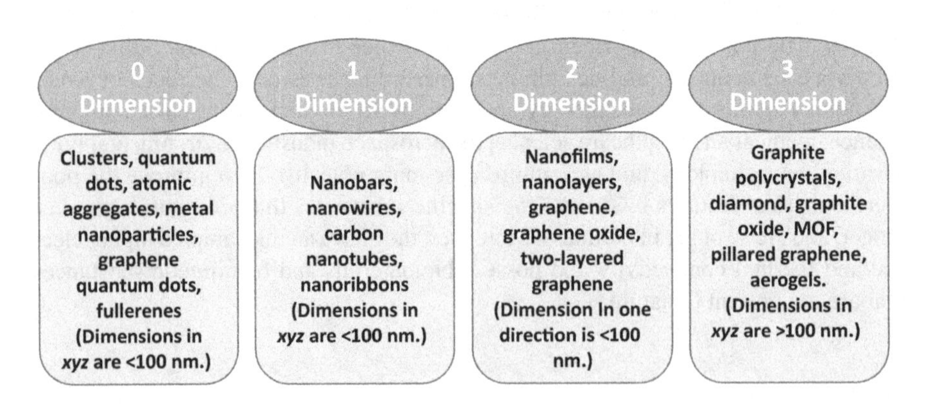

FIGURE 1.1 Schematic diagram of solid nanostructured materials with different dimensions.

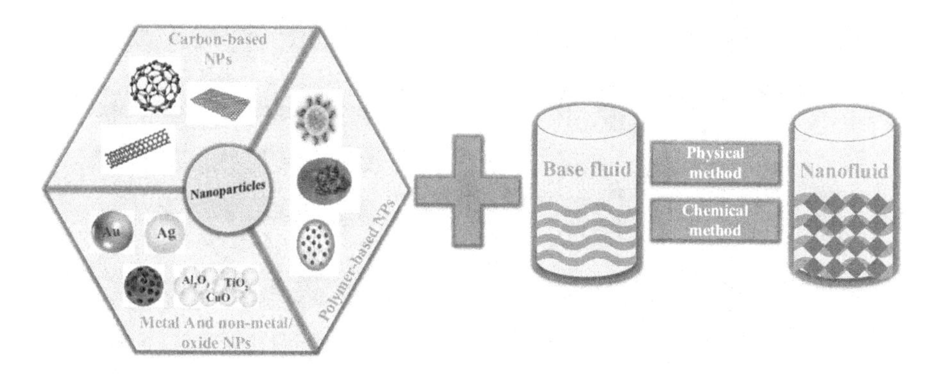

FIGURE 1.2 Schematic diagram of nanoparticles classification and a brief preparation process for nanofluids.

domains of nanomaterials. Metal and nonmetal oxides (e.g., Al_2O_3 and SiO_2), metals (e.g., Au and Ag), carbides (SiC), carbon-based nanomaterials (carbon nanotubes and fullerene), or polymers might be used to make nanofluids. Nanoparticles' categorization and a summary of nanofluids' preparation procedures are depicted in Figure 1.2. Water, oil, and ethylene glycol are the most frequent base fluids with low k [23,24]. Based on many pieces of research, in comparison to fluid heat transfer, nanofluids have a higher conductivity (k) and coefficient of convective heat transfer (h) [25–27]. Density, the capacity of isobaric specific heat, and dynamic viscosity completed the thermophysical features, and they are the three significant thermophysical parameters [28–30].

1.3 THE BENEFITS AND DRAWBACKS OF NANOSTRUCTURED MATERIALS

Generally, nanostructured materials outperform traditional materials in terms of characteristics, giving them a wide range of applications. The benefits and drawbacks of various common solid nanostructured materials and nanofluids are summarized in Figure 1.3. Carbon nanoparticles, for example, have a high surface-to-volume ratio, a high electrical conductivity, mechanical strength, a high k, and proper chemical stability in solid nanostructured materials. Aerogels are the most typical nano-porous materials due to their high porosity exhibition (95% or more) and deficient density (0.02–0.32 g/cm^3). The aerogels' interior pore size presence on the nanoscale is lower than the energy transformers' mean free path. Thus, they exhibit deficient k (0.02–0.036 W/m°K). As a result, one of the most efficient and promising insulating materials is aerogels [31].

Furthermore, the properties of the nano-porous high surface area of material are determinants of their outstanding gas storage [32].

Solid nanostructured materials also have some drawbacks concerning cost and procedure. Nanostructured materials' synthesis is costlier than conventional materials. Some specific nanomaterials' synthesizing and processing technology is not appropriate for large-scale production. Thus, commercial usage is restricted.

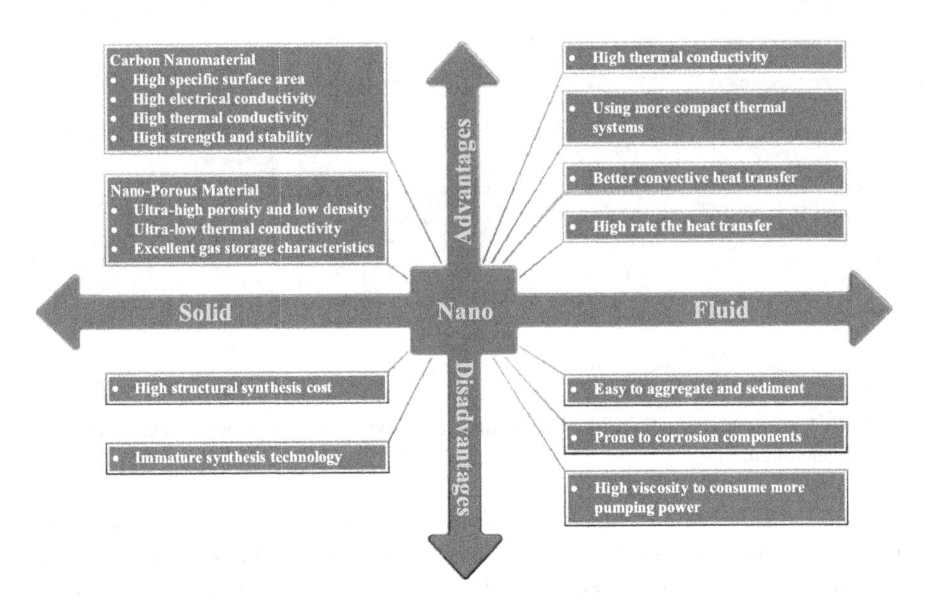

FIGURE 1.3 Advantages and disadvantages of some typical solid and colloidal nanofluids.

Nanofluids have some other drawbacks, as follows:

- Costly preparation and a somewhat complex preparation procedure.
- Preventing nanoparticles' aggregation and sedimentation.
- High consumption of nanofluids' pumping power during transport due to their higher dynamic viscosity than conventional fluids.
- Nanofluids components' corrosion in their contact.

1.4 NANOSTRUCTURED MATERIALS' APPLICATION

The differences between nanostructured materials and conventional engineering materials were stated in previous sections. Now, what is the application and use of these materials? Generally, nanostructured materials are helpful for different fields, including the energy field, aerospace technologies, environment, microelectronics field, bioengineering high-tech, medical science, and other fields [3,33,34]. Nanofluids are utilized in heating systems [35], solar energy systems [36–40], engine cooling devices [41], freezers, as porous media [42], and several flow procedures requiring cooling or heating [43]; thus, they are all of primary scientific and industrial importance. A high-level summary of the application sectors for nanomaterials is depicted in Figure 1.4.

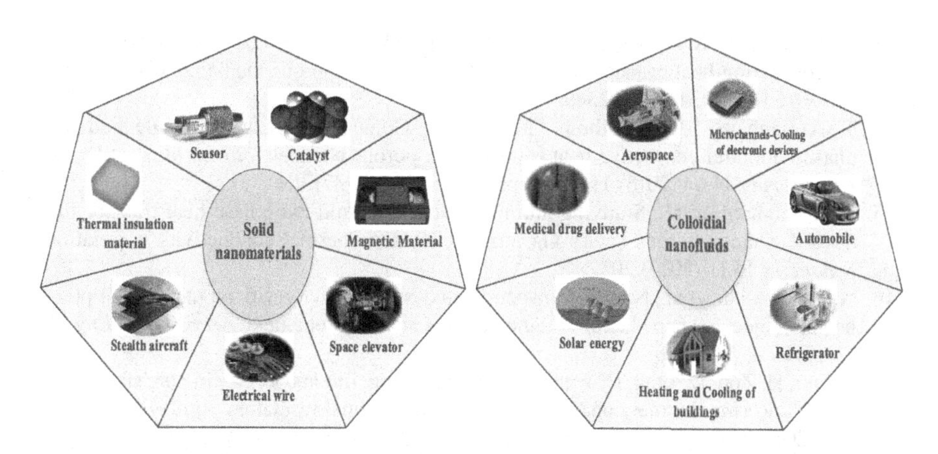

FIGURE 1.4 Widespread application fields for nanomaterials: From solid state to colloids.

REFERENCES

1. S. Iijima, Helical microtubules of graphitic carbon, *Nature* 354 (1991) 56–58. https://doi.org/10.1038/354056a0.
2. A.L. Porter, Y. Jan, Where does nanotechnology belong in the map of science? *Nat. Nanotechnol.* 4 (2009) 534–536. https://doi.org/10.1038/nnano.2009.207.
3. Y.J. Chen, B. Groves, R.A. Muscat, G. Seelig, DNA nanotechnology from the test tube to the cell, *Nat. Nanotechnol.* 10 (2015) 748–760. https://doi.org/10.1038/nnano.2015.195.
4. Y. Huang, X. Dong, Y. Shi, C.M. Li, L.J. Li, P. Chen, Nanoelectronic biosensors based on CVD grown graphene, *Nanoscale* 2 (2010) 1485–1488. https://doi.org/10.1039/c0nr00142b.
5. A. Shafiei, et al., Nanoparticles migration effects on enhancing the cooling process of triangular electronic chips using novel E-shaped porous cavity, *Comput. Part. Mech.* 10.4 (2023) 793–808.
6. H.A. Nabwey, et al., A comprehensive review of nanofluid heat transfer in porous media, *Nanomaterials* 13.5 (2023) 937.
7. A. Chamkha, et al., Effects of magnetic field inclination and internal heat sources on nanofluid heat transfer and entropy generation in a double lid driven L-shaped cavity, *Therm. Sci.* 25.2 Part A (2021) 1033–1046.
8. T. Armaghani, et al., Effects of particle migration on nanofluid forced convection heat transfer in a local thermal non-equilibrium porous channel, *J. Nanofluids* 3.1 (2014) 51–59.
9. T. Armaghani, M.A. Ismael, A.J. Chamkha, Analysis of entropy generation and natural convection in an inclined partially porous layered cavity filled with a nanofluid, *Can. J. Phys.* 95.3 (2017) 238–252.
10. T. Armaghani, et al., Numerical investigation of water-alumina nanofluid natural convection heat transfer and entropy generation in a baffled L-shaped cavity, *J. Mol. Liq.* 223 (2016) 243–251.
11. A. Chamkha, et al., Entropy generation and natural convection of CuO-water nanofluid in C-shaped cavity under magnetic field, *Entropy* 18.2 (2016) 50.
12. A.M. Rashad, et al., Entropy generation and MHD natural convection of a nanofluid in an inclined square porous cavity: Effects of a heat sink and source size and location, *Chin. J. Phys.* 56.1 (2018) 193–211.

13. M. Nazarahari, R. Ghasemi Asl, T. Armaghani, Experimental study of nanofluids natural convection heat transfer in various shape pores of porous media, *J. Therm. Anal. Calorim* 149 (2024) 2331–2349.
14. T. Armaghani, M.J. Maghrebi, M. Nazari, Comparison between single and two phase nanofluid convective heat transfer in a porous channel *Journal of Modeling in Engineering* 14.44 (2016) 11–20. https://sid.ir/paper/173713/en.
15. T. Armaghani, et al., Studying alumina-water nanofluid two-phase heat transfer in a novel E-shaped porous cavity via introducing new thermal conductivity correlation, *Symmetry* 15.11 (2023) 2057.
16. A.S. Dogonchi, et al., Natural convection analysis in a cavity with an inclined elliptical heater subject to shape factor of nanoparticles and magnetic field, *Arab. J. Sci. Eng.* 44 (2019) 7919–7931.
17. L. Qiu, H. Zou, D. Tang, D. Wen, Y. Feng, X. Zhang, Inhomogeneity in pore size appreciably lowering thermal conductivity for porous thermal insulators, *Appl. Therm. Eng.* 130 (2018) 1004–1011. https://doi.org/10.1016/j.applthermaleng.2017.11.066.
18. I. Vukovic, G.T. Brinke, K. Loos, Block copolymer template-directed synthesis of well-ordered metallic nanostructures, *Polymer* 54 (2013) 2591–2605. https://doi.org/10.1016/j.polymer.2013.03.013.
19. M.M. Mariscal, J.A. Olmos-Asar, C. Gutierrez-Wing, A. Mayoral, M.J. Yacaman, On the atomic structure of thiol-protected gold nanoparticles: A combined experimental and theoretical study, *Phys. Chem. Chem. Phys.* 12 (2010) 11785–11790. https://doi.org/10.1039/c004229c.
20. H. Masuda, A. Ebata, K. Teramae, N. Hishinuma, Alteration of thermal conductivity and viscosity of liquid by dispersing ultra-fine particles. Dispersion of Al_2O_3, SiO_2 and TiO_2 ultra-fine particles, *Netsu Bussei.* 7 (1993) 227–233. https://doi.org/10.2963/jjtp.7.227.
21. S.U.S. Choi, Nanofluids: From vision to reality through research, *J. Heat Transfer.* 131 (2009) 033106. https://doi.org/10.1115/1.3056479.
22. Y. Xuan, Q. Li, Heat transfer enhancement of nanofuids, *Int. J. Heat Fluid Flow.* 21 (2000) 58–64. https://doi.org/10.1016/S0142-727X(99)00067-3.
23. S. Aman, I. Khan, Z. Ismail, M.Z. Salleh, Q.M. Al-Mdallal, Heat transfer enhancement in free convection flow of CNTs Maxwell nanofluids with four different types of molecular liquids, *Sci. Rep.* 7 (2017) 2445. https://doi.org/10.1038/s41598-017-01358-3.
24. J. Sui, L. Zheng, X. Zhang, Y. Chen, Z. Cheng, A novel equivalent agglomeration model for heat conduction enhancement in nanofluids, *Sci. Rep.* 6 (2016) 19560. https://doi.org/10.1038/srep19560.
25. M.U. Sajid, H.M. Ali, Thermal conductivity of hybrid nanofluids: A critical review, *Int. J. Heat Mass Transf.* 126 (2018) 211–234. https://doi.org/10.1016/j.ijheatmasstransfer.2018.05.021.
26. O. Mahian, L. Kolsi, M. Amani, P. Estellé, G. Ahmadi, C. Kleinstreuer, J.S. Marshall, M. Siavashi, R.A. Taylor, H. Niazmand, S. Wongwises, T. Hayat, A. Kolanjiyil, A. Kasaeian, I. Pop, Recent advances in modeling and simulation of nanofluid flows-Part I: Fundamentals and theory, *Phys. Rep.* 790 (2019) 1–48. https://doi.org/10.1016/j.physrep.2018.11.004.
27. L. Yang, J. Xu, K. Du, X. Zhang, Recent developments on viscosity and thermal conductivity of nanofluids, *Powder Technol.* 317 (2017) 348–369. https://doi.org/10.1016/j.powtec.2017.04.061.
28. G. Żyła, J.P. Vallejo, L. Lugo, Isobaric heat capacity and density of ethylene glycol based nanofluids containing various nitride nanoparticle types: An experimental study, *J. Mol. Liq.* 261 (2018) 530–539. https://doi.org/10.1016/j.molliq.2018.04.012.

29. A. Mariano, M.J. Pastoriza-Gallego, L. Lugo, L. Mussari, M.M. Piñeiro, Co_3O_4 ethylene glycol-based nanofluids: Thermal conductivity, viscosity and high pressure density, *Int. J. Heat Mass Transf.* 85 (2015) 54–60. https://doi.org/10.1016/j.ijheatmasstransfer.2015.01.061.
30. S.N. Shoghl, J. Jamali, M.K. Mostafa, Electrical conductivity, viscosity, and density of different nanofluids: An experimental study, *Exp. Therm. Fluid Sci.* 74 (2016) 339–346. https://doi.org/10.1016/j.expthermflusci.2016.01.004.
31. Y. He, T. Xie, Advances of thermal conductivity models of nanoscale silica aerogel insulation material, *Appl. Therm. Eng.* 81 (2015) 28–50. https://doi.org/10.1016/j.applthermaleng.2015.02.013.
32. D.P. Broom, K.M. Thomas, Gas adsorption by nanoporous materials: Future applications and experimental challenges, *MRS Bull.* 38 (2013) 412–421. https://doi.org/10.1557/mrs.2013.105.
33. L. Dai, D.W. Chang, J.B. Back, W. Lu, Carbon nanomatcrials for advanced energy conversion and storage, *Small* 8 (2012) 1130–1166. https://doi.org/10.1002/smll.201101594.
34. H. Jiang, Chemical preparation of graphene-based nanomaterials and their applications in chemical and biological sensors, *Small* 7 (2011) 2413–2427. https://doi.org/10.1002/smll.201002352.
35. S. Rashidi, O. Mahian, E.M. Languri, Applications of nanofluids in condensing and evaporating systems: A review, *J. Therm. Anal. Calorim.* 131 (2018) 2027–2039. https://doi.org/10.1007/s10973-017-6773-7.
36. S.S. Meibodi, A. Kianifar, O. Mahian, S. Wongwises, Second law analysis of a nanofluid-based solar collector using experimental data, *J. Therm. Anal. Calorim.* 126 (2016) 617–625. https://doi.org/10.1007/s10973-016-5522-7.
37. O. Mahian, A. Kianifar, S.Z. Heris, D. Wen, A.Z. Sahin, S. Wongwises, Nanofluids effects on the evaporation rate in a solar still equipped with a heat exchanger, *Nano Energy* 36 (2017) 134–155. https://doi.org/10.1016/j.nanoen.2017.04.025.
38. O. Mahian, A. Kianifar, A.Z. Sahin, S. Wongwises, Performance analysis of a minichannel-based solar collector using different nanofluids, *Energy Convers. Manag.* 88 (2014) 129–138. https://dx.doi.org/10.1016/j.enconman.2014.08.021.
39. O. Mahian, A. Kianifar, S.A. Kalogirou, I. Pop, S. Wongwises, A review on the applications of nanofluids in solar energy field, *Int. J. Heat Mass Transf.* 57 (2013) 582–594. https://dx.doi.org/10.1016/j.ijheatmasstransfer.2012.10.037.
40. O. Mahian, A. Kianifar, A.Z. Sahin, S. Wongwises, Heat transfer, pressure drop, and entropy generation in a solar collector using SiO2/water nanofluids: Effects of nanoparticle size and pH, *J. Heat Transfer.* 137 (2015) 061011. https://doi.org/10.1115/1.4029870.
41. N.A.C. Sidik, R. Mamat, A review on the application of nanofluids in vehicle engine cooling system, *Int. Commun. Heat Mass Transf.* 68 (2015) 85–90. https://doi.org/10.1016/j.icheatmasstransfer.2015.08.017.
42. A. Kasaeian, R. Daneshazarian, O. Mahian, L. Kolsi, A.J. Chamkha, S. Wongwises, I. Pop, Nanofluid flow and heat transfer in porous media: A review of the latest developments, *Int. J. Heat Mass Transf.* 107 (2017) 778–791. https://doi.org/10.1016/j.ijheatmasstransfer.2016.11.074.
43. O. Mahian, L. Kolsi, M. Amani, P. Estellé, G. Ahmadi, C. Kleinstreuer, J.S. Marshall, R.A. Taylor, H. Niazmand, S. Wongwises, T. Hayat, A. Kolanjiyil, A. Kasaeian, I. Pop, Recent advances in modeling and simulation of nanofluid flows – Part II: Applications, *Phys. Rep.* 791 (2019) 1–59. https://doi.org/10.1016/j.physrep.2018.11.003.

2 Solid State

The main topics covered in this section are the thermophysical properties of solid nanomaterials and mostly carbon fibers, nanotubes, nanofilms, nanowires, graphene, and nanocomposites. To measure such properties, the most recently developed techniques are used such as the T-type probe method, the 3ω method, and the combination form known as 3ω–T-type method, the comprehensive T-type method, the laser flash Raman spectroscopy, and the H-type method. The main energy carriers of crystalline nanomaterials are crystal vibrations and phonons that are featured with characteristic lengths of the order of nanometers. Conduction of heat in semiconductors and dielectric materials is controlled by collective lattice vibrations. Technological management has been achieved in heating devices like thermal diodes that are capable of controlling the flow of heat in the preferred direction. In addition, thermal coverage that is capable of hiding the object from heat and thermal crystal and capable of dissipating heat by technology are other examples. Moreover, manipulated lightning bolts can convert heat into electricity and it is directly used as a Peltier refrigerator [1].

A variety of factors influence k, $c_{p,}$ and α, like the intrinsic electron–phonon and phonon–phonon scatterings, surface interactions, grain boundaries, temperature, and grain size [1]. It is imperative to study the transfer of heat in nanostructures to understand and design the next-generation electronic and energy devices. This has resulted in deep studies on the causes and capability to create considerable reductions in thermal conductivity in crystalline materials. This is mostly through realizing the fact that the free phonon pathway (MFP) is featured with the characteristic size of nanostructures. Crystalline bound is also called Casimir boundary scattering. Using an extremely sensitive measurement system, we demonstrate that crystalline Si(c-Si) nanotubes (NTs) with shell thickness as thin as ~5 nm have a low thermal conductivity of ~1.1 W/m K. It is notable that this value is less than the desired boundary scattering limit and it is even nearly 30% less than the measured value for amorphous Si (a-Si) NTs with identical geometries [1].

Along with advances in many unprecedented disordered systems, there has been a convergence of the domains of low-dimensional materials (low D) and disordered materials. There is no clear picture of the heat transfer properties in these systems. Recent studies have used amorphous graphene and fiberglass nanodiamonds as prototype systems and demonstrated the way structural disturbance influences the transmission of vibrational energy in low D and irregular materials. Molecular dynamics simulation, mode localization analysis, and a general model demonstrated that the properties of heat transfer in these materials are featured with similarities and differences with irregular 3D materials. In comparison with 3D materials, low D erratic systems demonstrated vibrational propagation and propagation modes. In addition, and despite 3D, the contribution of diffusion to heat transfer in low D systems is almost negligible, which can be due to the intrinsic difference in low D random walking nature. In spite of the absence of diffusions, studies have shown that the

DOI: 10.1201/9781032664118-2

suppression of thermal conductivity because of disorder in low D systems is mild or comparable to 3D. The mild suppression is rooted in the presence of low-frequency vibrational modes that have a well-defined polarization and keep thermal conductivity despite disorder [2].

Amorphous nanomaterials have a completely disordered structure in which the phonons do not meet the intrinsic thermal transport mechanism. Every phonon vibration mode is localized, and the pertinent disordered diffusion between high-energy localized vibrational model (LVMS) is the main mechanism of transporting heat. Thus, the disorder and temperature of amorphous materials are the key factors in their thermal conductivity [2]. To have an accurate description of the thermal transport processes, we need to know the phonon transport mechanism and adjacent LVMs in solid nanomaterials. The theoretical basis for studying the thermophysical specifications of solid nanomaterials (e.g., correlated scattering, phonon dispersion, ballistic transport, low-frequency propagating, and diffusive vibration modes) is a broader range of phonon transport mechanisms and LVM [3]. Figure 2.1 illustrates a schematic diagram of the study on thermophysical specifications of solid materials.

2.1 THERMAL CONDUCTIVITY

Generally, the top objective of studies on the nanomaterial's thermal conductivity is to increase or decrease conductivity as much as feasible given the application. To study thermal conductivity, our concentration is on theory, experimental measurement techniques, Molecular Dynamics (MD) simulation, and the main factors affecting thermal conductivity value.

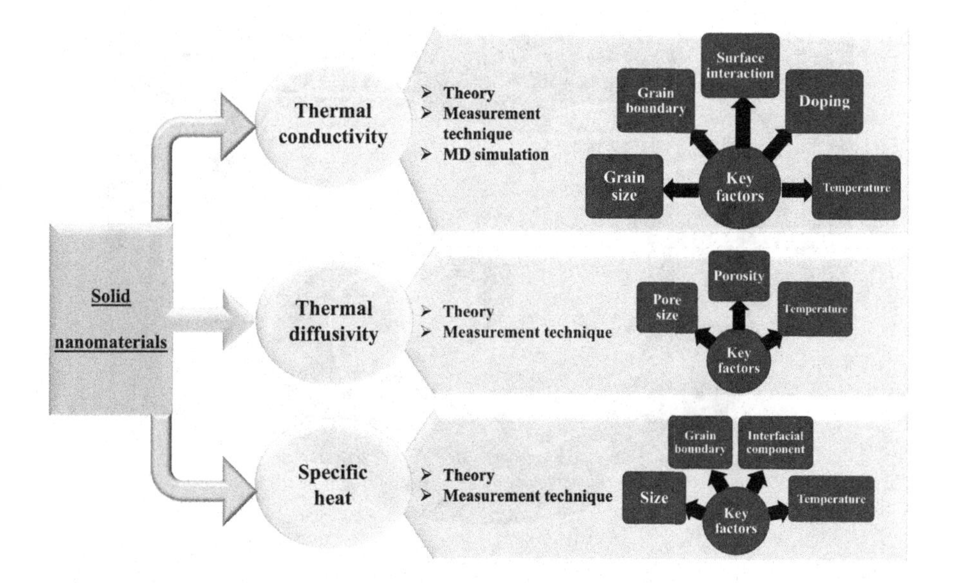

FIGURE 2.1 A schematic view of the general framework of thermophysical properties of solid nanomaterials.

2.1.1 Theory

To have a deeper perception of the thermophysical specifications and thermal transport mechanisms in nanoscale materials, more theoretical studies are needed [4–18]. Needless to say, electrons and phonons are carriers of energy that transport thermal energy [4]. Thus, it is important to study the phonon MFP distribution to understand the internal thermal transport mechanism of crystalline materials [10]. As to nanostructured materials, the wave vectors and MFP of phonons carry a considerable sum of energy at the nanoscale. Thus, it is important to understand the MFP distribution of phonons to realize the thermal transport of phonons as carriers. A flowchart of the theoretical study of k for solid nanostructured materials is illustrated in Figure 2.2.

The thermal conductivity of bulk materials and nanostructures has been a topic in Yang and Dames's study [4]. They used MFP spectra and thermal conductivity accumulation function (k_{acc}). As to bulk materials, the main common formula to calculate the k_{bulk} is on the basis of kinetic theory, which is obtainable through the Boltzmann Transport Equation (BTE) with an approximate relaxation time and the assumption that the dispersion relation under the relaxation time approximation is isotropic [5]:

$$k_{bu} = \sum_{S} \int_{0}^{\infty} \frac{1}{3} c_{\omega} v \Lambda_{bu} d\omega \tag{2.1}$$

where ω stands for the phonons frequency, c_{ω} refers to the volumetric isobaric specific heat capacity for each phonon frequency unit, v refers to the group velocity, Λ_{bulk} refers to the bulk MFP and S indexes the polarizations. To determine a direct

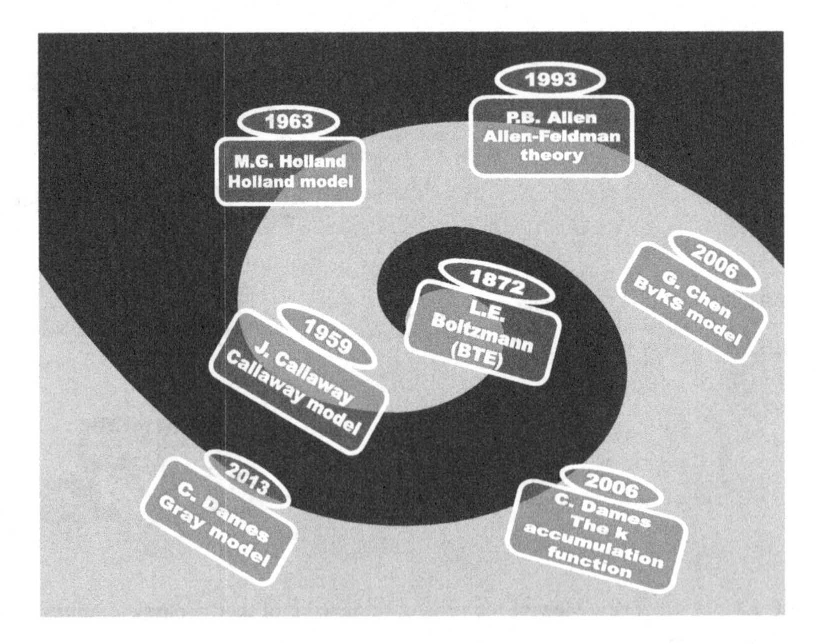

FIGURE 2.2 Chart of the theoretical study of k of solid nanostructured materials.

relationship between the Λ_{bu} and k_{bu}, the integration variable should be changed from ω to Λ_{bu}.

$$k_{bu} = -\sum_{S} \int_{0}^{\infty} \frac{1}{3} c_{\omega} v \Lambda_{bu} \left(\frac{d\Lambda_{bu}}{d\omega} \right)^{-1} d\Lambda_{bu} \tag{2.2}$$

Because of the convergence of the integral of Eq. (2.2), the order of exchanging integration and summation is obtained as follows:

$$k_{bu} = \int_{0}^{\infty} k_{\Lambda_{bu}} d\Lambda_{bu} \tag{2.3}$$

where $k_{\Lambda_{bu}} = -\sum_{S} \frac{1}{3} c_{\omega} v \Lambda_{bu} \frac{d\omega}{d\Lambda_{bu}}$ stands for the contribution to k per MFP, which is also the differential MFP distribution or MFP spectrum for k_{bu}. Therefore, Yang and Dames [4] reviewed the concept of bulk materials MFP distribution and used it to obtain k of nanostructured materials (k_{nano}). By assuming a negligible wave confinement effect, the v and c_{ω} of the nanostructure become the same as that of the bulk. Therefore, the reason for the difference between bulk and nanostructure thermal transport is the remarkable MFP reduction due to scattering at boundaries and interfaces, and the k_{nano} is given as [6–8]:

$$k_{nano,\ type} = \sum_{S} \int_{0}^{\infty} \frac{1}{3} c_{\omega} v \Lambda_{nano} d\omega \tag{2.4}$$

The subscript type indicates the geometry type, like nanowire or nanofilm. The scattering from boundaries and interfaces leads to $\Lambda_{nano} < \Lambda_{bu}$, which alters the integration variable from ω to Λ_{bu}:

$$k_{nano,\ type} = \int_{0}^{\infty} \left[-\sum_{S} \frac{1}{3} c_{\omega} v \Lambda_{bu} \left(\frac{d\Lambda_{bu}}{d\omega} \right)^{-1} \right] \frac{\Lambda_{nano}}{\Lambda_{bu}} d\Lambda_{bu} \tag{2.5}$$

$$k_{nano,\ type} = \int_{0}^{\infty} k_{\Lambda_{bu}} \frac{\Lambda_{nano}}{\Lambda_{bu}} d\Lambda_{bu} \tag{2.6}$$

Because k_{nano} depends on the structural characteristic length (L_c) and the nanostructure type, the results showed that the conversion of Λ_{nano} between bulk and strongly constrained behavior is mostly determined by L_c and Λ_{bu}. Generally, Eq. (2.7) represents the relationship between these three parameters (Λ_{nano}, L_c, and Λ_{bu}):

$$B_{type}(Kn) = \frac{\Lambda_{nano}}{\Lambda_{bu}} = B_{type}\left(\frac{\Lambda_{bu}}{L_c} \right) \tag{2.7}$$

where the Knudsen number (Kn) represents Λ_{bu}/L_c. B_{type} stands for a function of the ratio Λ_{bu}/L_c, which is obtained through the nanostructure type and Kn. Given

Eq. (2.7) and assuming that the function $B_{\text{type}}(Kn)$ is known, Eq. (2.6) is rewritten as follows:

$$k_{\text{nano, type}} = \int_0^\infty k_{\Lambda_{\text{bu}}} B_{\text{type}}(Kn)\, d\Lambda_{\text{bu}} \tag{2.8}$$

In addition, there is a different perspective from the bulk MFP spectrum known as k_{acc}. Based on the normalized k_{acc}, it is defined as:

$$k_{\text{acc}}(\Lambda_c) = \frac{1}{k_{\text{bu}}} \int_0^{\Lambda_c} k_{\Lambda_{\text{bu}}}\, d\Lambda_{\text{bu}} \tag{2.9}$$

Which gives us the fraction of the total k because of carriers with MFPs less than Λ_c. The Eq. (2.9) was introduced by Dames and Chen [9]. Equation (2.8) is featured with a counterpart related to the k_{acc}, which can be obtained using Eqs. (2.8 and 2.9):

$$k_{\text{nano, type}} = -k_{\text{bu}} \int_0^\infty k_{\text{acc}}(\Lambda_{\text{bu}}) \frac{dB_{\text{type}}}{d\Lambda_{\text{bu}}}\, d\Lambda_{\text{bu}} \tag{2.10}$$

where $\Lambda_{\text{bulk}} = 0$, $B_{\text{type}} \to 0$ as $\Lambda_{\text{bulk}} \to \infty$, $k_{\text{acc}}(\Lambda_{\text{bulk}}) = 0$, since the boundary scattering yields considerable thermal resistance, it is among the main reasons for the sharp decrease in k.

There is another mainstream simple approximation method to examine the influence of the MFP spectrum of k. it is featured with a good ability to approximate when the distribution of real MFPs is narrow. Still, when the distribution is broad, such as a strong frequency-dependent scattering system, the gray model is no longer valid. In general, the effect of MFPs on heat conduction is mostly described using single lumped "gray" or effective MFP.

In addition, the gray model is another common simple approximation approach to investigate the effect of the MFP spectrum on k. It has a good approximation in systems with a narrow distribution of real MFPs. However, in other systems where the distribution is broad, such as a strong frequency-dependent scattering system, the gray model is invalid. Traditionally, the distribution of MFPs contribution to heat conduction is described primarily by a single lumped "gray" or effective MFP:

$$\Lambda_{\text{Gr}} = k_{\text{bu}} \Big/ \left(\sum_S \int_0^\infty \frac{1}{3} c_p v\, d\omega \right) \tag{2.11}$$

The gray model is based on the assumption that the bulk MFP distribution utilizes identical gray values regardless of the given temperature. Thus, the k spectrum depends on Dirac δ, $k_{\Lambda_{\text{bu}}} = k_{\text{bu}}\delta(\Lambda_{\text{bu}} - \Lambda_{\text{Gr}})$, and κ_a is a Heaviside step function given as $\kappa_a = H(\Lambda_{\text{bu}} - \Lambda_{\text{Gr}})$. Thus, the formula of k_{nano} of the gray model can be determined through Eqs. (2.10 and 2.11):

$$k_{\text{nano, Gr}} = k_{\text{bu}} B_{\text{type}}\left(\frac{\Lambda_{\text{Gr}}}{L_c}\right) = k_{\text{bu}} B_{\text{type}}(Kn_{\text{Gr}}) \tag{2.12}$$

where Kn_{Gr} represents the Knudsen number of the gray medium.

In addition, the decrease in the thermal conductivity because of grain boundary scattering is extensively explained using scattering length, which is assumed to be the same as grain size and independent from the phonon frequency (gray). To examine these assumptions and decouple the contributions of porosity and grain size, five samples of undoped nanocrystalline silicon were measured with average grain sizes ranging from 550 to 64 nm. In addition, porosities range from 17% to less than 1% at 310 to 16 K temperature. To prepare the samples, current-activated pressure-assisted densification (CAPAD) was used. With low-range temperatures, the thermal conductivity of all samples had a T2 dependence that may not be explained using the traditional gray model. For the whole temperature range, measurements are explained using a new model that depends on the frequency in which the mean free path for grain boundary scattering is proportional to the frequency of phonon. This is in agreement with the asymptotic analysis of atomic simulations in the literature. In every case, the recommended length of scattering is below the average grain size. The results can help us to integrate nanocrystalline materials into devices like advanced thermos electrics [10]. Callaway [10] introduced a theoretical model to calculate the k of the lattice at lower temperatures. The model was named Callaway model and it is based on the assumption that the phonon scattering process can be represented using a relaxation time of ω. In addition, the crystal vibrational spectrum has no dispersion and isotropy.

The phonon scattering mechanisms in (i) normal three-phonon processes (N processes), (ii) boundary scattering, (iii) point impurities, and (iv) Umklapp processes (U processes) were taken into account by assuming identical contributions from longitudinal and transverse phonons to the heat conduction. Based on such assumptions and approximations, Callaway writes an integral expression for k as follows:

$$k = c_p T^3 \int_0^{\theta_D/T} \frac{x^4 e^x \left(e^x - 1\right)^{-2} dx}{v_b / L + Ix^4T^4 + \left(\varsigma_1 + \varsigma_2\right)x^2T^5} + k_2 \tag{2.13}$$

where T represents temperature, $c_p = \left(\frac{k}{2\pi^2 v_s}\right)\left(\frac{k}{\hbar}\right)^3$, v_s stands for an average phonon velocity, $I = A\left(\frac{k}{\hbar}\right)^4$ – isotope scattering contribution, A refers to a coefficient, subscript $i = 1$ and 2 stand for the N and U processes, respectively. $\varsigma_i = \varsigma_i\left(\frac{k}{\hbar}\right)^2$ refers to three phonons scattering, $\hbar$ stands for the reduced Planck constant, θ_D stands for Debye temperature, k_2 refers to a correction term, and $x = \frac{\hbar\omega}{k_B T}$.

Given the analysis above, Allen [11] revised the Callaway model to make improvements to it. He analyzed the Deby-type phonon model and the relative roles of the N and U processes. He also argued that the Callaway model fails to perfectly estimate the suppression of the N processes in relaxing thermal current. Thereby, he introduced a new result for k by elongating the relaxation time. Given the Debye approximation, the Callaway model is as follows:

$$k = k_{\mathrm{RTA}}\left(1+\frac{\overline{\tau_{\mathrm{C}}(\omega,T)/\tau_{\mathrm{N}}(\omega,T)}}{\overline{\tau_{\mathrm{C}}(\omega,T)/\tau_{\mathrm{U}}(\omega,T)}}\right) \tag{2.14}$$

where k_{RTA} represents the k corresponding to the relaxation-time approximation (RTA), where τ stands for the relaxation time.

Holland [12] introduced an analytical model of crystal k, which was not the same as the Callaway model. The model covers the effect of longitudinal and transverse phonons on conducting heat so a new theoretical model is proposed for k with two parts:

$$k = k_{\mathrm{T}} + k_{\mathrm{L}} \tag{2.15}$$

Where,

$$k_{\mathrm{T}} = \frac{2}{3}\int_0^{\theta_{\mathrm{T}}/T} \frac{c_{\mathrm{pT}}\ T^3x^4e^x\left(e^x-1\right)^{-2}dx}{\tau_{\mathrm{T}}^{-1}} \tag{2.15a}$$

$$k_{\mathrm{L}} = \frac{1}{3}\int_0^{\theta_{\mathrm{L}}/T} \frac{c_{\mathrm{pT}}\ T^3x^4e^x\left(e^x-1\right)^{-2}dx}{\tau_{\mathrm{L}}^{-1}} \tag{2.15b}$$

The subscripts T represents transverse and L represents longitudinal phonons.

Therefore, this method was adopted to achieve a high level of fit to the data on silicon from 1.7 to 1300°K and on germanium from 1.7 to 1000°K. The method was also utilized to fit the data on isotropic pure germanium. Our comparison of the analysis with Callaway indicates that the same results are achieved in the impurity scattering and boundary scattering regions. The approximations used in different analyses are also discussed. In addition, a more thorough explanation of the Umklapp scattering relaxation time, which is valid for materials that are featured with a highly dispersed transverse acoustic phonon spectrum is given in an appendix. The issue of the validity of adding inverse relaxation time and the coupling due to normal three-phonon processes is given in another appendix. Moreover, the results of the thermal conductivity of a suspended single-layer graphene are reported. The temperatures at room for thermal conductivity are in the range ~(4.84 ± 0.44) × 103 to (5.30 ± 0.48) × 103 W/m K, which were based on a single-layer graphene from the dependence of the Raman G peak frequency on the excitation laser power and measures G peak temperature coefficient independently. The notably high thermal conductivity indicates that graphene can have better heat conduction compared to carbon nanotubes. The considerably high thermal conductivity of graphene is good for the proposed electronic applications and makes it a good choice for thermal management.

Slack and Galginaitis [13] was the first study to examine the scattering of Born-von Karman dispersion in 1964. According to the theory, Dames and Chen introduced the Bornvon Karman-Slack (BvKS) model [9]. In this model, the frequency (ω) is given as:

$$\omega = \omega_0 \sin\left(\frac{\pi q}{2q_0}\right) \tag{2.16}$$

where q_0 stands for a wave vector parameter, which is a constant through employing the number density of original unit cells and S_s, and ω_0 stands for the frequency parameter. The BvKS dispersion theory is featured with lowered wavevector group velocity near the first Brillouin zone boundary. The results of the BvKS model are highly in agreement with the results of the numerical assessments [4]. This means that the BvKS model outperforms the Holland model, the modified Callaway model, and the gray model in terms of accuracy and phonon boundary scattering.

Given the theoretical concepts mentioned above, Yang and Dames [4] determined the phonon contributed k of silicon using six models including the three top common analytical models (Callaway model [10], Holland model [12], and Bornvon Karman-Slack (ByKS) model [9]), MD simulation, first-principle (1stP) calculation, and simple gray model. In the case of bulk Si, the results of the calculation of the three analytical models have a good consistency with the experimental data (relative error > 10%). In the case of a single silicon nanowire that is 115 nm in diameter, the results by ByKS model have a better fit with experimental data compared to other models. In addition, given the diffusion boundary scattering assumption, the normalized k of silicon nanowires and the normalized in-plane k of silicon thin films were determined on the basis of the integral transformation mentioned above.

A theoretical and experimental study by Cuffe et al. [14] was conducted on the contributions of reconstructing the MFP distribution to k for Si nanofilms. To check the accuracy of the tests, the authors used the phonon BTE to determine k for Si nanofilms by assuming an estimated isotropic dispersion:

$$k_{\text{mem}} = \sum_{S} \int_{0}^{\infty} \frac{1}{3} c_{\omega} v S\left(\frac{\Lambda_{\text{bu}}}{d}\right) \Lambda_{\text{bu}} d\omega \tag{2.17}$$

where d stands for the thickness of the nanofilm and S refers to a function that shows the contribution to the heat flux of a phonon with a particular MFP and it is as follows:

$$S\left(\frac{\Lambda_{\text{bu}}}{d}\right) = 1 - \frac{3}{8}\frac{\Lambda_{\text{bu}}}{d} + \frac{3}{2}\frac{\Lambda_{\text{bu}}}{d} \int_{1}^{\infty} \left(\frac{1}{t^3} - \frac{1}{t^5}\right) e^{-\frac{d}{\Lambda_{\text{bu}}}t} dt \tag{2.18}$$

and the c_ω is defined as:

$$c_{\omega} = \frac{\partial}{\partial T} D(\omega) g_{\text{BE}} \hbar\omega = \frac{q(\omega)^2}{2\pi^2} \frac{dq(\omega)}{d\omega} \hbar\omega \frac{\partial g_{\text{BE}}}{\partial T} \tag{2.19}$$

Where $D(\omega)$ refers to the density of states, g_{BE} stands for the Bose-Einstein distribution, and $q(\omega)$ stands for the polarization-dependent phonon wave vector. Given the normalized accumulated MFP distribution in Eq. (2.9), it is possible to determine the relationship between k_{mem} and k_{bu}, as follows:

$$\frac{k_{\text{mem}}}{k_{\text{bu}}} = \int_{0}^{\infty} k_{\text{acc}}(\Lambda_{\text{bu}}) \frac{dS\left(\frac{\Lambda_{\text{bu}}}{d}\right)}{d\Lambda_{\text{bu}}} d\Lambda_{\text{bu}} \tag{2.20}$$

Still, it is not easy to invert Eq. (2.2) to find k_{acc} as a function of d based on experimental measurements (k_{mem}). It is possible to reconstruct the k_{acc} by introducing a few limitations on it. By employing an algorithm through a convex optimization procedure, we can obtain the results of reconstruction, which is consistent with the distribution obtained from the first principles and the MD calculations.

In the case of amorphous nanomaterials, the affecting factor in k is disorder degrees (ε). It is clear that the most common materials like silica, silicon, and graphene have a variety of different stable structures such as single-crystal polycrystalline, and amorphous. The main difference between these structures is in ε, which is a key structural parameter with a notable indicator of the thermal transport properties of materials. In particular, an ultra-low ε represents a highly orderly internal structure of the material. This structure is taken as a single-crystal structure with different outstanding properties such as extensively high k (5300 W/m:K) [15]. The high value of ε is indicative of a disordered internal structure, which is called an amorphous structure. The value of k is close to an insulator which is normally less than W/m:K [16]. In the case of amorphous nanomaterials with high ε, the velocity of phonon and polarization direction within the structure are not given and it is not possible to use the pertinent concept of a phonon to examine thermophysical properties. With ε between 0 and 1, the conductivity of heat mostly depends on phonons and diffusions. The latter has a role in k through harmonic coupling with other modes because of the disorder. Thus, with nanostructures that have a specific level of disorder, the total vibrational k is comprised of two parts:

$$k = k_P + k_D \tag{2.21}$$

where k_P stands for the k contributed by phonons from propagating modes, and k_D stands for the k contributed by diffusions from non-propagating modes given by the Allen-Feldman (AF) theory [17]. The k_P can be obtained from the c_p in different modes and vibrational density of states (VDOS) of phonons [16]. For deriving Eq. (2.22), it is assumed that the system is isotropic with a single polarization, so that mode properties are related only to ω:

$$k_P = \frac{1}{V_s}\int_0^{\omega_{cut}} D(\omega)c_p(\omega)\alpha(\omega)d\omega \tag{2.22}$$

where V_s refers to the system volume, and ω_{cut} stands for the phonon cutoff frequency. With the frequency below ω_{cut}, phonons affect transport and contribute at higher frequencies diffusions. $D(\omega)$, $c_p(\omega)$ and $\alpha(\omega) = s_s^2\tau(\omega)/3$ stand for the VDOS, mode-specific heat capacity, and mode thermal diffusivity of the vibration mode related to ω respectively. $\tau(\omega)$ stands for the phonon lifetime and s_s is the sound speed. In the case of phonons, the DOS of the material can be explained using the Debye model:

$$D(\omega) = \frac{3V\omega^2}{2\pi^2 {s_s}^3} \tag{2.23}$$

Therefore, it is possible to decide about the ω_{cut} and s_s of the material depending on the fact that the VDOS of the material meets the Debye model requirement. The quantum expression for the $c_p(\omega)$ is as follows:

$$c_p(\omega) = k_B \left[\frac{\hbar\omega / 2k_B T}{\sin h(\hbar\omega / 2k_B T)} \right]^2 \tag{2.24}$$

where k_B refers to the Boltzmann constant. Like the contribution that phonons make to k, it is possible to derive the contribution of diffusions to k from the c in different modes and α [17,18]:

$$k_D = \frac{1}{V} \sum_{\omega > \omega_{cut}} c_p(\omega_i) \alpha(\omega_i) \tag{2.25}$$

where ω_i stands for the vibration frequency of the ith localized mode, $c_p(\omega_i)$ refers to the diffusion isobaric specific heat capacity, and $\alpha(\omega_i)$ refers to the diffusion diffusivity of the *i*th localized mode. Based on AF theory, the thermal diffusivity of the localized mode is derived as follows [17]:

$$\alpha(\omega_i) = \frac{\pi V^2}{\hbar^2 \omega_i^2} \sum_{j \neq i} \left| M_{ij} \right|^2 \delta(\omega_i - \omega_j) \tag{2.26}$$

where M_{ij} refers to the thermal coupling between *i*th modes and *j*th modes, which is obtainable through the modal frequencies and the spatial overlap of mode vectors.

2.1.2 Experimental Measurement Techniques

2.1.2.1 T-Type Probe Method (for Nanotubes, Nanowires)

To measure the thermal conductivity of one carbon fiber with a diverse production process, the steady-state short-hot-wire method is used. This method relies on the transfer phenomena of a pin fin connected to a short hot wire. The short hot wire is fed with a fixed direct current to produce a uniform heat flux. In addition, both ends are attached to lead wires and kept at the initial temperature. In addition, the test fiber is connected as a pin fin to the center of the hot wire at one end and the other end is attached to a heat sink. It is assumed that 1D steady-state heat conduction is along the hot wire and test fiber. In addition, the basic equations are solved analytically. Given the solutions, the relationships of the average temperature increase of the hot wire, the heat generation rate, the flux from the hot wire to the fiber, and the temperature at the end connected to the fiber are obtained accurately. Given these relations, it is easy to estimate the thermal conductivity of a single carbon fiber when the average temperature increases and the heat generation rate of the hot wire is determined for one system. In addition, the electrical conductivity of one carbon fiber is determined with the identical condition that is used with thermal conductivity using a four-point contact method. Using crystal microstructure, the relationship between thermal and electrical conductivity is discussed in more detail in [19].

One efficient technique to measure the k of individual micro/nanoscale fibrous materials is the T-type probe technique. It can be used to measure nanotubes and nanowires. This technique was first introduced by Zhang et al. [19] using the heat conduction principle of the pin fin when connected to a hot wire. A platinum wire with electrical and thermal properties is utilized as the heating wire so that the two ends of the wire are attached to the heat sink. By inducing a direct current to the two ends, the uniform heat flux forms a parabolic temperature distribution in the metal wire. By connecting one end of the individual fibrous material on the heat sink and leaving the other end overlapped at the middle of the heating wire, heat flows from the heating wire into the fibrous material. Consequently, a parabolic temperature distribution forms inside the test structure. Based on the temperature distribution evolution, the axial k of the fibrous material is obtained [20]. It is notable that the uncertainty of measurement in this method is below 7% in vacuum conditions. A major issue that affects the measurement accuracy is the thermal contact resistance at the point that fibrous material and the hot wire are connected. By measuring the fibrous samples with a set of different lengths while keeping the same contact condition, Wang et al. [20] separated the axial thermal conductivity of the fibrous sample from the thermal contact resistance.

Along with measuring one fibrous sample with a micrometer diameter, Fujii et al. [21] measured the k using the T-type probe method for a single CNT with a diameter of 10 nm. A unique sample holder containing suspended platinum film nanosensors was built using electron beam etching on a multilayer film. Using a scanning electron microscope (SEM) a single CNT was grown using the arc discharge evaporation method and was suspended on the electrode of the nanosensors using the two ends fixed using locally focused electron beam irradiation. To determine the value of k of the CNT, the average temperature rise and the heat generation rate of the nanosensors were measured. The results showed that the k of the single CNT declined along with an increase in diameter. In particular, the value of k increased above 2000 W/m:k with a single CNT with a diameter of 9.8 nm (CNT I). In the case of a single CNT, which was grown like CNT I, with a diameter of 16.1 nm (CNTII), the value of k declined to 1600 W/m:K. In the case of one CNT that was grown like CNT I and II, with a diameter of 28.2 nm (CNT III), the value of k declined to 500 W/m:K. In addition, the k does not decrease with an increase in temperature when the temperature hits 320 k. To explain this, the phonon scattering is the main mechanism of thermal transport.

2.1.2.2 3ω Method (for Nanotubes, Nanowires, Nanofilms)

Performance of several newly designed devices like phase change memory devices, thermal management of high power and nanoscale electronic and optoelectronic devices, heat-assisted magnetic recording, and delivery vehicles for molecules in medical applications, and next-gen long-distance electricity transport wires depend on nanoscale electrical and thermal transport. Carbon nanotubes (CNTs) have drawn a great deal of attention for these applications given their outstanding thermal and electrical properties. Thanks to the notable long mean free path (MFP) and ballistic transport nature, theoretically speaking, the electrical and thermal conductivity of each CNT is of great importance.

The old 3ω method uses a micrometer-sized metal strip with four pads that are placed on the surface of the material to be examined. The strip functions as a heater and a heat sensor. Based on a sinusoidal alternating current with an angular frequency of 1ω, the strip creates 2ω temperature fluctuation. With a small range of temperature, the resistance of the strip changes with 2ω frequency. The obtained third harmonic voltage ($U_{3\omega}$) has information about the thermophysical specifications of the sample. This method was first introduced by Cahill [22,23] to measure k for different materials at micro and nanoscale ranges [17–28].

The value of k of an ultrathin nanocrystalline diamond film (16 μm) was first measured by Ahmed et al. [24] equal to (~26 W/m·K) using a frequency range up to 1 MHz. Given the uniquely high thermal conductivity of polycrystalline diamond (>2000 W/m k), it is a highly attractive material to optimize the thermal management of devices of high power. Here, the thermal conductivity of a diamond sample capturing grain size evolution was examined from nucleation toward the growth surface. To this end, an optimized 3ω technique was used. The results indicated a decrease in thermal conductivity along with a decrease in grain size, which is consistent with the theory. These findings indicate that the minimum film thickness and polishing thickness from nucleation are required to reach a single-crystal diamond performance to produce an optimal polycrystalline diamond for applications of spreading heat [25]. Along with one-layer film, Olson et al. [25] developed the 3ω method for structures with several layers. They used a non-approximate analytical solution to find the k of multilayer film structures as thermal impedance. In addition, the α value of the material is obtainable at the same time. Qiu et al. [26] shear-pressed and polished the top of a vertically aligned CNT array VACNT array to a horizontally aligned CNT (HACNT) film to examine microfilms with anisotropic k values. Afterward, they deposited a group of sensors of different widths and directions. By analyzing the different temperature changes compared to frequency curves, the anisotropic k value was obtained along x (along the CNT axes), y (perpendicular to CNT axes), and z (out of the plane) equal to 127, 42, and 4 W/m:K, respectively.

Afterward, Qiu et al. [29] used an identical test structure to measure the k of one conductive carbon fiber (CF) and CNT fiber. Both cases showed accurately the axial thermal transport performance of each fibrous sample. Because this method uses the fiber as the heater and thermometer, it depends on the highly conductive properties of the materials [30,31] mentioned earlier. In practice, CNTs must be assembled into macroscopic architectures. Still, the bulk electrical and thermal conductivities of CNT ensembles were notably less than that of individual values. A general example of this is CNT spun fiber of which axial electrical conductivity has been supported by experiments in the 104–105 S/m range and thermal conductivity is about 3–380 W/m K. These reductions of on the order of magnitude in electrical and thermal conductivities specifically rate efficient applications of CNTs. Studies have also shown that in the case of light weighted high conductive CNT fiber designed to fulfill the European Space Agency's stringent space thermal control requirement throughout their mission, accumulation of heat may increase the fiber temperature above 500°C. This can result in self-ignition of which the risk is more than blow out of conventional metal wires. Thus, it is essential to develop CNT fibers with better electrical and thermal transport for future uses in pertinent energy conservation and conversion, IT, and thermal management systems. To deal with the problem of measuring non-conductive fiber, Qiu et al. [32] changed the 3ω method by depositing a platinum conductive layer on the surface of the non-conductive fiber to pass electrical current through the fiber and achieving k measurement of each porous polyimide fiber.

Another major challenge in developing the path of the 3ω method is the free-standing sensor-based 3ω technique. This technique makes sure that material with an uneven surface is measurable even in operation [33]. A key feature of this method is that it can realize simultaneous measurement of k and contact thermal resistance between the sample-sensor interfaces [34]. In addition, the nanoscale interfacial thermal transport characteristics can be measured in nanoparticle-decorate graphene layers [35], which is a top issue in micro/nanoscale thermal transport research.

2.1.2.3 3ω-T Type Method (for Nanotubes, Nanowires)

In the case of the majority of materials and thermoelectric materials in particular, we need to extract multiple physical parameters (k, α, c_p, σ, temperature coefficient of resistance, Seebeck coefficient) at the same time to examine their major specifications. At micrometer and nanometer scales, two major issues have to be dealt with; one is the major difference between each sample. Therefore, it is more reasonable to measure all the properties of one sample. The other issue is the multiple transfer of samples in each measurement for diverse physical specifications that create damage to the morphologies of the microscale samples. This also affects intrinsic physical specifications. A reliable technique to examine the thermoelectric performance of 1D nanostructures is the T-type technique. The thermoelectric specifications such as the Seebeck coefficient, thermal conductivity, and electrical conductivity of an individual freestanding single-crystal Bi_2S_3 nanowire were first examined by using the T-type technique. The obtained figure is notably lower than the reported values of nanostructure bulk Bi_2S_3 samples. In addition, the Seebeck coefficient in the mechanism is almost zero in the temperature range 300–420 k and changes sign at 320 k.

Therefore, to overcome these problems, Ma et al. [36] introduced a comprehensive 3ω-T type method for in-situ measurement of several parameters for each Bi_2S_3 nanowire. In one study, a simple one-step hydrothermal method was introduced for the large-scale synthesis of ultra-long single-crystalline Bi_2S_3 nanowires. The study managed to characterize the nanowires comprehensively. The diameter of the nanowires was about 60 nm and in length they ranged from tens to microns to millimeters. The mechanism of growth was examined using high-resolution transmission electron microscopy. Based on optical examinations, the Bi_2S_3 nanowires were narrow-based semiconductors with a band gap of $Eg \approx 1.33$ eV. Electrical transport measurements on each nanowire yielded a resistivity of around 1.2 Ω cm and an emission current of 3.5 μA at a bias field of 35 V/μm. This current is indicative of a current density of around 105 A/cm^2, so the Bi_2S_3 nanowire is a candidate for use in field-emission electronic devices [36].

It is possible to use the standard DC T-type technique to measure the thermal contact resistance and thermal impedance of nanowires. In addition, it can be used to measure the thermal diffusivity (b) of the nanowire, and the value of k is measured using the $k = b^2/\rho c$ formula. Compared to the 3ω method, the 3ω-T type technique can measure the thermophysical specifications of metallic and non-metallic nanowires regardless of conductive properties. In comparison with the standard DC T-type method, the thermal penetrability depth of the hot wire in 3ω T type method is adjustable using frequency sinusoidal alternating current. Thus, it is possible to separate the thermal resistance of the test wire from the function [37].

In the case of the 3ω-T method, the relative uncertainty of α is 6% and 18% for microscale and nanoscale samples respectively.

2.1.2.4 H-Type Method (for Nanofilms)

One of the key phenomena for active heat flow control is thermal rectification. Asymmetric nanostructures like nanowires and thin films are expected to have a significant thermal rectification. Graphene is a one-atom-thick membrane, and it has drawn a great deal of attention for achieving thermal rectification as illustrated by several MD simulations. One of the main standard techniques to measure the k of nanofilms is the H-type technique. The principle of measurement is to place the nanofilm between two metallic sensors and by regulating the heating power of the sensor, control the temperature of the metallic sensor. The temperature difference between the sensor and the nanofilm contains the k of the nanofilm. In general, the temperature difference between the film and the metal sensor means a lower k of the film. Wang et al. [38] found thermal rectification in different asymmetric monolayer graphene nanostructures using the H-type method. They performed the experiment with five nanopores on one side, graphene with nanoparticles deposited on one side, and graphene with a tapered width. The obtained thermal rectification factor η in the first structure was about 26% and this figure in the other two structures was about 10%. According to the theoretical and MD analysis, there are two different mechanisms for different thermal ratification behavior in diverse asymmetric graphene nanostructures.

An asymmetric single-layer defect-engineered graphene rectifier was built using electron beam lithography and ion etching. To measure the k of the thermal rectifiers, the temperature change of the graphene sandwich between the metal sensors was used. In addition, a 2D beat conduction model was formed for the H-type method. In addition, the COMSOL multiphysics software was used for thermodynamic process analysis. The value of k for the defect-engineered graphene thermal rectifier was notably declined along with the average thermal rectification factor ξ_{ave} of 26%. In addition, the asymmetric structure of a one-layer graphene thermal rectifier demonstrated a higher k value and less thermal rectification factor compared to the defect-engineered graphene thermal rectifier.

2.1.2.5 Raman Spectroscopy Method (for Nanofilms)

While the ideal 2-D graphene is expected to have a notably high k, the results of experiments do not support this because conventional thermal measurement methods like the laser flash method and the thermal bridge method cannot measure the thermal conductivity of graphene. Because of the strong dependence of temperature on Roman G peak frequency, Balandin et al. [15] recommended measuring the k for suspended single-layer graphene (SLG) using Raman spectroscopy. As shown by the results, the k of SLG was 4800–5300 W/m:k at ambient temperature (AT), which was a little higher than single-wall CNT (SWCNT) [39,40]. In addition, Ghosh et al. [41] found a notably high k of a set of graphene flakes suspended across trenches in a Si/SiO_2 wafer using the Raman spectroscopy technique. Rendering k by the graphene flakes was in 3080–5150 W/m:K, while the phonon MFP was around 775 nm. This study also examined the mechanism of high thermal transport. Additionally, the authors showed that graphene multilayer film can be used for micro/nano thermal management as they have a high chance of appearance given the good contact of graphene to heat sinks.

The temperature-dependent thermal conductivity $\kappa_{(T)}$ of crystalline ropes of one-walled carbon nanotubes was measured from 350 K to 8 K. There was a gentle decline of $\kappa_{(T)}$ along with temperature so there was a linear temperature decrease under 30 K. Compared with electrical conductivity experiments, it was found that the AT thermal conductivity of one nanotube can be comparable to that of diamond or in-plane graphite. In addition, the ratio of thermal to electrical conductance shows that thermal conductivity is controlled by phonons at different temperatures. Under 30 K, the linear temperature dependence and estimated magnitude of $\kappa_{(T)}$ indicate that there is an energy-independent phonon MFP of ~0.5–1.5 μm [39].

To have a reliable measurement of graphene thermal conductivity, the Raman spectroscopy method has received a great deal of attention to realize non-contact, non-destructive, and fast thermal property measurement. By measuring monolayer graphene with a large area, Faugeras et al. [42] obtained the value of k equal to 5000 W/m:K. The authors argued that this notable deviation was mostly because of diverse assumptions that are made for the optical absorbance efficiency of graphene samples. Taking into account the effect of the environment, Ruoff et al. [43] placed the SLG in vacuum, air, and CO_2 gas and compared the differences in k of SLG using Raman spectroscopy. The notable difference in k between environments means that the heat loss to the surrounding gas must be taken into account to prevent measurable errors in k. To remove the effect of substrate, Ruoff et al. [44] grew the graphene monolayer over a hole substrate and reported k around 630 W/m:K at 669 K for the suspended graphene with no substrate. Molybdenum disulfide (MoS_2) was added by Yan et al. [45] and they measured different types of monolayer MoS_2 including suspended, Si_3N_4-supported and sapphire-supported with k equal to 34.5 ± 4 W/m K, which is in good agreement with the results of simulations. By determining the characteristics of single-layer MoS_2, valuable thermal data was obtained for optical and electronic uses. Additionally, following Balandin's work, Maboudian et al. [46] used Raman thermography to measure single nanowire, which was a great expansion of material categories based on Raman measurements.

2.1.2.6 Time-Domain Thermoreflectance

A reliable technique to measure thermal conductivity is time-domain thermoreflectance (TDTR) as it is featured with high accuracy and wide applicability for different materials and micro/nanoscale materials in particular. Still, due to the long platform displacement, it needs a temporal delay. It is not easy to keep the size and position of the laser spot while it is focused on the sample. This is a limitation to its application and development. Cahill et al. [47] discovered the measurement error because of small spot sizes so that the unchanged position can be lowered once the ratio of the in-phase signal to the inverted signal is utilized as the fitting signal. Researchers have used this technique to measure the thermal conductance (G) at a solid interface. Cahill [48] tried to use TDTR for measuring layered structure. They reported an accurate heat transfer model for the metal interface, which was more complicated because of the extract effect of electrons. To remove the limitations of the type of materials, Cahill et al. [49] increased the sensitivity and accuracy through the ratio of in-phase signal to out-of-phase signal and replaced the in-phase signal. They managed to extend the G measurement to typical metal interfaces like Cu–Al interfaces. Cahill and Lyeo [50] also examined G of an intriguing an-harmonic scattering interface developed by the interface of low and high Debye temperature materials including a bismuth/hydrogen-terminated diamond interface. Surprisingly, they found that G was low and it exceeded the radiation limit significantly. These results showed an extra thermal transport channel by the three-phonon processes because of highly dissimilar lattice vibrations. Cahill's works on the TDTR method have given us a good starting point for using TDTR in micro and nanoscale thermophysical measurement and also new ideas for more advances in the TDTR method.

In addition, researchers have extensively worked on the TDTR technique. Anisotropic thermal properties are important fundamentally and practically; however, they still pose a challenge to characterize using conventional methods. Here, a novel methodology based on asymmetric beam time-domain thermoreflectance (AB-TDTR) was introduced for measuring 3D anisotropic thermal transport by expanding the conventional TDTR technique. With an elliptical laser beam and controlled elliptical ratio and spot size, it is possible to use experimental signals based on measuring thermal conductivity along the cross-plane or any specific in-plane directions. It is possible to derive an analytical solution for a multilayer system for AB-TDTR signal in response to the periodical pulse, elliptical laser beam, and heating geometry to extract the anisotropic thermal conductivity based on actual measurements. There are examples of experimental data for different materials with in-plane thermal conductivity ranging from 5 to 2000 W/m K such as isotropic materials (boron phosphide, silicon, and boron nitride), transversely isotropic materials (graphite, quartz, and sapphire), and transversely anisotropic materials (black phosphorus). In addition, an accurate sensitivity analysis was carried out to perform the optimal setting of experimental configuration for various materials. The obtained AB-TDTR metrology gives us a novel method to measure anisotropic thermal phenomena for rational materials design and thermal applications [51].

The Asymmeteric-Beam Time-Domain Thermoreflectance (AB-TDTR) technique was introduced by Hu et al. [51] on the basis of the controllable elliptical ratio and spot size. Using the significantly various sensitivity of out-of-plane or in-plane signals, k can be obtained in the intended direction. The technology can be employed for measuring k for isotropic (boron phosphide boron nitride, and silicon), transversely isotropic (sapphire, quartz, and graphite), and anisotropic materials (black phosphorus). An accurate k measurement method (<10%) was introduced by Koh et al. [52] for thin films (400 nm) using the dual-frequency TDTR. Sun and Koh [53] showed that it is possible to measure rough surfaces once the artificial signals are removed by avoiding the pump beam leaking into the photodetector. Cahill et al. [54] measured k in diverse directions of black phosphorus utilizing the low frequency for improving the depth of thermal penetration of the laser for out-of-plane measurement and the beam-offset method for in-plane measurement. A novel TDTR technique was reported by Yang et al. [55] based on a changing spot site approach which

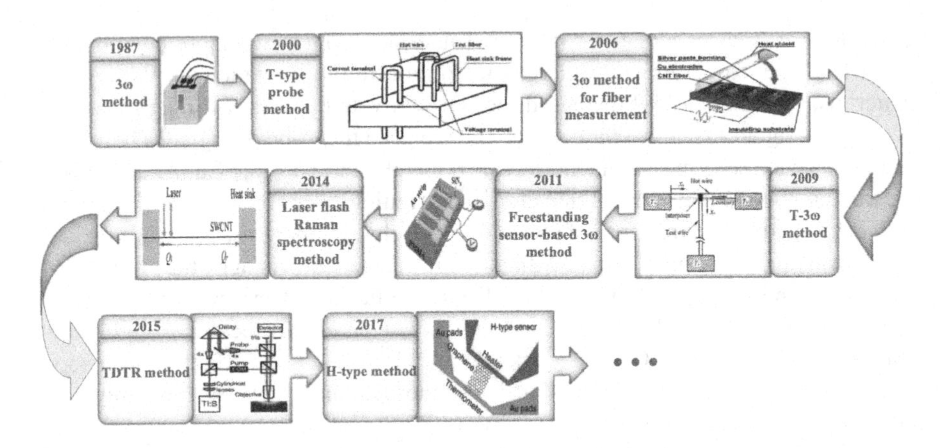

FIGURE 2.3 Schematic diagrams of measurement techniques.

solves the measuring issues in strongly anisotropic materials. This technique has been tested on different materials such as molybdenum disulfide, rutile titania, and highly ordered pyrolytic graphite. A schematic view of the *k* measurement methods for nanostructured materials is illustrated in Figure 2.3 [19,30,34,38,55–57].

2.1.3 Experimental Studies

In the section above, different techniques for measurement of *k* of solid nanostructured materials were introduced such as measurement principles and selecting nanomaterials. Because specific nanomaterials like CNT, graphene, fullerene, nanofilms, and nanowires have a specific crystal structure, grain size and boundary, defect concentration, interfacial interaction, and doping degree along with temperature; these features have a major role in the value of *k* [58]. These factors and their effect on *k* are further discussed in the following.

2.1.3.1 Effect of Grain Size

Clearly, the effect of grain size (l) on the *k* of polycrystalline solid materials is undeniable. Researchers in 1980 initiated studies on grain size and its effect on thermophysical specifications of polycrystalline materials. One of these notable researchers from Australia is Savvides [59], who showed that the *k* of polycrystalline $Ge_{30}Si_{70}$ is negatively and linearly correlated with $L^{-1/2}$ where L exceeds 10 μm, this is in good agreement with theoretical predictions. In addition, Zhang et al. [60] showed by experiments that the *k* of polycrystalline platinum nanofilms is mostly lowered by grain boundary scattering rather than surface scattering. According to Dong et al. [61], when L is big enough in nanocrystal materials, the internal thermal resistance becomes closer to the single crystal and the thermal resistance of crystals converges to zero. With *k* for the nanocrystal material, the size effect is greater than the grain boundary effect. Wang et al. [7] showed the same finding. Similarly, Soyez et al. [62] concluded that an increase in the size of nanoparticles was a way to increase *k*.

It is notable that these studies were mainly on nanofilm materials, which represent nanocrystalline materials. It is notable that Qiu et al. [63] showed that the effect of grain size on k was also visible in CFs following graphitization treatment. This report is based on the fact that the fiber size is not at the nanoscale, while the inner structure is made of nanocrystals. They were the first group of researchers who showed macroscopic materials also render size effects on K. In short, the reciprocal of in-plane coherence length (L_a^{-1}) can illustrate that the number of grain boundaries is linearly related to the k ($k \propto -L_a^{-1}$). On the other hand, some studies have shown that k can be controlled by defects rather than L. For instance, Watari et al. [64] argued that k of polycrystalline Si_3N_4 was dominated by internal defects like dislocation, while L apparently had no effect on k.

2.1.3.2 Effect of Grain Boundary

As noted earlier, boundaries of grain are key factors that affect k of nanocrystalline materials. Many nanostructures are known for man and require a great deal of research work on grain boundary scattering to achieve thermoelectric performance [7]. In general, nanocrystalline boundaries can scatter short-wave phonons effectively, while lowering the lattice k of the material in a low-temperature range. To control this feature, researchers have chosen nanocrystalline silicon rather than one-crystal silicon as an ideal thermoelectric material. The reason for this is that the k of nanocrystalline silicon is two or three orders of magnitude less than that of single-crystal silicon. This indicates a high figure of merit (ZT) value [65]. A great deal of attention has been attracted by nanostructured thermoelectric materials. Still, in spite of this trend, there is a limited amount of information is available on the thermoelectric properties of single-crystal for high doping concentrations at higher temperatures in particular. Here, the temperature ranges of study for the thermoelectric properties of heavily doped (1018–100/cm^3) n- and p-type single-crystal Si were from AT to 1000 k. To determine the value of ZT, the data of electrical conductivity, Seebeck coefficient, and thermal conductivity were used. The highest ZT values were 0.015 for n-type and 0.008 for p-type Si at AT. To have a clear picture of carrier and phonon transport and predict the thermoelectric properties of Si, a simple theoretical model was introduced using the BTE with approximate relaxation time [65].

The effect of grain boundaries on the k of nanodiamond films was examined by Mohr et al. [66]. They reported that the value of k depends on the cohesive energy of the grain boundaries. In addition, the structure, width, and chemical composition of the grain boundaries affect the k of the nanodiamond film. In the case of grain boundaries with a higher sp3/sp2 ratio, the value of k was higher. Additionally, the effect of the elastic modulus of grain boundary on k was higher than the elastic behavior. According to Anaya et al. [67], the in-plane k of polycrystalline diamond film was notably inhomogeneous, which is due to the limitation of the MFP of phonons at the interface of grains. These findings give us new insights into controlling heat transfer through controlling grain boundaries.

2.1.3.3 Effect of Surface Interactions

A top research question in thermal functional materials is the way of using surface interactions to control thermophysical properties. Qiu et al. [30] experimentally showed that decorating noble nanoparticles (Au, Pd, Pt) onto CNT surfaces can create vibrations of low frequency of C atoms at the CNT interface, which facilitates the transport of heat through the interface. Through decorating 10% (wt.%) Au nanoparticles between CNTs it is possible to boost the axial *k* of the assembled fibers by 70%. In the same way, it is possible to improve interfacial thermal transport by adding small halogen molecules into the CNT interface by introducing low-frequency phonons and creating extra thermal channels [31].

Another way to modify surface interaction is to functionalize the surface. Through reverse Non-Equilibrium Molecular Dynamics (NEMD), Li et al. [68] simulated the *k* of hybrid graphene nanoribbons. They found that graphene with gradient hydrogen arrangement had a notable thermal rectification with no dependence on the chirality and length. Qui et al. [30] managed to prepare CNT fibers using an inter-tube functionalized surface. They reported that the *k* of fibers was diametrically increased due to the higher interfacial heat transfer.

Given that non-bonded van der Waals force control surface interaction and this usually results in a weak energy transport, there are reports of novel interfacial structures with more powerful bonding and improved thermal transfer. The interfacial force from van der Waals to covalent bounding by the self-assembled monolayer was reported by Losego et al. [69]. They achieved about an 80% increase in the interfacial thermal conductance (G) between Au and z-cut quartz surfaces. Given the works mentioned, Ramanath et al. [70] managed to bond copper and silica surfaces strongly using the organic nanomolecular monolayer (NML). To control G different terminated-groups of NMLs were selected like CH3-NML and SH-NML with the highest value of *k* equal to 430 MW/m^2k. This new approach created a great deal of attention in researchers. Majumdar et al. [71] studied the phonon mismatch level between the bonded Self-assembled monolayer (SAM) and metal leads and uncovered the mismatch level and thermal transport relationship. They reported that the higher the phonon matching level, the stronger the interfacial thermal transport capacity. The samples were extended to polymers by Tang et al. [72]. They found a linear dependence between the interfacial thermal transfer and solubility between polymer and SAM. There has been a surge of attention on interfacial thermal conductance (ITC) due to its importance in determining the thermal performance of hybrid materials like polymer-based nanocomposites. Here, we systematically examined the ITC between sapphire and polystyrene (PS) through the time-domain thermoreflectance (TDTR) technique. SAMs Silane bases with different groups –NH2, –Cl, –SH, and –H were added to the sapphire/PS interface to examine the effect on ITC. The results showed that ITC was improved by a factor of 7 by functionalization of the sapphire surface with SAM, which ends with a chloride group (–Cl). As shown by the results, the improvement of thermal transport across the SAM functionalized interface is rooted in strong covalent bonding between sapphire and silane-based SAM and also the high level of compatibility between the SAM and PS. The results showed that ITC among other SAMs, had a linear dependence on solubility parameters, which can be a key factor in ITC compared with wettability and adhesion. As an intermediate layer, SAMs connect the sapphire and PS. This feature can be used for ceramic-polymer immiscible interfaced through functionalizing the ceramic surface using molecules that are miscible with the polymer materials. The results help us to design critical heat transfer materials like composite and nanofluids for managing heat [72].

Studies give us three major approaches to improve the transfer of heat and modulation including surface functionalization, interface decoration, and surface bonding strength. The results showed that k is notably controlled by the surface functionalization. Studies in this field have led to great findings about heat transfer mechanisms and the design of micro/nanoscale thermal management materials.

2.1.3.4 Effect of Doping

Because of limitations in the preparation techniques of CNT, short CNTs are preferred. Still, studies have shown that with short CNTs assembled into CNT fibers, the thermophysical specifications are notably less than those theoretically predicted. On the other hand, long CNTs demonstrate better thermophysical specifications in comparison with short CNTs. To examine the contradictions of the length and performance of CNTs, Bahabtu et al. [73] dealt with this problem. Using the high-throughput wet spinning method, ultra-long CNTs were prepared. Afterwards, the CNTs were added to CNT fibers featured excellent thermophysical properties. To examine the effect of doping on the thermal transport, they examined the effect of acid doping, annealing, and iodine doping on the *k* of CNT fibers. The average *k* of the acid-doped fibers was obtained at 380 W/m:K, and the *k* value of the iodine doped CNT fibers was 635 W/m:K. This notable increase in *k* is due to the higher inter-tube contact and notably lowered impurity and defect scattering caused by the annealing process of iodine doping.

One of the major issues in the electronic industry is heat removal so thermal conduction in low-dimension structures has shown interesting features. Carbon allotropes and their derivatives have received a great deal of attention given their ability for heat conduction. Carbon materials have a notably large range capability of heat transfer at AT including the lowest in amorphous carbon to the highest in graphene and carbon nanotubes. The main focus here is on graphene, carbon nanotubes, and nanostructure carbon materials with diverse levels of disorder. The unusual size dependence of heat conduction in 2D crystals and graphene in particular are under focus. The opportunities to use graphene and carbon materials for the thermal management of electronics are also discussed [74].

Another case of a defect affecting thermophysical properties is from ideal 2-D structured graphene. There are several reports that graphene with structural defects renders a notably lowered *k*, mostly because of increased phonon scattering [74]. Because of the fact that substitutional doping of atoms does not destroy the hexagonal lattice structure of graphene, the thermophysical properties of graphene have drawn a great deal of attention. As shown by studies, the *k* of doped graphene is strongly related to the mass and concentration of the dopants and the difference in mass between the adopted and the C atom (ΔM) in particular. Hu et al. [75] examined the influence of isotope doping (^{13}C, $\Delta M = 1$) on the *k* of graphene. As the findings showed, the isotope doping decreases phonon relaxation time and lowers the *k* dependence of isotope-doped graphene on temperature. It is notable that the normalized accumulative *k* of doped graphene is slightly different from pristine graphene thanks to the fact that the relationship time of high-frequency phonons is trivially lowered and that of low-frequency phonons does not change notably.

The effects on k of replacing the carbon atom in one-layered graphene with boron atoms were examined by Mortazavi et al. [76] (B, $\Delta M = 1.2$). As they showed with a concentration of doped boron atoms in SLG equal to 0.75%, the k of graphene along both chiral directions is lowered by more than 60%. Moreover, when the concentration of boron atoms doped in graphene increased to 1%, the effect of chirality on the k of graphene was nullified. Phonon scattering is considered the main key factor for decreasing the k for graphene and removing chirality from the structure. Goharshadi et al. [77] and Zhang et al. [78] argued that a very low nitrogen-doped concentration (0.5–1%) is strong enough to decrease the k of graphene. In addition, Zhang et al. [78] argued that other authors have only focused on the effect of nitrogen configurations (graphite-N) on the k of graphene. The point is that most nitrogen atoms are pyridinic-like structures (pyridinic-N) in practice). The thermophysical specifications of graphite-N and pyridinic-N-doped graphene were examined by them. They showed that along with the effect of nitrogen concentration on the k of graphene, the type of nitrogen atoms doped is important. Moreover, the inhibition of the k by pyridinc-N-doping is more important than the graphite-N doping for graphene.

One common feature of all these studies is that they all are theoretical work on substitutional dopants with a small ΔM. To examine the effect of heavy dopants (high ΔM) on the k of graphene, the same group of silicon dopant (Si, $\Delta M = 16$) was used by Lee et al. [79] as the substitutional dopants for C. Using a low pressure chemical vapor deposition (CVD) method, high quality Si-doped graphene was synthesized. The findings indicated the higher inhibitory effect of 1–2 orders of magnitude, of heavy dopants on the k of graphene. The top cause of this phenomenon is the immense mass difference between the C atom and Si atoms. In addition, the present Si atoms lower the phonon MFP.

2.1.3.5 Effect of Defects

The defect of nanomaterials is inevitable and appear as vacancies, impurity atoms, and rough edges. Nanomaterials with low dimensions like graphene and nanowires are ideal platforms for examining the influence of defects on k [75,80–87]. In the case of vacancy defect, Li et al. [80] built a perfect crystal β-SiC using a single point defect structure. The local density of states (LDOS) showed that the presence of the point defect created a change in the thermal transport mechanism from one dictated by phonon–phonon collisions to one dictated by phonon-defect scattering. Malekpour et al. [81] supported this conclusion by using electron beam irradiation to create defects. In addition, Kaskins [82] highlighted that one-point vacancy is a high-energy defect, which leads to the development of a double-coordination of three carbon atoms and thus destroys the sp2 bond of the surrounding crystal lattice. Based on the Green-Kubo method, qualitative analysis has indicated that with 0.42% vacancy defect in graphene, a notable decrease in k happens from 2903 W/m:K to 118.1 W/m:K [83]. Given that low-frequency phonons with a long MFP increase heat conduction and they are notably inhibited by defect scattering [75], the increase in the defect concentration lowers the dependence of k on temperature. In addition, it suppresses k as well. The most commonly observed vacancies include Stone-Thrower-Wales vacancy, double vacancy, and monovacancy. According to Feng et al. [84], graphene with monovacancy and long MFP renders the lowest k. This can be described based on the vacancies that increase the effect of long-wavelength phonon scattering on k.

With the same concentration, the total cross-sectional area of the monovacancy is bigger than the other two defects, which causes scattering of long-wavelength phonons, which results in the largest decrease in *k*.

Another structural defect is the impurity atom. A mixture of graphene nanoribbons (GNR) and isotopes was studied by Hu et al. [85] who showed that impurity atoms created the inter-atomic bond length and bond energy to be distorted, which destroyed the lattice symmetry and created phonon scattering. Chien et al. [86] used 25% functional coverage to GNRs so that it decreased *k* by 50%. The reason for this is that functionalized sp3 bonds are not bound and only the surface of GNR supports them. Moreover, the surface support and the rotation of the functional group interact with the low-energy phonons around and create phonon scattering.

In addition, Chen et al. [87] examined the role of defects in nanowire materials. With low temperature, the zero acoustic mode phonon has a major role in thermal transport. The results of computations of the scattering matrix method showed that with an increase in vacancy defect, the zero acoustic mode scattering increases, and the effect of the defect on the *k* decreases. The reason for this is extra phonon scattering because of the localization of phonons close to the defect. Thus, it is important to control defects in materials to regulate *k*.

2.1.3.6 Effect of Temperature

Needless to say, the effect of temperature on the k is important for studying the thermophysical specifications of nanoscale materials at high temperature in particular. There have been extensive studies on the effect of temperature on nanoscale materials. Studies have examined carbon nanomaterials, aerogels, and composite phase change materials (PCM). Results have demonstrated notably different temperature dependences and their potential for thermal management.

Given its unique thermal transport properties from several aspects, graphene is a hot nanomaterial. According to studies, the temperature dependence of *k* is indicative of a new microscopic thermal transport mechanism. According to Fang et al. [88], the value of *k* decreases with an increase in temperature from 300 k to 1000 k. To explain this, the fact that the higher temperatures decrease the phonon MFP is notable. Pop et al. [89] determined the trend of *k n* graphene at low temperatures and showed that low-frequency phonons control heat transfer at low temperatures. Additionally, the level of dispersion of the low-frequency phonons of zigzag graphene is higher than that of armchair graphene [90], which leads to a higher *k* of armchair graphene compared to zigzag graphene at 100 k. According to Renteria et al. [91], in-plane *k* of freestanding reduced graphene oxide (rGO) films that was grown from 3 W/m:K at AT to 61 W/m:K at 1000 k. In addition, out-of-plane *k* was 0.09 W/m:k at 1000 k. Based on this strong anisotropy of *k*, a new idea was developed for using graphene as a thermal management material for electronic devices.

One of the good thermal insulation materials is silica aerogel, which is relatively transparent to 3–8 μm near infrared radiation. Therefore, the thermal insulation properties of silica aerogel at high temperature are strongly affected by thermal radiation. Xie et al. [92] examined the relationship between *k* and temperature of silica aerogel. They showed that the value of *k* increased gradually with temperature at low temperatures and the rate of increase was higher at higher temperatures. This is because of radiation heat transfer at high temperatures.

One of the common uses of carbon nanomaterials is as core materials for composite PCMs to improve thermal transport. This is based on their high k and low ρ [93,94]. According to Fan et al. [95], the temperature dependence of k for paraffin-based composite PCM filled with different carbon nanomaterials (CNTs, carbon nanofibers, and graphene). They showed that k remained relatively constant when the temperature was less than the solid phase transition point. In addition, the value of k increased notably with the temperature at a temperature close to the solid-liquid phase change point (melting point). It is notable that the k of the composite phase graphene-filled materials is the weakest in temperature dependence.

Based on these discussions, a few meaningful conclusions can be made. In the case of nanocrystalline materials, the grain size effect is higher than the grain boundary effect on the k. Nanocrystalline boundaries can scatter short-wave phonons, and this lowers the lattice k of the material at low temperature. By decorating a noble metal (Au Pd, Pt) between the interfaces, it is possible to trigger low-frequency vibration of the C atoms. In addition, adding small molecules of halogen between the interfaces creates an extra hot channel that improves the transport of heat. Not all doping and defect concentrations have the same effect on the k of nanomaterials, which is due to an increase in the phonon scattering. In addition, the difference in the mass of doping atoms is a key factor for the sharp decline in k. The value of k of different nanomaterials depends differently on temperature. Figure 2.4 illustrates how different factors affect the k of nanomaterials [30,61,96]. Clearly, the effect of different factors on the crystal structure is not the same so it may increase or decrease k.

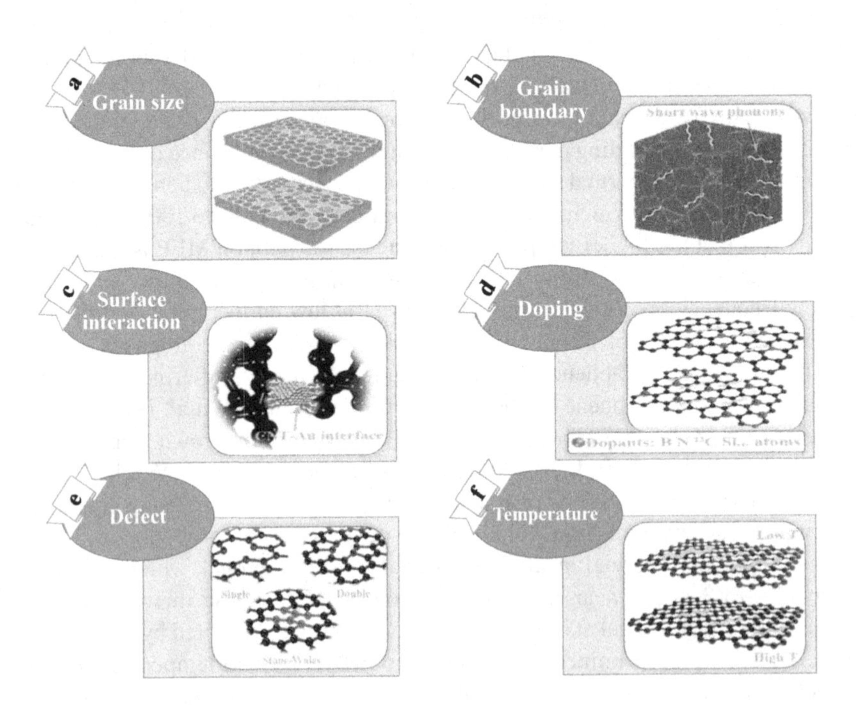

FIGURE 2.4 Schematic diagram of the effect of different factors on the k of nanomaterials.

2.1.4 MD Simulations

While there have been achievements in the characterization of k on the micro-scale and even nanoscale materials like porous polyimide fiber [32] and an individual CNT [97], it is not possible to explain the thermal transport mechanism of the heat carriers using the available experimental observations. Thereby, molecular-level simulation is a valuable tool to examine the thermal transport mechanism in simple materials [98]. Among the simulation methods are MD, BTE, thermal diffusion equation, and non-equilibrium Green's function. The MD approach uses Newton's law that describes the position and momentum space trajectories of particles in the system. It is based on the ergodic hypothesis and calculates a large number of particles. As to the computation domain, MD is the most common method for atomic-level simulation that gives a clear picture of the thermal behavior of materials.

Thanks to the nanoscale nature and excellent thermal transport properties, several research works have been conducted on carbon nanomaterials. To uncover the novel mechanisms in charge of exceptional thermal transport phenomena, several simulation methods have been used. According to Xu et al. [99], there is a notably boosted G between individual CNTs through altering the molecular structure with higher phonon mode coupling. They showed that the electron–phonon coupling effect of heterogeneous carbon-metal interfaces causes notably high G compared with different nanoscale interfaces. A new 3D carbon network was proposed by Farmer et al. [100], which is a mixture of graphene sheets and CNTs so that the CNTs were between layers of graphene as the pillars. Because of the scattering of phonon at the carbon-graphene junctions, the transfer of heat was controlled by the length of the CNT pillars.

Moreover, Li et al. [101] examined the thermal transport properties of silicon nanowires (SiNW) to check the low-dimensional phonon transport mechanisms. They showed that the phonon–phonon interaction is negligible if the SiNW length is below its MFP, while the phonon–phonon interaction controls the thermal transport when the SiNW length is too higher than the MFP. Their results also indicated that Fourier's law is not valid for low-dimensional thermal transport. Hu et al. [102] studied the relationship between the diameter and k. Through a combination of the normal scattering process and the Umklapp scattering processes of acoustic phonons, they showed the dependence of the nonmonotonic diameter of k for thin SiNW, so that the critical diameter was 2–3 nm. The thermal transport of SiNW was studied by Alibakhshi et al. [103] and they examined phonon transport by comparing Equilibrium Molecular Dynamics (EMD) and NEMD methods. They argued that the k depends on the length, cross-section width, and temperature. The improvement of k can be through increasing the length, increasing the cross-section, and lowering the temperature.

Along with the fast growth of nanoscale devices and nanocomposite materials, the interest in interfacial thermal transport has been growing. Studies on the thermal transport mechanisms at the materials' interface try to develop efficient micro/nanoscale structure for heat transfer in different applications. Stevens et al. [104] studied the influence of temperature and the disorder of solid interfaces. They showed a good agreement between the diffuse mismatch model (DMM) for high mismatch interfaces at a temperature about half of the melting point. Still, there is an undeniable inconsistency in highly matched interfaces which is due to the significant temperature dependence in the calculations of MD by Stevens et al. This dependence is not the case for the DMM, which only covers elastic scattering. In addition, defects create a considerable effect on highly matched interfaces; while their effect on highly mismatched interfaces is negligible. Therefore, as an effective approach to increase interfacial heat transfer, interface mixing was introduced, which is in agreement with the results in a doubling of G.

A method was introduced to make a notable improvement in the thermal transport across the interface between two solids with identical phonon spectra. If the two solids have identical crystal structure and lattice constant, it can be expected from the molecular dynamic modeling to achieve more than a 50% decrease in the thermal boundary resistance by adding a three-unit-cell-thick interlayer of which the Debye temperature is about the square root of the product of the Debye temperature of the two solids. In addition, if the two solids are largely different in terms of lattice constant, the interfacial atomic restructuring can have a key role in thermal transport. To reduce the thermal boundary resistance in an efficient way, the interlayer's lattice constant must be close to the average of the lattice constants of the two solids. In the case under study, more than a 60% decrease in the thermal boundary resistance is achievable when the Debye temperature of the interlayer is the same or a bit higher than the square root of the product of the Debye temperature of the two solids. Improvement in thermal transport is mostly achieved by more phonon states taking part in the boundary transport by adding an interlayer. Thermal management is a major issue in using nanostructures. In the case of micro/nanodevices like processors or semiconductor lasers, we need to transport heat from the core to an external heat sink as efficiently as possible. Given the higher surface-to-volume ratio in nanostructure components, thermal transport between interfaces is usually controlled by the overall thermal behavior. Decreasing thermal boundary resistance is one solution to improve the overall heat transport efficiency. Materials like Si. Ge. SiO_2 and GaAs are extensively utilized in semiconductor devices. They have different crystal structures and Debye temperature. It was illustrated that thermal boundary resistance can be notably lowered by inserting an interlayer between the two solids [105].

Thus, the thermal transport between phonon mismatch solids was studied by Tsai and Liang [105]. They showed that adding intercalation within a high mismatched solid-solid interface can improve G up to 60% at a specific Debye temperature. Merabia and Termentzidis [106] compared the simulation results and classified the model. They showed that the EMD results were in agreement with DMM, while NEMD results were consistent with AMM. Their results were a great contribution to studies on finding the simulation method. Additionally, the value of k of superlattices is another hotspot in nanomaterial thermophysical specifications. According to Keblinski et al. [107], interfacial geometry depends on k for superlattices obtained by EMD and NEMD simulations. The height and roughness control the thermal transport of superlattices when the MFP and superlattice period are of the same order of magnitude. The in-plane k, in particular, is controlled by the height to roughness ratio and superlattice period. On the other hand, the out-of-plane k constantly increases because of interfacial phonon scattering.

Along with the effects of phonon transport, the role of electrons in transporting heat cannot be neglected in specific metal-containing structures. The point is that the effect of electron–phonon (*e–p*) coupling is supported in [30]. The first in-depth study on e-p coupling for thermal transport at mental-nonmetal interfaces was in [108] by Majumdar and Reddy who used DMM named two-temperature model (TTM). They showed that G is considerably dependent on the *e–p* coupling factor and the lattice *k* of the metal. After writing this report, there has been a consensus that in studying nanoscale heat transfer MD, the results must be taken into account along with the TTM for obtaining G [109]. In addition, the *e–p* coupling was also taken into account for metal-dielectric interfaces [110], which differs from normal metal-nonmetal interfaces. It is widely accepted that the contribution of *e–p* coupling to interfacial thermal transport is very limited for metal-dielectric interfaces. The outcomes of this simulation about thc mechanisms of heat transport at the micro/nanoscale still are not consistent with the experimental studies. Along with the development of computation power, the similarity between simulations and experiments is increasing.

In conclusion, *k* of some nanomaterials like individual CNTs [40,111–115], CNT fibers [29–31], CNT arrays [26,34,116], CNT bundles [117–120], CNT films [121–123], graphene [15,38,44,124], CFs [19,20,63,125], BN [125–127], silicon nanomaterials [128–131], and polymers [33,125,132–135] was summarized and compared with the *k* of traditional materials like metal, glass, and water [125,136–138]. Through this, the superiority of nanomaterials for the transfer of heat and thermal insulation was demonstrated. Clearly, nanomaterials demonstrate outstanding thermophysical properties in comparison with mainstream materials. This superiority is not limited to *k* and covers thermal insulation as well (Figure 2.5).

2.2 THERMAL DIFFUSIVITY

Clearly, *k* indicates the ability of a material to transfer heat, and α stands for the rate of heat transfer in the material. That is, α is taken as the rate of diffusion from the hot end to the cold end of the material. It is independent of the energy used in heat conduction so it is a transient value. Given the specification relationship between *k* and α in some cases, the value of *k* of the material is determined by measuring α. Therefore, it is important to examine α of solid nanostructure materials in theory and practice. The following sections focus on theoretical and experimental methods of α for solid nanostructure materials. To measure α, laser flash Raman spectroscopy method and transient electrothermal (TET) techniques are used. Figure 2.6 illustrates a high-level overview of studying α of solid nanomaterials.

2.2.1 THEORY

The relation between *k*, ρ, and c_p is considered as:

$$\alpha = k / \rho \cdot c_p \tag{2.27}$$

With an increase in α, the capability of the medium to balance internal temperature (o) increases. Therefore, α indicates the ability of the medium for energy propagation.

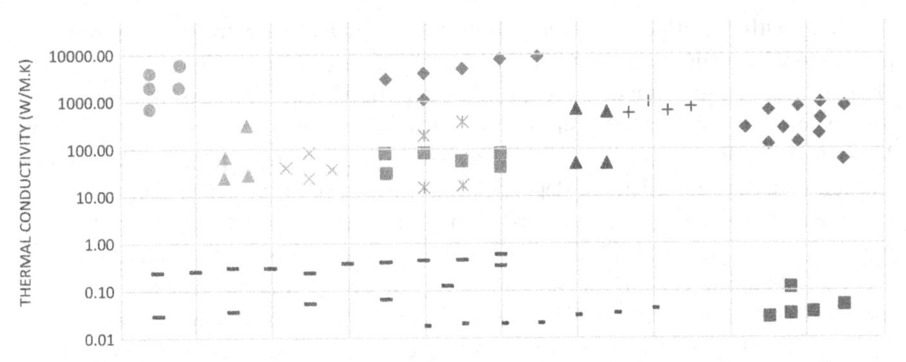

FIGURE 2.5 Schematic diagram of the reported k values of some representative nanomaterials versus traditional materials.

It is not easy to calculate α only theoretically. Theoretically, k and c_p are calculated to achieve α through Eq. (2.27). Using BET for theoretical calculation was discussed earlier. The specific heat capacity is mostly comprised of phonon specific heat capacity and electron specific heat capacity. Still, with specific nanomaterials, c_p is mostly controlled by phonons (see Section 2.3).

There are few completely theoretical research works on thermal diffusivity; therefore, α and the transport mechanisms are mostly studied using experimental techniques. In Eq. (2.28), direct measurements of α using another form of α which represents the simple relationship between time and temperature is given:

$$\frac{\partial T}{\partial t} = \alpha \nabla^2 T \tag{2.28}$$

Given the fact that α is representative of the rate of transfer of heat from the source to the heat sink, there is a need for a fast heating technique for measurement. The first character of α was determined using laser lash technology for heating the sample

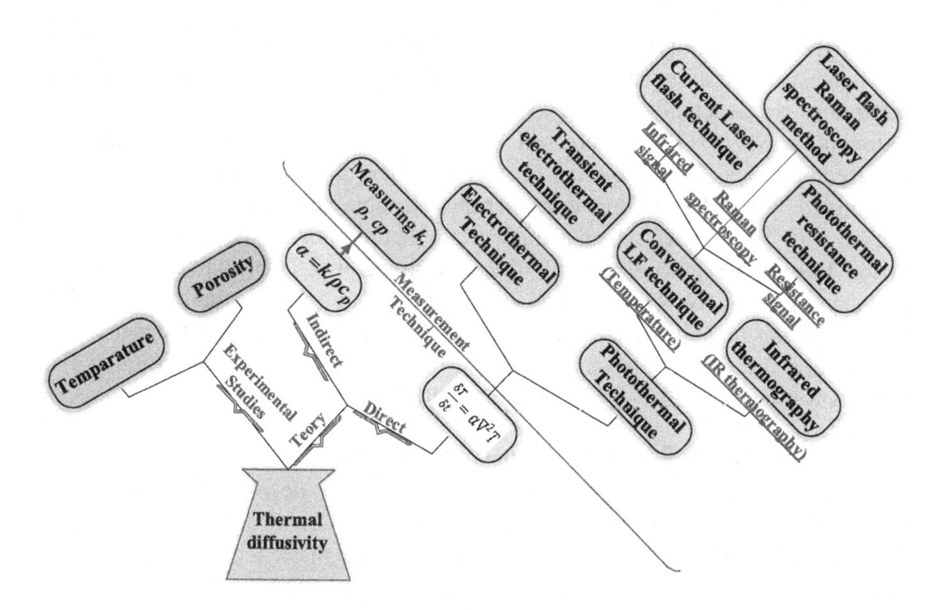

FIGURE 2.6 Schematic diagram of research on α of solid nanomaterials. ([142,147,157,184].)

using a laser pulse. Then, the temperature change was analyzed from a spot close to the heating location [139]. Based on 1D adiabatic condition, α is given as:

$$\alpha = 0.1388 \cdot d^2/t_{0.5} \tag{2.29}$$

where d represents the thickness and $t_{0.5}$ stands for the half maximum of time. Given that α needs rapid heating laser heating techniques are mostly used for measurement including laser flash Raman spectroscopy. Some of the general measurement techniques are introduced in the following sections.

2.2.2 Experimental Measurement Techniques

2.2.2.1 Laser Flash Technique

As a standard method to measure the thermophysical property, the laser flash technique was used [139]. The technique is featured with high sensitivity, transient, and non-contact measurement so that it is extensively used for measuring α for homogeneous materials [140]. To perform the test, a short laser pulse is emitted to the front side of the sample, and temperature change on the back surface is measured in real time. To obtain α, the temporal temperature profile is fitted. A key specification of this technique is the energy and width of the pulse that is adjusted easily. Using this technique, Lin et al. [141] determined the thermophysical properties of vertically aligned MWCNTs. Their study was based on a few assumptions. Firstly, the thin gold coating used to facilitate rapid planar distribution of heat is assumed to undergo ideal 1D heat transfer when the laser pulse is radiated to the front surface.

Secondly, the surface of the sample is opaque because of the gold coating. Therefore, the slightest temperature change was needed during the measurements. It is known that α does not depend on the thickness of the sample and the metal coating of the bottom of the MWCNT film increased the accuracy of measurement. This technique was used by Xie et al. [142] for non-contact heating of the MWCNT array and α was determined by fitting for temperature vs. time. The obtained α for the MWCNT array was $4.6 \times 10^{-4} m^2/s$ at RT, which is higher than that of known excellent thermal conductors like copper and silver. In addition, Akoshima et al. [143] synthesize a highly pure super growth SWNT forest of 1 mm length. They reported that α of SWCNT forests was not notably different from isotropic graphite with notable dependence. Specifically, the value of α was within the range of $0.47–0.77 \times 10^{--4} m^2/s$ at RT, which was higher than $0.2–0.6 \times 10^{-4} m^2/s$ at 1000 K.

2.2.2.2 Laser Flash Raman Spectroscopy Method

To measure thermal diffusivity, a laser flash device was developed, especially for the condition imposed by the necessity to measure thermal diffusivity of highly radioactive reactor-irradiated nuclear fuel. Some of the major requirements to measure irradiated samples are the ability to determine specifications of sample platelets or irregular contours and different sizes, to perform measurement as fast as possible, and to keep a good experimental accuracy along with preserving pulse laser energies at low levels.

The Raman spectroscopy method is a non-contact and non-destructive method to measure temperature rapidly. The method is recommended for measuring microscale materials like graphene [41], bilayer MoS_2 [144], and 3D nanofoams [145]. The method relies on temperature dependent Raman band shift and it mostly defines the thermal properties using a mixture of absorbed laser power and the steady-state heat conduction model. It is not easy, to accurately determine the absorbed laser power of materials. Therefore, the use of the method on micro/nanoscale materials is not easy. To deal with this limitation, Zhang et al. [146] proposed a new method known as laser flash-Raman spectroscopy. Using a pulsed laser with two independent steady-state and unsteady-state heat conduction models, they managed to remove the effect of the absorbed laser power parameter in analyzing thermal properties. After showing the accuracy of the technique by comparing α and the results of the 3ω method, they measured α for CFs and CNTs. Another benefit of this method is its simple and non-electrical design which makes it suitable for the thermophysical characteristics of insulating materials. Zhang et al. [146] developed the materials to 2D nanomaterial and examined the difference between 2D nanomaterial measurements with and without supporting substrate. Using variable spot sizes, they reported the results of 2D van der Waals heterostructure and showed measures of k, α, and interfacial resistance at the same time [147].

2.2.2.3 Transient Electrothermal (TET) Technique

One of the effective techniques to measure α of fibers or narrow strips is the transient electrothermal technique (TET). It can overcome the drawbacks of other methods like long test times and weak signals [148]. In short, a direct current is utilized to a suspended fiber which adjusts the heat sinks. The caused joule heat increases the temperature of the fiber and the transient temperature response of every moment is utilized to find the α of the material.

Over the term of the thermal conduction model for the TET technique, Guo et al. [137] neglected the heat loss on the sample surface. In addition, it is assumed that the sample heating is constant. The stability of the sample heating power was ensured by increasing the impedance of the circuit in the experimental setup. The results indicated the average α of the SWCNT bundle was obtained equal to $2.73 \times 10^{-5} m^2/s$. In addition, the α of the non-conductive material was determined utilizing a polyester fiber coated and a gold layer, which was obtained equal to $5.26 \times 10^{-7} m^2/s$. Taking into account the effect of convection and radiation and non-constant power heating in low vacuum condition, Xing et al. [149] created a model without dimension under surface heat transfer and non-constant heating condition. To determine a, thc Levenberg-Marquardt algorithm was used by analyzing the temperature response in the full model. The results showed a linear relationship between the dimensionless parameter model and the error of measurement, which gives us a method to evaluate the effect of different conditions on the measurement error. In addition to fibrous material, Feng et al. [150] examined the α of film materials.

The thermal transport measurement in multi-wall carbon nanotube (MWCNT) bundles is reported at a higher temperature of 830 K using a new generalized electrothermal technique. In comparison with individual CNTs, the thermal conductivity (k) of MWCNT bundles is two-three orders of magnitude lower, which means that the thermal transport in MWCNT bundles is controlled by the tube-to-tube thermal contact resistance. It is not easy to measure the effective density for the two MWCNT bundles using other techniques, and it is obtained at 116 and 234 kg/m^3. With a small decrease in thermal diffusivity with temperature, the value of k increases slightly with temperature up to 500 k followed by a decrease. This is the first time that the behavior of specific heat for MWCNTs at a temperature higher than AT is clarified. The specific heat is close to graphite at 300–400 k; however, it is below that for graphite higher than 400 k. This indicates that the behavior of phonons in MWCNT bundles is controlled by boundary scattering, not the three-phonon Umklapp process [151]. According to the radiation heat loss analysis, it should be taken into account for measuring the thermophysical specifications of micro/nanowires with high aspect ratios at higher temperature. This is more important for individual MWCNTs given the extremely small diameters [151]. It is notable that the α in the TET method is obtained from the transition state before the temperature becomes steady. By extending the electrothermal method, Huang et al. [151] determined a form of the initial transient change to the final steady state. The benefit of this method is that it does not need the value of c_p and ρ of the material. It is possible to obtain a more accurate α using global fitting of the normalized temperature curve in time. Huang et al. [152] utilized the TET technique to examine the thermophysical properties of a natural fiber (spider silk). They found that α and k increased under stretching. The value of α increased specifically by 50% with a 20% strain. Xing et al. [148] determined the value of α for spider silk using TET and 3ω techniques to compare the two techniques. As their results showed, the TET method was significantly better than the other technique in terms of time to measure α of long fibers with large diameters.

Still, in terms of other variables, the TET method has no superiority over the 3ω technique in terms of the accuracy of measurements.

2.2.2.4 Photothermal Resistance Technique

Another name of the photothermal resistance technique is the optical heating and electrical thermal sensing (OHETS) technique. It was introduced in 2006 by Hou et al. [153]. The technique can characterize the thermophysical properties of one-dimensional micro/nanoscale conductive and non-conductive materials. In short, the samples are suspended on two electrodes and irradiated using a periodic laser beam. Because the laser beam is heated, the sample temperature changes periodically. Therefore, the electrical resistance of the sample also creates periodic changes. Afterwards, the changes in electrical resistance are reflected through voltage changes by using a direct current to the sample, which derives the thermophysical property parameters of the sample. Using this principle, Hou et al. [154] supported the reliability of the technique by measuring a platinum wire with a diameter of 25.4 μm as the reference. Therefore, they managed to determine α of three SWNT bundles, clothing fibers, and human hair. Because SWNT bundles are conductors, and the two others are not, a metal fill was added to the non-conductive sample surface to make sure that the temperature is obtained accurately by employing an electrical voltage signal. As shown by the results, α of the three SWNT bundles was 2.98×10^{-5} m^2/s, 4.41×10^{-5} m^2/s and 6.64×10^{-5} m^2/s respectively. In the case of human hair and clothing fibers, that are not conductive, the value of α was on the order of 10^{-6} m^2/s, which is indicative of the excellent thermophysical properties of CNTs in comparison with traditional materials.

Additionally, Hou et al. [155] measured the value of α of one submicron (about 800 nm) polyacrylonitrile (PAN) fibers. Because PAN fibers are not conductive, an Au film with nanometer thickness was coated on the surface of the PAN fiber irradiated with a periodic laser beam pulse on the suspended individual fibers. The variation of the temperature was periodic and derived by changing the electrical resistance of the nanoscale Au film. By measuring the value of α for three Au films coated with different thicknesses, they found that by adopting a reasonable thickness for Au film (about nanometer level), the technique was successfully measured α at low k, non-conductive 1-D micro/nanoscale materials.

2.2.2.5 Infrared Thermography

A technique to determine thermal diffusivity in thin plates uses infrared images of evolving thermal patterns irradiated with a laser beam. The technique is non-contacting, one-sided, and remote and there is no need for independent estimates of either the emissivity of the sample or thickness of the sample. Using a line-segment pattern for thermal input, the technique gives the in-plane component of the diffusivity tensor in anisotropic materials. In addition, the rate of heat loss to the environment of the plate is given. Two data analysis methods are available, one for a heating line of general cross-section and the other for Gaussian cross-section, which saves a great deal of time. The two methods generate a statistical examination of the quality of measurement and determine diffusivity and loss rate. The results for plates of metals and graphite-epoxy composite materials are available. In addition, in the anisotropic graphite-epoxy sample, principal components and orientation for the diffusivity tensor are determined [156].

Thus, another useful technique to measure α is infrared thermography proposed by Welch et al. [156]. The technique measures the α of metal and graphite-epoxy plates. The measurement relies on changes in temperature due to photothermal excitation and temperature distribution is determined using an infrared camera. The measurement process is featured with a sinusoidal laser beam irradiated on the surface of the sample. Along with the propagation of heat along the radial direction, the amplitude and the phase of the thermal wave are modulated by adjusting the frequency of the laser to make sure that the sample stays thermally thin. Then, the thermal response can be given using the 1-D heat conduction equation. The value of α indicates the amplitude and phase change of the heat wave; while the phase signal is more effective for the local change of the surface temperature and the emissivity of the sample to be tested. Thus, a new technique was proposed by Giri et al. [157] to determine α using the phase information. They found that the active infrared thermography technique can characterize α perpendicular to the axial direction of the template Bi_2Te_3 nanowires determined using electrodeposition on an anodized aluminum (AAO) template. As recommended by the results, α perpendicular to the axial direction of the nanowire of the AAO template with different sizes of pore in 8.8–9.2 × 10^{-7} m^2/s range and the value of α for AAO/Bi_2Te_3 nanocomposite was a bit lower than 6.7 × 10^{-7} m^2/s.

In addition, the value of α was measured by Mendioroz et al. [158] for thin plates and filaments through a lock-in thermography technique. They analyzed the experimental factors affecting α and determined the experimental condition to make an accurate measurement. They showed that the infrared thermography method can measure α of different isotropic and anisotropic thin samples in a vacuum. It is notable that the sample should be large enough to remove the effect of boundary effects. They also showed that the 1-D heat flux loss model overestimates α of the sample; that is, some α values reported by other studies are overestimated. The α of polyester-ether-ketone (PEEK) films with thicknesses of 25, 75, 125, and 250 μm were measured in air and vacuum, respectively. Their findings indicated that α in vacuum was below α in air, which is due to the removal of heat flow loss in the vacuum.

2.2.3 Experimental Studies

2.2.3.1 Effect of Porosity

To examine the role of porosity (ζ) on α, the highly porous solids featured with good thermal insulation need to be taken into account. Generally, it is believed that the thermal insulation performance is mostly a function of the pore structure [159]. The α of nanoporous spinel/forsterite/zirconia composite ceramics was examined by Wahsh et al. [160]. They showed that the level of ζ increases with zirconia content, which leads to a decrease in α. This can be explained by the tetragonal to monoclinic phase transition of zirconia and the thermal expansion mismatch that different phases have in different ceramic composites. According to other studies, using density ρ instead of ζ to build the relation with α is possible. The reason for this is that

ρ and ζ are correlated ($\zeta = 1 - \rho/\rho_s$). Where ρ_s indicates the density of dense materials without ζ. The relationship between ρ and α of aerogel-based non-porous fibrous material was examined by Benkataraman et al. [161]. They showed that with a higher ρ_s the value of k increases. Hirata et al. [162] examined the α and k of porous mullite ceramics theoretically and experimentally. They showed that α and k were dependent ζ similarly. These findings are expected since with a higher ρ or lower ζ, the mass fraction of solid components will be higher, which is responsible for the heat conduction network in the porous material.

In the same way, Mayr et al. [163] determined the α of CF reinforced polymers (CFRP) through active IR and explained the effects of pore shape and ζ on α. They showed that the α of CF declined linearly with an increase of ζ, and it was less than that of spherical pores under the same ζ. Wu et al. [164] showed that the α of CFRP and the pore shape and ζ are closely related. Along with ζ increases, the α of CFs declined correspondingly. Moreover, the α of CFRP was examined in three directions and the results showed that it was largest along the fiber bundle, which indicated the remarkable anisotropy in thermophysical properties.

2.2.3.2 Effect of Temperature

Generally, two factors affect the temperature-dependent α, including the lattice wave the phonon transfer energy, and the free electron transfer energy. With an increase in temperature, the chance of collision between phonons increases, which leads to a higher lattice scattering and lower MFP. Through this, the phonons' effect on α decreases.

A typical anisotropic thermally conductive material is graphene, which is extensively utilized to dissipate heat in electronic devices. Authors have examined the effect of temperature on graphene and its thermophysical properties [165]. Pan et al. [166] manufactured reduced graphene oxide-multiwalled carbon nanotubes (rGO/CNT) film to examine the temperature effect on α. They showed that in-plane and cross-plane α values of rGO/CNT increased with an increase in temperature, which was due to the higher phonon changing frequency and large amplitude that resulted in faster transfer of heat at a higher temperature [167]. Still, others have shown an opposite trend, which is explainable by Umklapp phonon scattering [79]. To measure α, Tuan et al. [168] bonded graphene and an alumina substrate. They showed that α of graphene declined along with an increase in temperature and the in-plane α was equal to $7.5 \times 10^{-6} m^2/s$ at AT. This figure is 400 times higher than the cross-plane α of graphene. Given this, it is clear that graphene and composite films have a notable potential for thermal management.

Another nanocomposite with good thermal transport properties is CNT-based composites. According to Jackson et al. [169], the value of α decreases with an increase in temperature. With a higher CNT volume fraction, the value of α of the composites increases with temperature. Xie et al. [142] determined the value of α of MWCNT in temperatures ranging from –55°C to 200°C. They found consistent results that the α increases with temperature from –55°C to 70°C. This is the same trend obtained with the 3ω technique [119]; however, the results were one order of magnitude higher than α obtained with 3ω. To explain this, it is notable that CNTs used in [119] were longer, which have more defects that result in a higher phonon scattering probability so that the heat transfer capability decreases. The variation of α in the temperature range of 70°C –200°C is not much, which is due to the fact that the phonon scattering mechanism is not clear and needs further studies [119].

The effect of temperature on thermophysical properties has been extensively studied in nanopowders, which are featured as common matrix materials for nanocomposite thanks to the scale effect. The thermophysical properties of SiO_2/Al_2O_3/carbon black composite power tablets were studied by Voges et al. [170]. They showed that the thermophysical properties were comparable to those of aerogel so that α decreased with an increase in temperature. This is because of the higher Umklapp scattering of phonons with higher temperature, which leads to a decrease in α. According to Zhan and Mukherjee [171], α of CNT-reinforced Al_2O_3-based ceramics decrease with temperature, which is because of Umklapp scattering and lowered phonon MFPs.

Studies on α of solid-state nanomaterials were summarized. Given the notable difficulties in the theoretical calculation of α, the more general practices were determined indirectly using k, c_p, and ρ, which are determined using experimental measurement techniques.

The majority of techniques for measurement are transient so they can be classified as photothermal techniques and electrothermal techniques. One of the standard photothermal technologies is the traditional laser flash method. Along with technological advances, better techniques for measurement have been developed such as laser flash method, laser Raman flash method, photothermal resistance method, and infrared thermography. In the case of electrothermal techniques, the main focus is on TET. Moreover, most of the factors in α of solid nanomaterials are involved in ζ and temperature. The results showed that α and k depend on ζ in the same way, while the temperature-dependent α depends on two factors including the lattice wave and the phonon transfer energy and the free electron transfer energy. The α value of some nanomaterials is summarized by Figure 2.7 (individual CNTs [132,137,143], CNT fibers [29–31], CNT array [26,153,172,173], CNT films [117,174], CNT bundles [175,176], CNT composites [171,177], silicon nanomaterials [128], graphene [178–180], polymer [148,181–183], and aerogel [184–186]) and traditional materials (metal, air, foam, etc. [187]). The image demonstrates the superiority of the thermophysical properties of nanomaterials in comparison with traditional materials.

2.3 ISOBARIC SPECIFIC HEAT CAPACITY

2.3.1 Theory

By assuming the atoms as solid with simple harmonic oscillators and the vibrational degrees of freedom that are completely exited, the following formula gives the high temperature limit of c_p of a monoatomic solid.

Assuming that the atoms in the solid are simple harmonic oscillators and the vibrational degrees of freedom are fully excited, the high temperature limit of c_p of a monoatomic solid is defined as:

$$\overline{c_p} = 3\overline{R} = 3N_A k_B \tag{2.30}$$

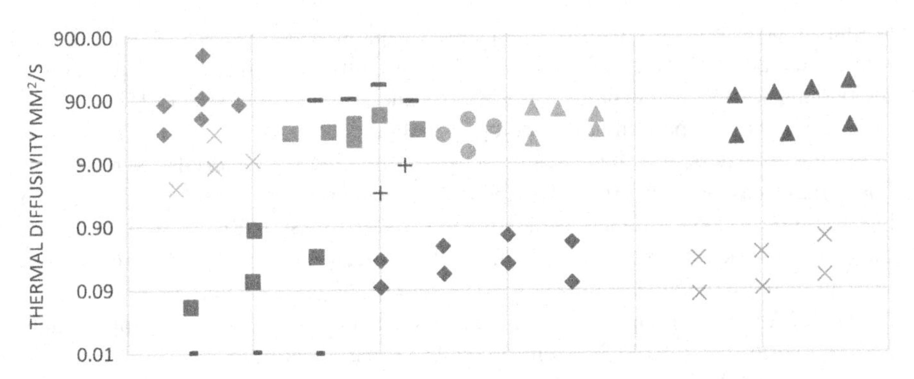

FIGURE 2.7 Schematic diagram of α values with some reported representative nanomaterials versus traditional materials from the literature.

This is the Dulong-Petit law that is explained using the energy equipartition theorem using classical statistical mechanics. The law is not able to predict the behavior of material when the temperature is low; however, it overestimates the c_p of different materials like graphite and diamond. There is evidence that the c_p of some solids declines along with temperature decline. This is not consistent with the Dulong-Petit law. Einstein used quantum theory to explain this trend along with harmonic oscillators to explain atomic vibrations and a few new assumptions: quantization of energy and atomic vibration at the same frequency [188]. In Einstein's model, c_p is defined as:

$$c_p = 3N_A k_B f_E\left(\frac{T_E}{T}\right) \tag{2.31}$$

where $T_E = \frac{\hbar\omega_E}{k_B}$ refers to the Einstein temperature and $f_E\left(\frac{T_E}{T}\right)$ refers to the Einstein heat capacity function. When $T >> T_E$, the Einstein c_p can is written as $c_{p,E} = 3N_A k_B$, which is consistent with the results of the Dulong-Petit law.

Afterwards, Debye [189] introduced the collective vibrations concept of lattice. The concept assumes that a crystal is an isotropic continuum so that the thermal motion of the atom happens as elastic waves. Elastic wave vibration mode is the same as the harmonic oscillator. By using the concept of Einstein's energy quantization, Debye modified the assumption of atomic single frequency vibration to cover all the vibrations under the Debye frequency. The Debye frequency is the upper limit of the given elastic wave frequency and the Debye mode is defined as follows:

$$c_p = 9N_A k_B\left(\frac{T}{T_{De}}\right)^3 \int_0^{T_D} \frac{x^4 e^x}{\left(e^x - 1\right)^2} dx \tag{2.32}$$

where $x = \frac{\hbar\omega_E}{k_B T}$. At high temperature $T >> T_{De}$, which is consistent with the Einstein model and the Dulong-Petit law. With low temperature $T << T_{De}$, the c_p of the Debye model is $c_p = \frac{12}{5} N_A k_B\left(\frac{T}{T_{De}}\right)^3$, which is also known as the Debye-T^3 law. This is in agreement with experimental curves as only the low-frequency modes are excited by thermal excitation (Figure 2.8).

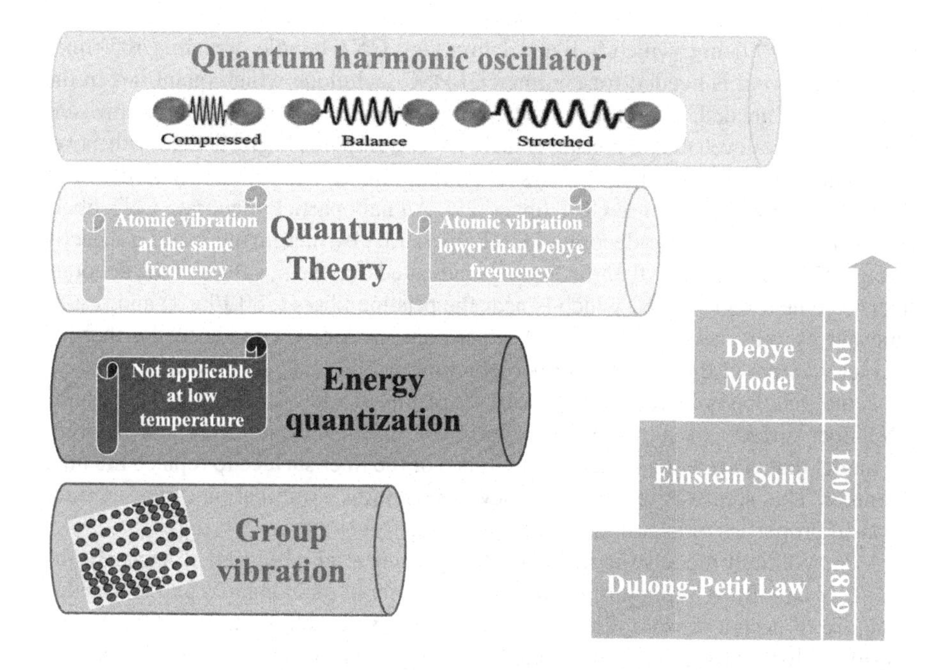

FIGURE 2.8 Schematic diagram of the theoretical study progress on c_p of solid nanomaterials.

2.3.2 Experimental Measurement Techniques

2.3.2.1 3ω Method

One way to measure the c_p for different nanomaterials is 3ω. The traditional 3ω technique has been used to measure millimeter-long aligned MWCNT arrays [119,190,191], CNT fibers [30,31], and nanofilms [192–194]. It is notable that these nanomaterials are mostly found in limited volume and therefore, standard c_p measurement methods like differential scanning calorimetry (DSC) are ineffective. Lu et al. [190] measured c_p of MWCNT arrays in 10–300 k temperature range in vacuum. They showed a linear dependence between c_p and temperature, which was about 500 J/kg:k at room temperature. This is mostly due to the behavior of phonons. Based on theoretical calculations, c_p and the integral of the phonon spectrum and temperature are related. The phonon spectrum does not change over the experimental temperature range (less than Debye temperature), which leads to a linear dependence of c_p on temperature. Hu et al. [191] showed that the c_p of a 13 μm-thick vertically aligned CNT array at room temperature is 312 J/kg:k, which increases to 375 J/kg:k with a temperature increase to 325 K. These figures are not comparable with other reports [19,31,190], which can be explained by the role of highly low c_p value of trapped air in the CNT array

In addition, Lu et al. utilized the 3ω technique to determine the c_p of standard filament samples like high-purity platinum wire of 2 μm diameter and 1 mm length [191]. They estimated the air convection-induced heat loss and the proper vacuum condition was determined to make sure that the measurement uncertainty of c_p was below 4%. In addition, the quantity of sample needed to measure c_p can be notably less than that needed by other methods. For instance, the filament sample can be as light as 0.0001 mg, which is like a 1 mm long CNT bundle, which is 105 times lighter than what is needed for commercial DSC technique which quantities in the mg range are needed. Therefore, one can say that the 3ω technique enables measuring c_p of trace-level nanomaterial, which is very important given that with novel nanomaterials, the quantity is always a limitation. Qui et al. determined the c_p of the functionalized and dense CNT fibers [29], Au nanoparticle-decorate CNT fibers [30], and iodine molecule-decorated CNT fibers [31] by measuring k and α directly through the 3ω method. All of the findings indicated that the c_p values of the decorate fiber were near 685 J/kg/K, which is near the pristine fibers (780 J/kg:k) and results reported by other studies [19,190]. A great contribution was made to the thermal characterization of the assembled fiber structure by Qui et al. [192], which indicated the actual relation is correlated only with the results of a few layers of bundles on the CNT fiber surface. As shown by the experiment results, they created a new theoretical model that utilizes the right circumference of the fiber surface to replace the fiber diameter. This step was needed to achieve an accurate examination of the thermophysical properties of assembled nanostructures.

A 3ω-Völklein model was introduced by Ftouni et al. [193,194] to measure the c_p of SiN nanofilms. Völklein geometry [195] is about an elongated and suspended structure to design sensors. This geometry was mixed with 3ω to achieve a higher sensitivity for measuring ultrathin nanofilms. Prior to measurement, they placed niobium nitride as a thermometer on SiN nanofilms with thicknesses of 50 and 100 nm.

There was no significant difference in c_p during the measurement of nanofilm that had different thicknesses. The results are consistent with the Debye-like c_p model for temperatures higher than 100 k. Given that the model is highly sensitive (4×10^{-6} J/kg·K) it is a good option to measure nanofilms.

It is important to determine the thermal properties of thin film materials to realize their structure and conduction mechanism as well as for technical applications. Two important parameters for thermal conductivity and diffusivity to design integrated devices are thermal microsensors and actuators. To have a reliable system simulation and optimized design of such devices, we need to pay attention to determining thermophysical properties. Thereby, it is better to have such data. Generally, there is a considerable difference between thin films and bulk values as the parasitic surface effects are notably stronger because of the smaller dimensions of aspect ratios. It is not easy to measure these properties. Here, the steady-state and transient methods are reported for determining the properties in-plane and cross-plane film properties using microelectrothermal chips [195].

2.3.2.2 AC-Calorimetric Technique

Note that c_p is of great value in research works to understand lattice vibration, superconductivity, and structural phase transition [196,197]. The AC calorimetry technique was introduced by Sullivan and Seidel in 1968 for measuring the c_p of samples lighter than 200 mg [198]. In short, the sample heating is done using an alternating current with an angular frequency of $\omega/2$ that passes through the heater, and the temperature increases and changes periodically with time. Afterwards, the current is amplified using a lock-in amplifier. Eventually, the c_p of the sample is determinable by measuring the temperature of the system in which the current frequency is a factor.

As a dynamic measurement method, the c_p of the sample in AC calorimetry is measured during the process of heating. Because the thermal equilibrium is not easy to establish in the dynamic process, the accuracy of measuring c_p is less than that of the adiabatic method [199]. Still, AC calorimetry is a good choice for measuring small samples so that the sensitivity of measurement can be 10^{-8} to 10^{-12} J/K.

The slightly abnormal c_p of superconducting Sn nanowires was measured by Zhang et al. [200] at the superconducting transition using the AC calorimetry method. The results indicated that the c_p of Sn nanowires is considerably different from the bulk Sn with two peaks at 3.7 and 5.5 K that are explained by the inner bulk contribution and the surface contribution respectively [200]. This method was used by Lee et al. [201] to study the structural phase transition of Zr/B multilayer nanofilms. This method can be performed at a low scan rate below 4000 K/s, which minimizes loss of heat [202]. Two exothermic peaks appeared in the c_p based on the temperature rise curve. The first peak is because of the non-uniform diffusion of Zr and B at low temperatures that yield amorphous Zr/B and the second peak is because of crystallization at high temperatures. The c_p of randomly oriented SWCNT and MWCNT was measured by Pradhan et al. [203]. They used bulk graphite powder as a reference sample. They found that the c_p values of the graphite powder and the MWCNT were the most stable readings. At the same time, the c_p of MWCNT had the lowest dependence on temperature, for which macroscopic arrangement is notable.

Taking into account that the AC calorimeter technique can measure samples less than 1 ng, Tress et al. [204] employed the method for measuring the glass transition of polystyrene nanofilms. Their method was compared to other techniques like broadband dielectric spectroscopy (BDS), spectroscopic vis-ellipsometry (SE), and DSC. Researchers have shown that in the case of multilayer molecular weight polystyrene, the glass transition temperature changes within ±3 K at a layer thickness of 5 nm. The results showed that the c_p of nanomaterials obtained by the AC calorimetry method can explain why the c_p increases compared to bulk materials.

2.3.3 Experimental Studies

2.3.3.1 Effect of Size

There are a few experimental works on the effect of the solid size of the c_p. One of these few studies is Novotny et al. [205], who showed that 27 and 37 Å lead particles (purity:99.9999%) had a higher c_p compared to bulk lead at ultra-low temperature (1.5–5 k). This can be elaborated on based on the fact that the specific surface area of small particles is high and this improves surface phonon softening. Because softening leads to a decrease of vibration frequency, c_p, which in turn depends on vibration of the lowest frequency of the lattice, improves consequently. In the same way, Rupp et al. [206] determined the values of c_p of nanocrystalline Cu and polycrystalline Cu. They found a size effect of around 10% increase in c_p. The c_p of ultrathin ferroelectric $BaTiO_3$ films was measured by Strukov et al. [207] using an ac-hot probe technique. With the BaTiO film thickness below 100 nm, the c_p did not show any increase in the ferroelectric phase transition temperature, which is seen with other sizes, due to the large diffusion of the transition at the nanoscale. Another study on ternary nitrate eutectic [208] examined the influence of adding SiO_2 nanoparticles of different diameters (5–60 nm) on c_p. They showed that the effect of all nanoparticles was the same. In all cases, the c_p was around 1800 J/kg/K, which means a 13% increase in comparison with the counterpart without adding nanoparticles. This means that the size of the added nanoparticle does not affect the c_p for the eutectic material.

Singh et al. [209] measured c_p values for metal nanowires and films. Their results showed that the atomic vibration of the nano-solid surface was higher than that of the bulk material. Along with the decrease in size, there was an increase in the atomic thermal vibration energy of the material and consequently, the c_p was affected. Wang et al. [210] examined the c_p of CuO nanoparticles of different sizes. The value of c_p increased from 0.49 KG/kg:K to 0.51 kJ/kg:K along with a decrease in particle size from 12.5 to 3.4 nm at 325 K. This observation was also supported by an experimental study that showed the softening of surface atom vibrations increased c_p due to size effects [209]. This finding was also supported by semiconductor materials [211]. Some reports indicate nanomaterials demonstrate specific physical and chemical properties with significant differences from the corresponding properties of bulk materials. Due to the high surface-area-to-volume ratio of nanomaterials, the atom energy of these nanomaterials is not the same as conventional bulk materials. Therefore, there is a size-dependent thermodynamic property in nanomaterials. In addition, cohesive energy, or the heat of sublimation, is a key physical quantity that is responsible for the strength of metallic bonds.

Results have shown that the melting temperature of nanomaterials declines with particle size. The root mean square amplitude model, the size-dependent Deby temperature model, and the size-dependent thermal conductivity model have been calculated using Lindemann's criterion of Mott's equation. Researchers have argued that the Debye temperature declines with the size of nanomaterials. The role of particle size and thermodynamic energy have been used to determine properties (e.g., surface tension and young's modulus of nanocrystals) based on surface thermodynamics and the atomic bond energy. The first thermodynamic specification to predict melting temperature, melting enthalpy, melting entropy, and specific heat of nanomaterials is the cohesive energy. Researchers have recommended diverse models such as the latent heat model, the liquid drop model, and the surface area difference model to estimate the cohesive energy of nanomaterials.

Here, a surface free energy model was introduced that does not rely on adjustable parameters and uses size and shape as to the cohesive energy of nanomaterials. It is possible to determine the expressions for size/shape-dependent specific heat, melting entropy, and enthalpy based on the relationship between melting temperature and cohesive energy. Theoretical works using these expressions have been used on Ag, Cu, In, Se, Au, and Al nanomaterials in spherical, nanowire, and nanofilm shapes [211].

2.3.3.2 Effect of Grain Boundary

Before discussing the role of grain boundaries, the definition of nanocrystalline (NC) materials needs to be clarified as they are the main reasons for grain boundaries. In general, NC materials are comprised of nanomaterial grains (1–10 nm). The interface of grains with the same structure and orientation is the grain boundary, which represents 50% or higher in nanocrystalline materials. There is an inverse relationship between fractional and average grain size of grain boundaries in NC materials [212]. The thermophysical properties of NC materials are notably different from conventional crystals because of the small grain size and large specific surface area. For instance, the value of c_p is higher in NC materials. Thus, nanopolycrystallization is a way to improve the thermal transport in solid-state nanomaterials.

The effect of grain boundaries on c_p has been studied by several researchers. In terms of nanopolycrystalline metals and alloys, Tang et al. [213] examined the cp in NC-Ni and reported that its c_p was higher than conventional crystals. This can be explained by the existence of grain boundaries. The excessive isobaric specific heat capacity (the isobaric specific heat capacity difference of NC materials and the bulk materials, Δc_p) decreases at low and high temperatures. This finding is consistent with the findings reported by other studies and MD simulation results [214,215]. Rojas et al. [216] demonstrated based on Raman spectroscopy that the values of c_p of NC-Fe, Cu, Ni, and $LaAl_2$ alloys were more than those reported for corresponding bulk. In addition to high temperature regions, Δc_p has been studied at low-temperature regions. The findings indicate that the peak of Δc_p happens with a low-temperature range of 20–65 K. The softening of the interface and the decline of the neighboring coordination number of the surface explains this phenomenon.

In addition, the interface atoms lower the frequency of phonons, which changes the spectrum of phonon and Debye temperature of the NC materials and leads to a decline in c_p. In addition, Li et al. [217] examined the effect of grain boundaries on the c_p of NC-AL at low temperatures and reported results consistent with other studies.

Moreover, the effect of grain boundaries on the c_p of ultrananocrystalline diamond (UNCD) was examined by Adiga et al. [218]. As shown by the results, the c_p of UNCP at AT was about 20% higher than that of single-crystal diamond. In addition, Δc_p increased to a maximum level in UNCD of around 350 K. To explain this, it is notable that the density of vibration of the atoms increases near the grain boundary. These changes take place in the low-frequency region of 5–20 Hz and the effect of low and intermediate frequency phonons to c_p decreases at 350 k.

2.3.3.3 Effect of Interfacial Component

As to the c_p of nanoscale solid materials, our attention is more focused on nanometer-sized crystalline materials that are usually polycrystals with the size of crystal of several nanometers. The polycrystals are normally made of interfacial components created by all atoms in the crystal lattice. In addition, all atoms are positioned in the interface between the crystallites (grain boundary). Thus, to examine the effect of interfacial elements on c_p in crystalline materials, Rupp and Birringer [206] examined the c_p of nanometer-sized crystalline and polycrystalline Pb (crystal size of 6 nm) and Cu (crystal size of 8 nm). They showed that the c_p value of both Cu and Pb nanocrystals are a bit higher (about 10% and 40%) than that of crystalline state at temperature 150–300 k. The main reason for this is the relatively low density of Pb and the weak mutual coupling of atoms between the interfaces. This indicates that the interface interaction is a factor in the value of c_p. The effect of the interfacial elements on the c_p of nanocrystalline Fe with a grain size of 49 nm in an ultra-low-temperature range (1.8–26 k) was examined by Bai et al. [219]. They reported that the increase in c_p at a relatively high temperature (about 26 k) is mostly because of the Einstein oscillators caused by the weak mutual coupling of atoms at the interface.

The role of surface components on the c_p of metal nanocrystalline materials was examined by Lei et al. [220]. Needless to say, the surface phonon softening improves the c_p of the metal nanocrystal materials at low temperatures. However, the role of surface oxidation creates an abnormal decline in c_p. The author employed face centered cubic Al (fcc-Al) and amorphous Al_2O_3 (a-Al_2O_3) nanocrystals as the surface oxide layer around the nanocrystalline Al. By calculating the VDOS of fcc-Al and a-Al_2O_3 nanocrystals, they explained the phenomenon of c_p reduction. The constituent items of a-Al_2O_3@Al nanoparticles are 1nm thick a-Al_2O_3 shells encapsulating of fcc-Al core. Based on the results of VDOS calculations, the surface and volume VDOSs ($D(\omega)$) of the α-Al_2O_3 layer are not the same as pure fcc-Al nanocrystals at low frequencies. In addition, $D(\omega)$ is scaled to $D(\omega) \sim \omega^{1.5}$. The c_p of a-Al_2O_3@Al nanoparticles showed a different temperature dependence compared to pure metal nanocrystals at low temperatures.

2.3.3.4 Effect of Temperature

Debye introduced a new formula to calculate the c_p of solids in 1912, the formula assumes that the c_p is the sum of the atomic vibrations of different frequencies [221]. In the case that the temperature is notably less than the Debye temperature, the c_p is proportional to T3 in the 3-D case, to T2 in the 2-D case, and to T in the 1-D case (Debye's law) [222]. In the case that the temperature is notably higher than the Debye temperature, the c_p of the solid meets Dulong-Petit's law, which indicates that c_p is not dependent on temperature, the thermal vibration of the crystal lattice grows in magnitude, and the c_p of the solid increases [223]. The following explains the role of temperature on the c_p of different nanoparticles like CNT, graphene, and nanocrystals.

Hone et al. [224] examined the temperature dependence of c_p on SWCNTs ropes for CNTs at low temperatures. With a temperature range of 100–200 k, there is a linear correlation between c_p and T. For temperatures less than 50 k, the c_p is relative to T2, which is consistent with Debye's law. Li et al. [225] reported similar findings. Moreover, other studies have shown that the c_ps of both SWCNTs and MWCNTs are linearly dependent on T at low temperatures [119,226]. This can be explained based on wall-to-wall coupling in MWCNTs, which is weak compared to graphite. The c_p of double-walled carbon nanotubes (DWCNTs) has been examined by other studies. For instance, Xiang et al. [227] found that the interaction between adjacent and tubes and layers is too strong to be neglected in the case of c_p of DWCNTs at a low-temperature range of 0.3–30 K. The role of electrons in SWCNTs is notably lower than that of phonons at low temperatures [225,228]. In addition, a cubic polynomial controls the relationship between c_p and T.

Alofi et al. [229] examined the c_p of graphene with different acoustic branches as a function of T like in-plane longitudinal mode (LA), the in-plane transverse mode (TA), and the out-of-plane or flexural mode (ZA). In the cases of temperatures less than 100 k, the cp of the LA branches and the TA branches are relative to T2 and the c_p of the ZA branches is relative to T. This can be explained by the quadratic dispersion in the ZA phonons. Thereby, the overall c_p of graphene is relative to T at low temperatures. With $T > 2500$ k, the c_p does not depend on T, which is consistent with Dulong-Petit's law. The phonon c_p of the single-layer, bilayer, and twisted bilayer graphene was studied by Nika et al. [230] and they reported results similar to [224]. In other words, with low-temperature blow 15 k, the c_p of T relations is $c_p \sim T^n$ so that for single-layer, bilayer, and twisted bilayer graphene $n = 1$, 1.6, and 1.3 respectively. By studying graphene c_p dependence on different layers, we can select thermal modulation based on graphene.

Several studies have been conducted on c_p and its relation with T for nanocrystals. For instance, Zhang et al. [231] examined the c_p and T of nanocrystalline Cu. Based on the Debye formula, the c_p and t relationship of ideal crystalline Cu is $c_p \sim T^3$. Still, they showed that the c_p of nanocrystalline Cu was higher than that of the bulk Cu. Therefore, the relation is not consistent with Debye law. To explain this, the atomic distribution in the grain boundary of nanocrystalline Cu is more chaotic, which leads to a higher entropy and larger c_p.

The theoretical analysis, experimental techniques, and the main factors in the thermophysical studies of solid nanomaterials were examined. Theoretical works have been mostly on defining the c_p of solid nanomaterials using the Dulong-Petit law, Einstein model, and Debye model. They have also focused on the relationship and differences between the three models and the development processes. The best experimental techniques for thermophysical properties measurement of solid nanomaterials include the 3ω method and the AC-calorimetric technique. In terms of the size effect of c_p, it was shown that particles have higher c_p compared to bulk under ultra-low temperature. This finding can be clarified based on the fact that small size particles have a large specific surface area, which improves surface phonon softening. In addition, because of the small grain size and large specific surface area, there is a significant difference between the thermophysical properties of nanocrystal materials and conventional crystals [206]. Because of differences between interface components, the value of c_p of nanocrystals is a bit higher than that of polycrystalline state in temperature range 150–300 k. The temperature dependence is different for c_p of different typical nanomaterials over different temperature ranges.

REFERENCES

1. L. Yang, J. Chen, N. Yang, B. Li, Significant reduction of graphene thermal conductivity by phononic crystal structure, *Int. J. Heat Mass Transf.* 91 (2015) 428–432. https://doi.org/10.1016/j.ijheatmasstransfer.2015.07.111.
2. M.C. Wingert, J. Zheng, S. Kwon, R. Chen, Thermal transport in amorphous materials: A review, *Semicond. Sci. Technol.* 31 (2016) 113003. https://doi.org/10.1088/0268-1242/31/11/113003.
3. T. Zhu, E. Ertekin, Phonons, localization, and thermal conductivity of diamond nanothreads and amorphous graphene, *Nano Lett.* 16 (2016) 4763–4772. https://doi.org/10.1021/acs.nanolett.6b00557.
4. F. Yang, C. Dames, Mean free path spectra as a tool to understand thermal conductivity in bulk and nanostructures, *Phys. Rev. B* 87 (2013) 035437. https://doi.org/10.1103/PhysRevB.87.035437.
5. G. Chen, *Nanoscale Energy Transport and Conversion: A Parallel Treatment of Electrons, Molecules, Phonons, and Photons*, Oxford University Press, New York, 2005.
6. J. Nakagawa, Y. Kage, T. Hori, J. Shiomi, M. Nomura, Crystal structure dependent thermal conductivity in two-dimensional phononic crystal nanostructures, *Appl. Phys. Lett.* 107 (2015) 023104. https://doi.org/10.1063/1.4926653.
7. Z. Wang, J.E. Alaniz, W. Jang, J.E. Garay, C. Dames, Thermal conductivity of nanocrystalline silicon: Importance of grain size and frequency-dependent mean free paths, *Nano Lett.* 11 (2011) 2206–2213. https://doi.org/10.1021/nl1045395.
8. Y. Hu, L. Zeng, A.J. Minnich, M.S. Dresselhaus, G. Chen, Spectral mapping of thermal conductivity through nanoscale ballistic transport, *Nat. Nanotechnol.* 10 (2015) 701–706. https://doi.org/10.1038/nnano.2015.109.
9. C. Dames, G. Chen, *Thermoelectrics Handbook: Macro to Nano*, Taylor & Francis, New York, 2005.
10. J. Callaway, Model for lattice thermal conductivity at low temperatures, *Phys. Rev.* 113 (1959) 1046–1051. https://doi.org/10.1103/PhysRev.113.1046.
11. P.B. Allen, Improved Callaway model for lattice thermal conductivity, *Phys. Rev. B* 88 (2013) 144302. https://doi.org/10.1103/PhysRevB.88.144302.

12. M.G. Holland, Analysis of lattice thermal conductivity, *Phys. Rev.* 132 (1963) 2461–2471. https://doi.org/10.1103/PhysRev.132.2461.
13. G.A. Slack, S. Galginaitis, Thermal conductivity and phonon scattering by magnetic impurities in CdTe, *Phys. Rev.* 133 (1964) A253–A268. https://doi.org/10.1103/PhysRev.133.A253.
14. J. Cuffe, J.K. Eliason, A.A. Maznev, K.C. Collins, J.A. Johnson, A. Shchepetov, M. Prunnila, J. Ahopelto, C.M. Sotomayor Torres, G. Chen, K.A. Nelson, Reconstructing phonon mean-free-path contributions to thermal conductivity using nanoscale membranes, *Phys. Rev. B* 91 (2015) 245423. https://doi.org/10.1103/PhysRevB.91.245423.
15. A.A. Balandin, S. Ghosh, W. Bao, I. Calizo, D. Teweldebrhan, F. Miao, C.N. Lau, Superior thermal conductivity of single-layer graphene, *Nano Lett.* 8 (2008) 902–907. https://doi.org/10.1021/nl0731872.
16. J.M. Larkin, A.J.H. McGaughey, Thermal conductivity accumulation in amorphous silica and amorphous silicon, *Phys. Rev. B* 89 (2014) 144303. https://doi.org/10.1103/PhysRevB.89.144303.
17. J.L. Feldman, M.D. Kluge, P.B. Allen, F. Wooten, Thermal conductivity and localization in glasses: Numerical study of a model of amorphous silicon, *Phys. Rev. B* 48 (1993) 12589–12602. https://doi.org/10.1103/PhysRevB.48.12589.
18. J.L. Feldman, P.B. Allen, S.R. Bickham, Numerical study of low-frequency vibrations in amorphous silicon, *Phys. Rev. B* 59 (1999) 3551–3559. https://doi.org/10.1103/PhysRevB.59.3551.
19. X. Zhang, S. Fujiwara, M. Fujii, Measurements of thermal conductivity and electrical conductivity of a single carbon fiber, *Int. J. Thermophys.* 21 (2000) 965–980. https://doi.org/10.1023/A:1006674510648.
20. J. Wang, M. Gu, X. Zhang, Y. Song, Thermal conductivity measurement of an individual fibre using a T type probe method, *J. Phys. D. Appl. Phys.* 42 (2009) 105502. https://doi.org/10.1088/0022-3727/42/10/105502.
21. M. Fujii, X. Zhang, H. Xie, H. Ago, K. Takahashi, T. Ikuta, H. Abe, T. Shimizu, Measuring the thermal conductivity of a single carbon nanotube, *Phys. Rev. Lett.* 95 (2005) 065502. https://doi.org/10.1103/PhysRevLett.95.065502.
22. D.G. Cahill, R.O. Pohl, Thermal conductivity of amorphous solids above the plateau, *Phys. Rev. B* 35 (1987) 4067–4073. https://doi.org/10.1103/PhysRevB.35.4067.
23. D.G. Cahill, Thermal conductivity measurement from 30 to 750 K: The 3ω method, *Rev. Sci. Instrum.* 61 (1990) 802–808. https://doi.org/10.1063/1.1141498.
24. S. Ahmed, R. Liske, T. Wunderer, M. Leonhardt, R. Ziervogel, C. Fansler, T. Grotjohn, J. Asmussen, T. Schuelke, Extending the 3ω-method to the MHz range for thermal conductivity measurements of diamond thin films, *Diam. Relat. Mater.* 15 (2006) 389–393. https://doi.org/10.1016/j.diamond.2005.08.041.
25. B.W. Olson, S. Graham, K. Chen, A practical extension of the 3ω method to multilayer structures, *Rev. Sci. Instrum.* 76 (2005) 053901. https://doi.org/10.1063/1.1896619.
26. L. Qiu, X. Wang, G. Su, D. Tang, X. Zheng, J. Zhu, Z. Wang, P.M. Norris, P.D. Bradford, Y. Zhu, Remarkably enhanced thermal transport based on a flexible horizontally-aligned carbon nanotube array film, *Sci. Rep.* 6 (2016) 21014. https://doi.org/10.1038/srep21014.
27. L. Qiu, P. Guo, Q. Kong, C.W. Tan, K. Liang, J. Wei, J.N. Tey, Y. Feng, X. Zhang, B.K. Tay, Coating-boosted interfacial thermal transport for carbon nanotube array nano-thermal interface materials, *Carbon* 145 (2019) 725–733. https://doi.org/10.1016/j.carbon.2019.01.085.
28. L. Qiu, N. Zhu, H. Zou, Y. Feng, X. Zhang, D. Tang, Advances in thermal transport properties at nanoscale in China, *Int. J. Heat Mass Transf.* 125 (2018) 413–433. https://doi.org/10.1016/j.ijheatmasstransfer.2018.04.087.

29. L. Qiu, X. Wang, D. Tang, X. Zheng, P.M. Norris, D. Wen, J. Zhao, X. Zhang, Q. Li, Functionalization and densification of inter-bundle interfaces for improvement in electrical and thermal transport of carbon nanotube fibers, *Carbon* 105 (2016) 248–259. https://doi.org/10.1016/j.carbon.2016.04.043.
30. L. Qiu, H. Zou, X. Wang, Y. Feng, X. Zhang, J. Zhao, X. Zhang, Q. Li, Enhancing the interfacial interaction of carbon nanotubes fibers by Au nanoparticles with improved performance of the electrical and thermal conductivity, *Carbon* 141 (2019) 497–505. https://doi.org/10.1016/j.carbon.2018.09.073.
31. L. Qiu, H. Zou, N. Zhu, Y. Feng, X. Zhang, X. Zhang, Iodine nanoparticle-enhancing electrical and thermal transport for carbon nanotube fibers, *Appl. Therm. Eng.* 141 (2018) 913–920. https://doi.org/10.1016/j.applthermaleng.2018.06.049.
32. L. Qiu, Y. Ouyang, Y. Feng, X. Zhang, Note: Thermal conductivity measurement of individual porous polyimide fibers using a modified wire-shape 3ω method, *Rev. Sci. Instrum.* 89 (2018) 096112. https://doi.org/10.1063/1.5052692.
33. L. Qiu, X. Zheng, P. Yue, J. Zhu, D. Tang, Y. Dong, Y. Peng, Adaptable thermal conductivity characterization of microporous membranes based on freestanding sensor-based 3ω technique, *Int. J. Therm. Sci.* 89 (2015) 185–192. https://doi.org/10.1016/j.ijthermalsci.2014.11.005.
34. L. Qiu, K. Scheider, S.A. Radwan, L.A.S. Larkin, C.B. Saltonstall, Y. Feng, X. Zhang, P.M. Norris, Thermal transport barrier in carbon nanotube array nano-thermal interface materials, *Carbon* 120 (2017) 128–136. https://doi.org/10.1016/j.carbon.2017.05.037.
35. L. Qiu, P. Guo, H. Zou, Y. Feng, X. Zhang, S. Pervaiz, D. Wen, Extremely low thermal conductivity of graphene nanoplatelets using nanoparticle decoration, *ES Energy Environ.* 2 (2018) 66–72. https://doi.org/10.30919/esee8c139.
36. W. Ma, T. Miao, X. Zhang, K. Takahashi, T. Ikuta, B. Zhang, Z. Ge, A T-type method for characterization of the thermoelectric performance of an individual free-standing single crystal Bi_2S_3 nanowire, *Nanoscale* 8 (2016) 2704–2710. https://doi.org/10.1039/C5NR05946A.
37. X. Zhang, X. Shi, W. Ma, Development of multi-physical properties comprehensive measurement system for micro/nanoscale filamentary materials, *Sci. Sin. Technol.* 48 (2018) 403–414. https://doi.org/10.1360/N092018-00018.
38. H. Wang, S. Hu, K. Takahashi, X. Zhang, H. Takamatsu, J. Chen, Experimental study of thermal rectification in suspended monolayer graphene, *Nat. Commun.* 8 (2017) 15843. https://doi.org/10.1038/ncomms15843.
39. J. Hone, M. Whitney, C. Piskoti, A. Zettl, Thermal conductivity of single-walled carbon nanotubes, *Phys. Rev. B* 59 (1999) R2514–R2516. https://doi.org/10.1103/PhysRevB.59.R2514.
40. E. Pop, D. Mann, Q. Wang, K. Goodson, H. Dai, Thermal conductance of an individual single-wall carbon nanotube above room temperature, *Nano Lett.* 6 (2006) 96–100. https://doi.org/10.1021/nl052145f.
41. S. Ghosh, I. Calizo, D. Teweldebrhan, E.P. Pokatilov, D.L. Nika, A.A. Balandin, W. Bao, F. Miao, C.N. Lau, Extremely high thermal conductivity of graphene: Prospects for thermal management applications in nanoelectronic circuits, *Appl. Phys. Lett.* 92 (2008) 151911. https://doi.org/10.1063/1.2907977.
42. C. Faugeras, B. Faugeras, M. Orlita, M. Potemski, R.R. Nair, A.K. Geim, Thermal conductivity of graphene in corbino membrane geometry, *ACS Nano* 4 (2010) 1889–1892. https://doi.org/10.1021/nn9016229.
43. S. Chen, A.L. Moore, W. Cai, J.W. Suk, J. An, C. Mishra, C. Amos, C.W. Magnuson, J. Kang, L. Shi, R.S. Ruoff, Raman measurements of thermal transport in suspended monolayer graphene of variable sizes in vacuum and gaseous environments, *ACS Nano* 5 (2010) 321–328. https://doi.org/10.1021/nn102915x.

44. W. Cai, A.L. Moore, Y. Zhu, X. Li, S. Chen, L. Shi, R.S. Ruoff, Thermal transport in suspended and supported monolayer graphene grown by chemical vapor deposition, *Nano Lett.* 10 (2010) 1645–1651. https://doi.org/10.1021/nl9041966.
45. R. Yan, J.R. Simpson, S. Bertolazzi, J. Brivio, M. Watson, X. Wu, A. Kis, T. Luo, A.R. Hight Walker, H.G. Xing, Thermal conductivity of monolayer molybdenum disulfide obtained from temperature-dependent Raman spectroscopy, *ACS Nano* 8 (2014) 986–993. https://doi.org/10.1021/nn405826k.
46. G.S. Doerk, C. Carraro, R. Maboudian, Single nanowire thermal conductivity measurements by Raman thermography, *ACS Nano* 4 (2010) 4908–4914. https://doi.org/10.1021/nn1012429.
47. R.M. Costescu, M.A. Wall, D.G. Cahill, Thermal conductance of epitaxial interfaces, *Phys. Rev. B* 67 (2003) 054302. https://doi.org/10.1103/PhysRevB.67.054302.
48. D.G. Cahill, Analysis of heat flow in layered structures for time-domain thermoreflectance, *Rev. Sci. Instrum.* 75 (2004) 5119–5122. https://doi.org/10.1063/1.1819431.
49. B.C. Gundrum, D.G. Cahill, R.S. Averback, Thermal conductance of metal-metal interfaces, *Phys. Rev. B* 72 (2005) 245426. https://doi.org/10.1103/PhysRevB.72.245426.
50. H.K. Lyeo, D.G. Cahill, Thermal conductance of interfaces between highly dissimilar materials, *Phys. Rev. B* 73 (2006) 144301. https://doi.org/10.1103/PhysRevB.73.144301.
51. M. Li, J.S. Kang, Y. Hu, Anisotropic thermal conductivity measurement using a new Asymmetric-Beam Time-Domain Thermoreflectance (AB-TDTR) method, *Rev. Sci. Instrum.* 89 (2018) 084901. https://doi.org/10.1063/1.5026028.
52. P. Jiang, B. Huang, Y.K. Koh, Accurate measurements of cross-plane thermal conductivity of thin films by dual-frequency time-domain thermoreflectance (TDTR), *Rev. Sci. Instrum.* 87 (2016) 075101. https://doi.org/10.1063/1.4954969.
53. B. Sun, Y.K. Koh, Understanding and eliminating artifact signals from diffusely scattered pump beam in measurements of rough samples by time-domain thermoreflectance (TDTR), *Rev. Sci. Instrum.* 87 (2016) 064901. https://doi.org/10.1063/1.4952579.
54. H. Jang, J.D. Wood, C.R. Ryder, M.C. Hersam, D.G. Cahill, Anisotropic thermal conductivity of exfoliated black phosphorus, *Adv. Mater.* 27 (2015) 8017–8022. https://doi.org/10.1002/adma.201503466.
55. P. Jiang, X. Qian, R. Yang, Time-domain thermoreflectance (TDTR) measurements of anisotropic thermal conductivity using a variable spot size approach, *Rev. Sci. Instrum.* 88 (2017) 074901. https://doi.org/10.1063/1.4991715.
56. J. Wang, M. Gu, X. Zhang, G. Wu, Measurements of thermal effusivity of a fine wire and contact resistance of a junction using a T type probe, *Rev. Sci. Instrum.* 80 (2009) 076107. https://doi.org/10.1063/1.3159863.
57. J. Liu, H. Wang, Y. Hu, W. Ma, X. Zhang, Laser flash-Raman spectroscopy method for the measurement of the thermal properties of micro/nano wires, *Rev. Sci. Instrum.* 86 (2015) 014901. https://doi.org/10.1063/1.4904868.
58. E.E. Michaelides, Transport properties of nanofluids. A critical review, *J. Non-Equilibrium Thermodyn.* 38 (2013) 1–79. https://doi.org/10.1515/jnetdy-2012-0023.
59. N. Savvides, H.J. Goldsmid, Boundary scattering of phonons in fine-grained hot-pressed Ge-Si alloys. I. the dependence of lattice thermal conductivity on grain size and porosity, *J. Phys. C Solid State Phys.* 13 (1980) 4657–4670. https://doi.org/10.1088/0022-3719/13/25/009.
60. Q. Zhang, B. Cao, X. Zhang, M. Fujii, K. Takahashi, Size effects on the thermal conductivity of polycrystalline platinum nanofilms, *J. Phys. Condens. Matter.* 18 (2006) 7937–7950. https://doi.org/10.1088/0953-8984/18/34/007.
61. H. Dong, B. Wen, R. Melnik, Relative importance of grain boundaries and size effects in thermal conductivity of nanocrystalline materials, *Sci. Rep.* 4 (2014) 7037. https://doi.org/10.1038/srep07037.

62. G. Soyez, J.A. Eastman, L.J. Thompson, G.R. Bai, P.M. Baldo, A.W. McCormick, R.J. DiMelfi, A.A. Elmustafa, M.F. Tambwe, D.S. Stone, Grain-size-dependent thermal conductivity of nanocrystalline yttria-stabilized zirconia films grown by metal-organic chemical vapor deposition, *Appl. Phys. Lett.* 77 (2000) 1155–1157. https://doi.org/10.1063/1.1289803.
63. L. Qiu, X. Zheng, J. Zhu, G. Su, D. Tang, The effect of grain size on the lattice thermal conductivity of an individual polyacrylonitrile-based carbon fiber, *Carbon* 51 (2013) 265–273. https://doi.org/10.1016/j.carbon.2012.08.052.
64. K. Watari, K. Hirao, M. Toriyama, K. Ishizaki, Effect of grain size on the thermal conductivity of Si3N4, *J. Am. Ceram. Soc.* 82 (1999) 777–779. https://doi.org/10.1111/j.1151-2916.1999.tb01835.x.
65. Y. Ohishi, J. Xie, Y. Miyazaki, Y. Aikebaier, H. Muta, K. Kurosaki, S. Yamanaka, N. Uchida, T. Tada, Thermoelectric properties of heavily boron- and phosphorus-doped silicon, *Jpn. J. Appl. Phys.* 54 (2015) 071301. https://doi.org/10.7567/JJAP.54.071301.
66. M. Mohr, L. Daccache, S. Horvat, K. Brühne, T. Jacob, H.J. Fecht, Influence of grain boundaries on elasticity and thermal conductivity of nanocrystalline diamond films, *Acta Mater.* 122 (2017) 92–98. https://doi.org/10.1016/j.actamat.2016.09.042.
67. J. Anaya, S. Rossi, M. Alomari, E. Kohn, L. Tóth, B. Pécz, K.D. Hobart, T.J. Anderson, T.I. Feygelson, B.B. Pate, M. Kuball, Control of the in-plane thermal conductivity of ultra-thin nanocrystalline diamond films through the grain and grain boundary properties, *Acta Mater.* 103 (2016) 141–152. https://doi.org/10.1016/j.actamat.2015.09.045.
68. Y. Li, A. Wei, D. Datta, Thermal characteristics of graphene nanoribbons endorsed by surface functionalization, *Carbon* 113 (2017) 274–282. https://doi.org/10.1016/j.carbon.2016.11.067.
69. M.D. Losego, M.E. Grady, N.R. Sottos, D.G. Cahill, P. V. Braun, Effects of chemical bonding on heat transport across interfaces, *Nat. Mater.* 11 (2012) 502–506. https://doi.org/10.1038/nmat3303.
70. P.J. O'Brien, S. Shenogin, J. Liu, P.K. Chow, D. Laurencin, P.H. Mutin, M. Yamaguchi, P. Keblinski, G. Ramanath, Bonding-induced thermal conductance enhancement at inorganic heterointerfaces using nanomolecular monolayers, *Nat. Mater.* 12 (2013) 118–122. https://doi.org/10.1038/nmat3465.
71. S. Majumdar, J.A. Sierra-Suarez, S.N. Schiffres, W.L. Ong, C.F. Higgs, A.J.H. McGaughey, J.A. Malen, Vibrational mismatch of metal leads controls thermal conductance of self-assembled monolayer junctions, *Nano Lett.* 15 (2015) 2985–2991. https://doi.org/10.1021/nl504844d.
72. K. Zheng, F. Sun, J. Zhu, Y. Ma, X. Li, D. Tang, F. Wang, X. Wang, Enhancing the thermal conductance of polymer and sapphire interface via self-assembled monolayer, *ACS Nano* 10 (2016) 7792–7798. https://doi.org/10.1021/acsnano.6b03381.
73. N. Behabtu, C.C. Young, D.E. Tsentalovich, O. Kleinerman, X. Wang, A.W. Ma, E.A. Bengio, R.F. Waarbeek, J.J. Jong, R.E. Hoogerwerf, S.B. Fairchild, J.B. Ferguson, B. Maruyama, J. Kono, Y. Talmon, Y. Cohen, M.J. Otto, M. Pasquali, Strong, light, multifunctional fibers of carbon nanotubes with ultrahigh conductivity, *Science* 339 (2013) 182–186. https://doi.org/10.1126/science.1228061.
74. A.A. Balandin, Thermal properties of graphene and nanostructured carbon materials, *Nat. Mater.* 10 (2011) 569–581. https://doi.org/10.1038/nmat3064.
75. S. Hu, J. Chen, N. Yang, B. Li, Thermal transport in graphene with defect and doping: Phonon modes analysis, *Carbon* 116 (2017) 139–144. https://doi.org/10.1016/j.carbon.2017.01.089.
76. B. Mortazavi, S. Ahzi, Molecular dynamics study on the thermal conductivity and mechanical properties of boron doped graphene, *Solid State Commun.* 152 (2012) 1503–1507. https://doi.org/10.1016/j.ssc.2012.04.048.

77. E.K. Goharshadi, S.J. Mahdizadeh, Thermal conductivity and heat transport properties of nitrogen-doped graphene, *J. Mol. Graph. Model.* 62 (2015) 74–80. https://doi.org/10.1016/j.jmgm.2015.09.008.
78. T. Zhang, J. Li, Y. Cao, L. Zhu, G. Chen, Tailoring thermal transport properties of graphene by nitrogen doping, *J. Nanoparticle Res.* 19 (2017) 48. https://doi.org/10.1007/s11051-017-3749-2.
79. W. Lee, K.D. Kihm, H.G. Kim, W. Lee, S. Cheon, S. Yeom, G. Lim, K.R. Pyun, S.H. Ko, S. Shin, Two orders of magnitude suppression of graphene's thermal conductivity by heavy dopant (Si), *Carbon* 138 (2018) 98–107. https://doi.org/10.1016/j.carbon.2018.05.064.
80. J. Li, L. Porter, S. Yip, Atomistic modeling of finite-temperature properties of crystalline β-SiC II. Thermal conductivity and effects of point defects, *J. Nucl. Mater.* 255 (1998) 139–152. https://doi.org/10.1016/S0022-3115(98)00034-8.
81. H. Malekpour, P. Ramnani, S. Srinivasan, G. Balasubramanian, D.L. Nika, A. Mulchandani, R.K. Lake, A.A. Balandin, Thermal conductivity of graphene with defects induced by electron beam irradiation, *Nanoscale* 8 (2016) 14608–14616. https://doi.org/10.1039/C6NR03470E.
82. J. Haskins, A. Kinaci, C. Sevik, H. Sevinçli, G. Cuniberti, T. Çağin, Control of thermal and electronic transport in defect-engineered graphene nanoribbons, *ACS Nano* 5 (2011) 3779–3787. https://doi.org/10.1021/nn200114p.
83. H. Zhang, G. Lee, K. Cho, Thermal transport in graphene and effects of vacancy defects, *Phys. Rev. B* 84 (2011) 115460. https://doi.org/10.1103/PhysRevB.84.115460.
84. T. Feng, X. Ruan, Z. Ye, B. Cao, Spectral phonon mean free path and thermal conductivity accumulation in defected graphene: The effects of defect type and concentration, *Phys. Rev. B* 91 (2015) 224301. https://doi.org/10.1103/PhysRevB.91.224301.
85. J. Hu, S. Schiffli, A. Vallabhaneni, X. Ruan, Y.P. Chen, Tuning the thermal conductivity of graphene nanoribbons by edge passivation and isotope engineering: A molecular dynamics study, *Appl. Phys. Lett.* 97 (2010) 133107. https://doi.org/10.1063/1.3491267.
86. S.K. Chien, Y.T. Yang, C.K. Chen, Influence of chemisorption on the thermal conductivity of graphene nanoribbons, *Carbon* 50 (2012) 421–428. https://doi.org/10.1016/j.carbon.2011.08.056.
87. K. Chen, W. Li, W. Duan, Z. Shuai, B. Gu, Effect of defects on the thermal conductivity in a nanowire, *Phys. Rev. B* 72 (2005) 045422. https://doi.org/10.1103/PhysRevB.72.045422.
88. T.H. Fang, Z.W. Lee, W.J. Chang, C.C. Huang, Determining porosity effect on the thermal conductivity of single-layer graphene using a molecular dynamics simulation, *Phys. E Low-Dimens. Syst. Nanostruct.* 106 (2019) 90–94. https://doi.org/10.1016/j.physe.2018.10.017.
89. E. Pop, V. Varshney, A.K. Roy, Thermal properties of graphene: Fundamentals and applications, *MRS Bull.* 37 (2012) 1273–1281. https://doi.org/10.1557/mrs.2012.203.
90. Z.W. Tan, J.S. Wang, C.K. Gan, First-principles study of heat transport properties of graphene nanoribbons, *Nano Lett.* 11 (2011) 214–219. https://doi.org/10.1021/nl103508m.
91. J.D. Renteria, S. Ramirez, H. Malekpour, B. Alonso, A. Centeno, A. Zurutuza, A.I. Cocemasov, D.L. Nika, A.A. Balandin, Strongly anisotropic thermal conductivity of free-standing reduced graphene oxide films annealed at high temperature, *Adv. Funct. Mater.* 25 (2015) 4664–4672. https://doi.org/10.1002/adfm.201501429.
92. T. Xie, Y. He, Z. Hu, Theoretical study on thermal conductivities of silica aerogel composite insulating material, *Int. J. Heat Mass Transf.* 58 (2013) 540–552. https://doi.org/10.1016/j.ijheatmasstransfer.2012.11.016.
93. T.P. Teng, C.M. Cheng, C.P. Cheng, Performance assessment of heat storage by phase change materials containing MWCNTs and graphite, *Appl. Therm. Eng.* 50 (2013) 637–644. https://doi.org/10.1016/j.applthermaleng.2012.07.002.

94. M. Li, A nano-graphite/paraffin phase change material with high thermal conductivity, *Appl. Energy* 106 (2013) 25–30. https://doi.org/10.1016/j.apenergy.2013.01.031.
95. L. Fan, X. Fang, X. Wang, Y. Zeng, Y. Xiao, Z. Yu, X. Xu, Y. Hu, K. Cen, Effects of various carbon nanofillers on the thermal conductivity and energy storage properties of paraffin-based nanocomposite phase change materials, *Appl. Energy* 110 (2013) 163–172. https://doi.org/10.1016/j.apenergy.2013.04.043.
96. T. Ma, Z. Liu, J. Wen, Y. Gao, X. Ren, H. Chen, C. Jin, X.L. Ma, N. Xu, H.M. Cheng, W. Ren, Tailoring the thermal and electrical transport properties of graphene films by grain size engineering, *Nat. Commun.* 8 (2017) 14486. https://doi.org/10.1038/ncomms14486.
97. C. Yu, L. Shi, Z. Yao, D. Li, A. Majumdar, Thermal conductance and thermopower of an individual single-wall carbon nanotube, *Nano Lett.* 5 (2005) 1842–1846. https://doi.org/10.1021/nl051044e.
98. X.D. Din, E.E. Michaelides, Kinetic theory and molecular dynamics simulations of microscopic flows, *Phys. Fluids* 9 (1997) 3915–3925. https://doi.org/10.1063/1.869490.
99. Z. Xu, M.J. Buehler, Nanoengineering heat transfer performance at carbon nanotube interfaces, *ACS Nano* 3 (2009) 2767–2775. https://doi.org/10.1021/nn9006237.
100. V. Varshney, S.S. Patnaik, A.K. Roy, G. Froudakis, B.L. Farmer, P. Architectures, V. Varshney, S.S. Patnaik, A.K. Roy, G. Froudakis, B.L. Farmer, Modeling of thermal transport in pillared-graphene architectures, *ACS Nano* 4 (2010) 1153–1161. https://doi.org/10.1021/nn901341r.
101. N. Yang, G. Zhang, B. Li, Violation of Fourier's law and anomalous heat diffusion in silicon nanowires, *Nano Today* 5 (2010) 85–90. https://doi.org/10.1016/j.nantod.2010.02.002.
102. Y. Zhou, X. Zhang, M. Hu, Nonmonotonic diameter dependence of thermal conductivity of extremely thin Si nanowires: Competition between hydrodynamic phonon flow and boundary scattering, *Nano Lett.* 17 (2017) 1269–1276. https://doi.org/10.1021/acs.nanolett.6b05113.
103. A. Soleimani, H. Araghi, Z. Zabihi, A. Alibakhshi, A comparative study of molecular dynamics simulation methods for evaluation of the thermal conductivity and phonon transport in Si nanowires, *Comput. Mater. Sci.* 142 (2018) 346–354. https://doi.org/10.1016/j.commatsci.2017.10.024.
104. R.J. Stevens, L.V. Zhigilei, P.M. Norris, Effects of temperature and disorder on thermal boundary, *Int. J. Heat Mass Transf.* 50 (2007) 3977–3989. https://doi.org/10.1016/j.ijheatmasstransfer.2007.01.040.
105. Z. Liang, H.L. Tsai, Reduction of solid-solid thermal boundary resistance by inserting an interlayer, *Int. J. Heat Mass Transf.* 55 (2012) 2999–3007. https://doi.org/10.1016/j.ijheatmasstransfer.2012.02.019.
106. S. Merabia, K. Termentzidis, Thermal conductance at the interface between crystals using equilibrium and nonequilibrium molecular dynamics, *Phys. Rev. B* 86 (2012) 094303. https://doi.org/10.1103/PhysRevB.86.094303.
107. K. Termentzidis, S. Merabia, P. Chantrenne, P. Keblinski, Cross-plane thermal conductivity of superlattices with rough interfaces using equilibrium and non-equilibrium molecular dynamics, *Int. J. Heat Mass Transf.* 54 (2011) 2014–2020. https://doi.org/10.1016/j.ijheatmasstransfer.2011.01.001.
108. A. Majumdar, P. Reddy, Role of electron-phonon coupling in thermal conductance of metal-nonmetal interfaces, *Appl. Phys. Lett.* 84 (2004) 4768–4770. https://doi.org/10.1063/1.1758301.
109. Y. Wang, X. Ruan, A.K. Roy, Two-temperature nonequilibrium molecular dynamics simulation of thermal transport across metal-nonmetal interfaces, *Phys. Rev. B* 85 (2012) 205311. https://doi.org/10.1103/PhysRevB.85.205311.

110. P. Singh, M. Seong, S. Sinha, Detailed consideration of the electron-phonon thermal conductance at metal-dielectric interfaces, *Appl. Phys. Lett.* 102 (2013) 181906. https://doi.org/10.1063/1.4804383.
111. C.W. Chang, A.M. Fennimore, A. Afanasiev, D. Okawa, T. Ikuno, H. Garcia, D. Li, A. Majumdar, A. Zettl, Isotope effect on the thermal conductivity of boron nitride nanotubes, *Phys. Rev. Lett.* 97 (2006) 085901. https://doi.org/10.1103/PhysRevLett.97.085901.
112. L. Lindsay, D.A. Broido, N. Mingo, Diameter dependence of carbon nanotube thermal conductivity and extension to the graphene limit, *Phys. Rev. B* 82 (2010) 161402. https://doi.org/10.1103/PhysRevB.82.161402.
113. J. Che, T. Çagin, W.A. Goddard III, Thermal conductivity of carbon nanotubes, *Nanotechnology* 11 (2000) 65–69. https://doi.org/10.1088/0957-4484/11/2/305.
114. S. Berber, Y.K. Kwon, D. Tománek, Unusually high thermal conductivity of carbon nanotubes, *Phys. Rev. Lett.* 84 (2000) 4613–4616. https://doi.org/10.1103/PhysRevLett.84.4613.
115. D. Donadio, G. Galli, Thermal conductivity of isolated and interacting carbon nanotubes: Comparing results from molecular dynamics and the Boltzmann transport equation, *Phys. Rev. Lett.* 99 (2007) 255502. https://doi.org/10.1103/PhysRevLett.99.255502.
116. I. Ivanov, A. Puretzky, G. Eres, H. Wang, Z. Pan, H. Cui, R. Jin, J. Howe, D.B. Geohegan, Fast and highly anisotropic thermal transport through vertically aligned carbon nanotube arrays, *Appl. Phys. Lett.* 89 (2006) 2004–2007. https://doi.org/10.1063/1.2397008.
117. D. Yang, Q. Zhang, G. Chen, S.F. Yoon, J. Ahn, S. Wang, Q. Zhou, Q. Wang, J. Li, Thermal conductivity of multiwalled carbon nanotubes, *Phys. Rev. B* 66 (2002) 165440. https://doi.org/10.1103/PhysRevB.66.165440.
118. R. Jin, Z. Zhou, D. Mandrus, I.N. Ivanov, G. Eres, J.Y. Howe, A.A. Puretzky, D.B. Geohegan, The effect of annealing on the electrical and thermal transport properties of macroscopic bundles of long multi-wall carbon nanotubes, *Phys. B* 388 (2007) 326–330. https://doi.org/10.1016/j.physb.2006.06.135.
119. W. Yi, L. Lu, D. Zhang, Z. Pan, S. Xie, Linear specific heat of carbon nanotubes, *Phys. Rev. B* 59 (1999) R9015–R9018. https://doi.org/10.1103/PhysRevB.59.R9015.
120. A.E. Aliev, M.H. Lima, E.M. Silverman, R.H. Baughman, Thermal conductivity of multi-walled carbon nanotube sheets: Radiation losses and quenching of phonon modes, *Nanotechnology* 21 (2010) 035709. https://doi.org/10.1088/0957-4484/21/3/035709.
121. J.E. Fischer, W. Zhou, J. Vavro, M.C. Llaguno, C. Guthy, R. Haggenmueller, M.J. Casavant, D.E. Walters, R.E. Smalley, Magnetically aligned single wall carbon nanotube films: Preferred orientation and anisotropic transport properties, *J. Appl. Phys.* 93 (2003) 2157–2163. https://doi.org/10.1063/1.1536733.
122. T.S. Gspann, S.M. Juckes, J.F. Niven, M.B. Johnson, J.A. Elliott, M.A. White, A.H. Windle, High thermal conductivities of carbon nanotube films and micro-fibres and their dependence on morphology, *Carbon* 114 (2017) 160–168. https://doi.org/10.1016/j.carbon.2016.12.006.
123. J. Hone, M.C. Llaguno, N.M. Nemes, A.T. Johnson, J.E. Fischer, D.A. Walters, M.J. Casavant, J. Schmidt, R.E. Smalley, Electrical and thermal transport properties of magnetically aligned single wall carbon nanotube films, *Appl. Phys. Lett.* 77 (2000) 666–668. https://doi.org/10.1063/1.127079.
124. E. Muñoz, J. Lu, B.I. Yakobson, Ballistic thermal conductance of graphene ribbons, *Nano Lett.* 10 (2010) 1652–1656. https://doi.org/10.1021/nl904206d.
125. Z. Han, A. Fina, Thermal conductivity of carbon nanotubes and their polymer nanocomposites: A review, *Prog. Polym. Sci.* 36 (2011) 914–944. https://doi.org/10.1016/j.progpolymsci.2010.11.004.
126. X. Jiang, Q. Weng, X. Wang, X. Li, J. Zhang, D. Golberg, Y. Bando, Recent progress on fabrications and applications of Boron Nitride nanomaterials: A review, *J. Mater. Sci. Technol.* 31 (2015) 589–598. https://doi.org/10.1016/j.jmst.2014.12.008.

127. V. Guerra, C. Wan, T. McNally, Thermal conductivity of 2D nano-structured boron nitride (BN) and its composites with polymers, *Prog. Mater. Sci.* 100 (2019) 170–186. https://doi.org/10.1016/j.pmatsci.2018.10.002.
128. X. Zheng, L. Qiu, G. Su, D. Tang, Y. Liao, Y. Chen, Thermal conductivity and thermal diffusivity of SiO2 nanopowder, *J. Nanoparticle Res.* 13 (2011) 6887–6893. https://doi.org/10.1007/s11051-011-0596-4.
129. L.W. Hrubesh, Aerogel applications, J. Non. *Cryst. Solids*. 225 (1998) 335–342. https://doi.org/10.1016/S0022-3093(98)00135-5.
130. E.T. Afriyie, P. Karami, P. Norberg, K. Gudmundsson, Textural and thermal conductivity properties of a low density mesoporous silica material, *Energy Build.* 75 (2014) 210–215. https://doi.org/10.1016/j.enbuild.2014.02.012.
131. J. Wang, J. Kuhn, X. Lu, Monolithic silica aerogel insulation doped with TiO2 powder and ceramic fibers, *J. Non. Cryst. Solids*. 186 (1995) 296–300. https://doi.org/10.1016/0022-3093(95)00068-2.
132. F. Tian, B. Song, X. Chen, N.K. Ravichandran, Y. Lv, K. Chen, S. Sullivan, J. Kim, Y. Zhou, T.H. Liu, M. Goni, Z. Ding, J. Sun, G. Amila, G.U. Gamage, H. Sun, H. Ziyaee, S. Huyan, L. Deng, J. Zhou, A.J. Schmidt, S. Chen, C.W. Chu, P.Y. Huang, D. Broido, L. Shi, G. Chen, Z. Ren, High thermal conductivity in cubic boron arsenide crystals, *Science* 361 (2018) 582–585.
133. P. Yue, L. Qiu, X. Zheng, D. Tang, The effective thermal conductivity of porous polymethacrylimide foams, *Key Eng. Mater.* 609–610 (2014) 196–200. https://doi.org/10.4028/www.scientific.net/KEM.609-610.196.
134. J. Li, J. Yin, T. Ji, Y. Feng, Y. Liu, H. Zhao, Y. Li, C. Zhu, D. Yue, B. Su, X. Liu, Microstructure evolution effect on high-temperature thermal conductivity of LDPE/BNNS investigated by in-situ SAXS, *Mater. Lett.* 234 (2019) 74–78. https://doi.org/10.1016/j.matlet.2018.09.061.
135. G. Kalaprasad, P. Pradeep, G. Mathew, C. Pavithran, S. Thomas, Thermal conductivity and thermal diffusivity analyses of low-density polyethylene composites reinforced with sisal, glass and intimately mixed sisal/glass fibres, *Compos. Sci. Technol.* 60 (2000) 2967–2977. https://doi.org/10.1016/S0266-3538(00)00162-7.
136. F. Mashali, E.M. Languri, J. Davidson, D. Kerns, W. Johnson, K. Nawaz, G. Cunningham, Thermo-physical properties of diamond nanofluids: A review, *Int. J. Heat Mass Transf.* 129 (2019) 1123–1135. https://doi.org/10.1016/j.ijheatmasstransfer.2018.10.033.
137. J. Guo, X. Wang, T. Wang, Thermal characterization of microscale conductive and non-conductive wires using transient electrothermal technique, *J. Appl. Phys.* 101 (2007) 063537. https://doi.org/10.1063/1.2714679.
138. A.M. Papadopoulos, State of the art in thermal insulation materials and aims for future developments, *Energy Build.* 37 (2005) 77–86. https://doi.org/10.1016/j.enbuild.2004.05.006.
139. W.J. Parker, R.J. Jenkins, C.P. Butler, G.L. Abbott, Flash method of determining thermal diffusivity, heat capacity, and thermal conductivity, *J. Appl. Phys.* 32 (1961) 1679–1684. https://doi.org/10.1063/1.1728417.
140. J. Blumm, A. Lindemann, S. Min, Thermal characterization of liquids and pastes using the flash technique, *Thermochim. Acta* 455 (2007) 26–29. https://doi.org/10.1016/j.tca.2006.11.023.
141. W. Lin, J. Shang, W. Gu, C.P. Wong, Parametric study of intrinsic thermal transport in vertically aligned multi-walled carbon nanotubes using a laser flash technique, *Carbon* 50 (2012) 1591–1603. https://doi.org/10.1016/j.carbon.2011.11.038.
142. H. Xie, A. Cai, X. Wang, Thermal diffusivity and conductivity of multiwalled carbon nanotube arrays, *Phys. Lett. A* 369 (2007) 120–123. https://doi.org/10.1016/j.physleta.2007.02.079.

143. M. Akoshima, K. Hata, D.N. Futaba, K. Mizuno, T. Baba, M. Yumura, Thermal diffusivity of single-walled carbon nanotube forest measured by laser flash method, *Jpn. J. Appl. Phys.* 48 (2009) 05EC07. https://doi.org/10.1143/JJAP.48.05EC07.
144. X. Zhang, D. Sun, Y. Li, G.H. Lee, X. Cui, D. Chenet, Y. You, T.F. Heinz, J.C. Hone, Measurement of lateral and interfacial thermal conductivity of single- and bilayer MoS_2 and $MoSe_2$ using refined optothermal Raman technique, *ACS Appl. Mater. Interfaces* 7 (2015) 25923–25929. https://doi.org/10.1021/acsami.5b08580.
145. P. Thiyagarajan, Z. Yan, J.C. Yoon, M.W. Oh, J.H. Jang, Thermal conductivity reduction in three dimensional graphene-based nanofoam, *RSC Adv.* 5 (2015) 99394–99397. https://doi.org/10.1039/c5ra19130k.
146. Q. Li, W. Ma, X. Zhang, Laser flash Raman spectroscopy method for characterizing thermal diffusivity of supported 2D nanomaterials, *Int. J. Heat Mass Transf.* 95 (2016) 956–963. https://doi.org/10.1016/j.ijheatmasstransfer.2015.12.065.
147. Q. Li, X. Zhang, K. Takahashi, Variable-spot-size laser-flash Raman method to measure in-plane and interfacial thermal properties of 2D van der Waals heterostructures, *Int. J. Heat Mass Transf.* 125 (2018) 1230–1239. https://doi.org/10.1016/j.ijheatmasstransfer.2018.05.011.
148. C. Xing, T. Munro, C. Jensen, H. Ban, C.G. Copeland, R. V. Lewis, Thermal characterization of natural and synthetic spider silks by both the 3ω and transient electrothermal methods, *Mater. Des.* 119 (2017) 22–29. https://doi.org/10.1016/j.matdes.2017.01.057.
149. C. Xing, T. Munro, C. Jensen, H. Ban, Analysis of the electrothermal technique for thermal property characterization of thin fibers, *Meas. Sci. Technol.* 24 (2013) 105603. https://doi.org/10.1088/0957-0233/24/10/105603.
150. X. Feng, X. Wang, Thermophysical properties of free-standing micrometer-thick Poly(3-hexylthiophene) films, *Thin Solid Films* 519 (2011) 5700–5705. https://doi.org/10.1016/j.tsf.2011.03.043.
151. X. Huang, J. Wang, G. Eres, X. Wang, Thermophysical properties of multi-wall carbon nanotube bundles at elevated temperatures up to 830 K, *Carbon* 49 (2011) 1680–1691. https://doi.org/10.1016/j.carbon.2010.12.053.
152. X. Huang, G. Liu, X. Wang, New secrets of spider silk: Exceptionally high thermal conductivity and its abnormal change under stretching, *Adv. Mater.* 24 (2012) 1482–1486. https://doi.org/10.1002/adma.201104668.
153. J. Hou, X. Wang, C. Liu, H. Cheng, Development of photothermal-resistance technique and its application to thermal diffusivity measurement of single-wall carbon nanotube bundles, *Appl. Phys. Lett.* 88 (2006) 181910. https://doi.org/10.1063/1.2199614.
154. J. Hou, X. Wang, J. Guo, Thermal characterization of micro/nanoscale conductive and non-conductive wires based on optical heating and electrical thermal sensing, *J. Phys. D. Appl. Phys.* 39 (2006) 3362–3370. https://doi.org/10.1088/0022-3727/39/15/021.
155. J. Hou, X. Wang, L. Zhang, Thermal characterization of submicron polyacrylonitrile fibers based on optical heating and electrical thermal sensing, *Appl. Phys. Lett.* 89 (2006) 152504. https://doi.org/10.1063/1.2358952.
156. C.S. Welch, D.M. Heath, W.P. Winfree, Remote measurement of in-plane diffusivity components in plates, *J. Appl. Phys.* 61 (1987) 895–898. https://doi.org/10.1063/1.338140.
157. L.I. Giri, S. Tuli, M. Sharma, P. Bugnon, H. Berger, A. Magrez, Thermal diffusivity measurements of templated nanocomposite using infrared thermography, *Mater. Lett.* 115 (2014) 106–108. https://doi.org/10.1016/j.matlet.2013.10.042.
158. A. Mendioroz, R. Fuente-Dacal, E. Apianiz, A. Salazar, Thermal diffusivity measurements of thin plates and filaments using lock-in thermography, *Rev. Sci. Instrum.* 80 (2009) 074904. https://doi.org/10.1063/1.3176467.
159. K. Sakai, Y. Kobayashi, T. Saito, A. Isogai, Partitioned airs at microscale and nanoscale: Thermal diffusivity in ultrahigh porosity solids of nanocellulose, *Sci. Rep.* 6 (2016) 20434. https://doi.org/10.1038/srep20434.

160. M.M.S. Wahsh, R.M. Khattab, N.M. Khalil, F. Gouraud, M. Huger, T. Chotard, Fabrication and technological properties of nanoporous spinel/forsterite/zirconia ceramic composites, *Mater. Des.* 53 (2014) 561–567. https://doi.org/10.1016/j.matdes.2013.07.059.
161. M. Venkataraman, R. Mishra, J. Militky, L. Hes, Aerogel based nanoporous fibrous materials for thermal insulation, *Fibers Polym.* 15 (2014) 1444–1449. https://doi.org/10.1007/s12221-014-1444-9.
162. Y. Hirata, Y. Kinoshita, T. Shimonosono, T. Chaen, Theoretical and experimental analyses of thermal properties of porous polycrystalline mullite, *Ceram. Int.* 43 (2017) 9973–9978. https://doi.org/10.1016/j.ceramint.2017.05.009.
163. G. Mayr, B. Plank, J. Sekelja, G. Hendorfer, Active thermography as a quantitative method for non-destructive evaluation of porous carbon fiber reinforced polymers, *NDT E Int.* 44 (2011) 537–543. https://doi.org/10.1016/j.ndteint.2011.05.012.
164. E. Wu, M. Li, Q. Gao, A. Mandelis, Effect of porosity on thermal diffusivity of woven carbon fiber reinforced polymers, *Laser Opt. Prog.* 11 (2017) 326–331. https://doi.org/10.3788/LOP54.111601.
165. J. Renteria, D. Nika, A. Balandin, Graphene thermal properties: Applications in thermal management and energy storage, *Appl. Sci.* 4 (2014) 525–547. https://doi.org/10.3390/app4040525
166. T.W. Pan, W.S. Kuo, N.H. Tai, Tailoring anisotropic thermal properties of reduced graphene oxide/multi-walled carbon nanotube hybrid composite films, *Compos. Sci. Technol.* 151 (2017) 44–51. https://doi.org/10.1016/j.compscitech.2017.07.015.
167. R.E. Hummel, *Electronic Properties of Materials*, Springer Science & Business Media, New York, 2011.
168. W.H. Tuan, T.T. Chou, C.T. Kao, S.Y. Wang, B.J. Weng, Thermal diffusivity of graphite paper and its joint with alumina substrate, *J. Eur. Ceram. Soc.* 38 (2018) 187–191. https://doi.org/10.1016/j.jeurceramsoc.2017.07.029.
169. E.M. Jackson, P.E. Laibinis, W.E. Collins, A. Ueda, C.D. Wingard, B. Penn, Development and thermal properties of carbon nanotube-polymer composites, *Compos. Part B Eng.* 89 (2016) 362–373. https://doi.org/10.1016/j.compositesb.2015.12.018.
170. K. Voges, M. Vadala, D.C. Lupascu, Dense nanopowder composites for thermal insulation, *Phys. Status Solidi Appl. Mater. Sci.* 212 (2015) 439–442. https://doi.org/10.1002/pssa.201431551.
171. G. Zhan, A.K. Mukherjee, Carbon nanotube reinforced alumina-based ceramics with novel mechanical, electrical, and thermal properties, *Int. J. Appl. Ceram. Technol.* 1 (2004) 161–171. https://doi.org/10.1111/j.1744-7402.2004.tb00166.x.
172. T. Borca-Tasciuc, S. Vafaei, D.A. Borca-Tasciuc, B.Q. Wei, R. Vajtai, P.M. Ajayan, Anisotropic thermal diffusivity of aligned multiwall carbon nanotube arrays, *J. Appl. Phys.* 98 (2005) 054309. https://doi.org/10.1063/1.2034079.
173. Q. Gong, Z. Li, X. Bai, D. Li, Y. Zhao, J. Liang, Thermal properties of aligned carbon nanotube/carbon nanocomposites, *Mater. Sci. Eng. A* 384 (2004) 209–214. https://doi.org/10.1016/j.msea.2004.06.006.
174. Y. Yue, X. Huang, X. Wang, Thermal transport in multiwall carbon nanotube buckypapers, *Phys. Lett. A* 374 (2010) 4144–4151. https://doi.org/10.1016/j.physleta.2010.08.034.
175. Y. Xie, T. Wang, B. Zhu, C. Yan, P. Zhang, X. Wang, G. Eres, 19-Fold thermal conductivity increase of carbon nanotube bundles toward high-end thermal design applications, *Carbon* 139 (2018) 445–458. https://doi.org/10.1016/j.carbon.2018.07.009.
176. T. Wang, X. Wang, J. Guo, Z. Luo, K. Cen, Characterization of thermal diffusivity of micro/nanoscale wires by transient photo-electro-thermal technique, *Appl. Phys. A* 87 (2007) 599–605. https://doi.org/10.1007/s00339-007-3879-y.

177. L. Kumari, T. Zhang, G.H. Du, W.Z. Li, Q.W. Wang, A. Datye, K.H. Wu, Thermal properties of CNT-Alumina nanocomposites, *Compos. Sci. Technol.* 68 (2008) 2178–2183. https://doi.org/10.1016/j.compscitech.2008.04.001.
178. H. Lin, S. Xu, X. Wang, N. Mei, Significantly reduced thermal diffusivity of freestanding two-layer graphene in graphene foam, *Nanotechnology* 24 (2013) 415706. https://doi.org/10.1088/0957-4484/24/41/415706.
179. P. Goli, H. Ning, X. Li, C.Y. Lu, K.S. Novoselov, A.A. Balandin, Thermal properties of graphene-copper-graphene heterogeneous films, *Nano Lett.* 14 (2014) 1497–1503. https://doi.org/10.1021/nl404719n.
180. Z. Hou, W. Song, P. Wang, M.J. Meziani, C.Y. Kong, A. Anderson, H. Maimaiti, G.E. Lecroy, H. Qian, Y.P. Sun, Flexible graphene-graphene composites of superior thermal and electrical transport properties, *ACS Appl. Mater. Interfaces* 6 (2014) 15026–15032. https://doi.org/10.1021/am502986j.
181. S.N. Goyanes, J.D. Marconi, P.G. König, G.H. Rubiolo, C.L. Matteo, A.J. Marzocca, Analysis of thermal diffusivity in aluminum (particle)-filled PMMA compounds, *Polymer* 42 (2001) 5267–5274. https://doi.org/10.1016/S0032-3861(00)00877-6.
182. A. Prasad, A. Ambirajan, Criteria for accurate measurement of thermal diffusivity of solids using the Angstrom method, *Int. J. Therm. Sci.* 134 (2018) 216–223. https://doi.org/10.1016/j.ijthermalsci.2018.08.007.
183. B. Weidenfeller, M. Höfer, F.R. Schilling, Thermal conductivity, thermal diffusivity, and specific heat capacity of particle filled polypropylene, *Compos. Part A Appl. Sci. Manuf.* 35 (2004) 423–429. https://doi.org/10.1016/j.compositesa.2003.11.005.
184. M. Wiener, G. Reichenauer, S. Braxmeier, F. Hemberger, H.P. Ebert, Carbon aerogel-based high-temperature thermal insulation, *Int. J. Thermophys.* 30 (2009) 1372–1385. https://doi.org/10.1007/s10765-009-0595-1.
185. J. Feng, J. Feng, C. Zhang, Thermal conductivity of low density carbon aerogels, *J. Porous Mater.* 19 (2012) 551–556. https://doi.org/10.1007/s10934-011-9504-7.
186. Y. Xie, S. Xu, Z. Xu, H. Wu, C. Deng, X. Wang, Interface-mediated extremely low thermal conductivity of graphene aerogel, *Carbon* 98 (2016) 381–390. https://doi.org/10.1016/j.carbon.2015.11.033.
187. J.P. Holman, Heat Transfer, Tata McGraw-Hill Education, New York, 2002.
188. A. Einstein, Die plancksche theorie der strahlung und die theorie der spezifischen waerme, *Ann. Phys-Berlin* 327 (1907) 180–190. https://doi.org/10.1002/andp.19063270110.
189. P. Debye, Zur theorie der spezifischen wärmen, *Ann. Phys-Berlin* 344 (1912) 789–839. https://doi.org/10.1002/andp.19123441404.
190. L. Lu, W. Yi, D. Zhang, 3ω method for specific heat and thermal conductivity measurements, *Rev. Sci. Instrum.* 72 (2001) 2996–3003. https://doi.org/10.1063/1.1378340.
191. X.J. Hu, A.A. Padilla, J. Xu, T.S. Fisher, K.E. Goodson, 3-Omega measurements of vertically oriented carbon nanotubes on silicon, *J. Heat Trans-T ASME* 128 (2006) 1109–1113. https://doi.org/10.1115/1.2352778.
192. L. Qiu, P. Guo, X. Yang, Y. Ouyang, Y. Feng, X. Zhang, J. Zhao, X. Zhang, Q. Li, Electro curing of oriented bismaleimide between aligned carbon nanotubes for high mechanical and thermal performances, *Carbon* 145 (2019) 650–657. https://doi.org/10.1016/j.carbon.2019.01.074.
193. A. Sikora, H. Ftouni, J. Richard, C. Hébert, D. Eon, F. Omnès, O. Bourgeois, Highly sensitive thermal conductivity measurements of suspended membranes (SiN and diamond) using a 3ω-Völklein method, *Rev. Sci. Instrum.* 83 (2012) 054902. https://doi.org/10.1063/1.4704086.
194. H. Ftouni, D. Tainoff, J. Richard, K. Lulla, J. Guidi, E. Collin, O. Bourgeois, Specific heat measurement of thin suspended SiN membrane from 8 K to 300 K using the 3ω-Völklein method, *Rev. Sci. Instrum.* 84 (2013) 094902. https://doi.org/10.1063/1.4821501.

195. F. Völklein, H. Reith, A. Meier, Measuring methods for the investigation of in-plane and cross-plane thermal conductivity of thin films, *Phys. Status Solidi.* 210 (2013) 106–118. https://doi.org/10.1002/pssa.201228478.
196. J.M. Schliesser, B.F. Woodfield, Lattice vacancies responsible for the linear dependence of the low-temperature heat capacity of insulating materials, *Phys. Rev. B* 91 (2015) 024109. https://doi.org/10.1103/PhysRevB.91.024109.
197. Q. Shi, L. Zhang, M.E. Schlesinger, J. Boerio-Goates, B.F. Woodfield, Low temperature heat capacity study of Fe_3PO_7 and $Fe_4(P_2O_7)_3$, *J. Chem. Thermodyn.* 62 (2013) 86–91. https://doi.org/10.1016/j.jct.2013.02.023.
198. T.H.K. Barron, G.K. White, *Heat Capacity and Thermal Expansion at Low Temperatures*, Springer Science & Business Media, Berlin, 2012.
199. Q. Shi, Z. Tan, N Yin, Low temperature calorimetry and its application in material research, *Chin. Sci. Bull.* 61 (2016) 3100–3114. https://engine.scichina.com/doi/10.1360/N972016-00550.
200. Y. Zhang, C.H. Wong, J. Shen, S.T. Sze, B. Zhang, H. Zhang, Y. Dong, H. Xu, Z. Yan, Y. Li, X. Hu, R. Lortz, Dramatic enhancement of superconductivity in single-crystalline nanowire arrays of Sn, *Sci. Rep.* 6 (2016) 32963. https://doi.org/10.1038/srep32963.
201. D. Lee, G.D. Sim, K. Xiao, Y. Seok Choi, J.J. Vlassak, Scanning AC nanocalorimetry study of Zr/B reactive multilayers, *J. Appl. Phys.* 114 (2013) 214902. https://doi.org/10.1063/1.4833572.
202. K. Xiao, J.M. Gregoire, P.J. McCluskey, J.J. Vlassak, A scanning AC calorimetry technique for the analysis of nano-scale quantities of materials, *Rev. Sci. Instrum.* 83 (2012) 114901. https://doi.org/10.1063/1.4763571.
203. N.R. Pradhan, H. Duan, J. Liang, G.S. Iannacchione, The specific heat and effective thermal conductivity of composites containing single-wall and multi-wall carbon nanotubes, *Nanotechnology* 20 (2009) 245705. https://doi.org/10.1088/0957-4484/20/24/245705.
204. M. Tress, M. Erber, E.U. Mapesa, H. Huth, J. Müller, A. Serghei, C. Schick, K.J. Eichhorn, B. Voit, F. Kremer, Glassy dynamics and glass transition in nanometric thin layers of polystyrene, *Macromolecules* 43 (2010) 9937–9944. https://doi.org/10.1021/ma102031k.
205. V. Novotny, P.P.M. Meincke, J.H.P. Watson, Effect of size and surface on the specific heat of small lead particles, *Phys. Rev. Lett.* 28 (1972) 901–903. https://doi.org/10.1103/PhysRevLett.28.901.
206. J. Rupp, R. Birringer, Enhanced specific-heat-capacity (*c*p) measurements (150–300 K) of nanometer-sized crystalline materials, *Phys. Rev. B* 36 (1987) 7888–7890. https://doi.org/10.1103/PhysRevB.36.7888.
207. B.A. Strukov, S.T. Davitadze, S.N. Kravchun, S.A. Taraskin, M. Goltzman, V.V. Lemanov, S.G. Shulman, Specific heat and heat conductivity of $BaTiO_3$ polycrystalline films in the thickness range, *J. Phys. Condens. Matter.* 15 (2003) 4331–4340. https://doi.org/10.1088/0953-8984/15/25/304.
208. J. Seo, D. Shin, Size effect of nanoparticle on specific heat in a ternary nitrate ($LiNO_3$–$NaNO_3$–KNO_3) salt eutectic for thermal energy storage, *Appl. Therm. Eng.* 102 (2016) 144–148. https://doi.org/10.1016/j.applthermaleng.2016.03.134.
209. M. Singh, S. Lara, S. Tlali, Effects of size and shape on the specific heat, melting entropy and enthalpy of nanomaterials, *J. Taibah Univ. Sci.* 11 (2017) 922–929. https://doi.org/10.1016/j.jtusci.2016.09.011.
210. B. Wang, L. Zhou, X. Peng, Surface and size effects on the specific heat capacity of nanoparticles, *Int. J. Thermophys.* 27 (2006) 139–151. https://doi.org/10.1007/s10765-006-0022-9.

211. N. Arora, D.P. Joshi, U. Pachauri, Effect of size and dimension dependent specific heat on thermal conductivity of nanostructured semiconductors, *Mater. Chem. Phys.* 217 (2018) 235–241. https://doi.org/10.1016/j.matchemphys.2018.05.071.
212. A.I. Gusev, Effects of the nanocrystalline state in solids, *Uspekhi Fiz. Nauk.* 41 (1998) 49–76. https://doi.org/10.1070/PU1998v041n01ABEH000329.
213. J.F. Tang, X.S. Li, W.Y. Long, Y. Wang, Vibrational properties in nanocrystalline nickels: Temperature effects and composite model for thermodynamics, *Phys. Status Solidi.* 245 (2008) 1527–1533. https://doi.org/10.1002/pssb.200743160.
214. D. Wolf, J. Wang, S.R. Phillpot, H. Gleiter, Phonon-induced anomalous specific heat of a nanocrystalline model material by computer simulation, *Phys. Rev. Lett.* 74 (1995) 4686–4689. https://doi.org/10.1103/PhysRevLett.74.4686.
215. N. Sun, K. Lu, Heat-capacity comparison among the nanocrystalline, amorphous, and coarse-grained polycrystalline states in element selenium, *Phys. Rev. B* 54 (1996) 6058–6061. https://doi.org/10.1103/PhysRevB.54.6058.
216. D.P. Rojas, L. Fernández Barquín, J. Rodríguez Fernández, L. Rodríguez Fernández, J. Gonzalez, Phonon softening on the specific heat of nanocrystalline metals, *Nanotechnology* 21 (2010) 445702. https://doi.org/10.1088/0957-4484/21/44/445702.
217. Y. Li, J. Luo, Y. Yi, H. Zhu, Z. Gan, X. Ji, H. Lei, Study on low temperature specific heat capacity of Aluminum nanocrystalline, *J. Synth. Cryst.* 43 (2014) 676–681. https://doi.org/10.16553/j.cnki.issn1000-985x.2014.03.015.
218. S.P. Adiga, V.P. Adiga, R.W. Carpick, D.W. Brenner, Vibrational properties and specific heat of ultrananocrystalline diamond: Molecular dynamics simulations, *J. Phys. Chem. C* 115 (2011) 21691–21699. https://doi.org/10.1021/jp207424m.
219. H.Y. Bai, J.L. Luo, D. Jin, J.R. Sun, Particle size and interfacial effect on the specific heat of nanocrystalline Fe, *J. Appl. Phys.* 79 (1996) 361–364. https://doi.org/10.1063/1.4763571.
220. H. Lei, J. Luo, J. Li, F. Dai, M. Yang, J. Zhang, J. Zhang, Anomalous specific heats of metallic nanocrystals induced by surface oxidation, *Appl. Phys. Lett.* 109 (2016) 213106. https://doi.org/10.1063/1.4968815.
221. R.C. Zeller, R.O. Pohl, Thermal conductivity and specific heat of noncrystalline solids, *Phys. Rev. B* 4 (1971) 2029–2041. https://doi.org/10.1103/PhysRevB.4.2029.
222. D. Gerlich, B. Abeles, R.E. Miller, High-temperature specific heats of Ge, Si, and Ge–Si alloys, *J. Appl. Phys.* 36 (1965) 76–79. https://doi.org/10.1063/1.1713926.
223. J.M. Ziman, Electrons and Phonons: The Theory of Transport Phenomena in Solids, Oxford University Press, London, 2001.
224. J. Hone, B. Batlogg, Z. Benes, A.T. Johnson, J.E. Fischer, Quantized phonon spectrum of single-wall carbon nanotubes, *Science* 289 (2000) 1730–1733. https://doi.org/10.1126/science.289.5485.1730.
225. Y. Li, X. Qiu, Y. Yin, F. Yang, Q. Fan, The specific heat of carbon nanotube networks and their potential applications, *J. Phys. D. Appl. Phys.* 42 (2009) 155405. https://dx.doi.org/10.1088/0022-3727/42/15/155405.
226. V.N. Popov, Low-temperature specific heat of nanotube systems, *Phys. Rev. B* 66 (2002) 153408. https://doi.org/10.1103/PhysRevB.66.153408.
227. B. Xiang, C.B. Tsai, C.J. Lee, D.P. Yu, Y.Y. Chen, Low-temperature specific heat of double wall carbon nanotubes, *Solid State Commun.* 138 (2006) 516–520. https://doi.org/10.1016/j.ssc.2006.04.022.
228. J.C. Lasjaunias, K. Biljaković, Z. Benes, J.E. Fischer, P. Monceau, Low-temperature specific heat of single-wall carbon nanotubes, *Phys. Rev. B* 65 (2002) 113409. https://doi.org/10.1103/PhysRevB.65.113409.

229. A. Alofi, G.P. Srivastava, Phonon conductivity in graphene, *J. Appl. Phys.* 112 (2012) 013517. https://doi.org/10.1063/1.4733690.
230. D.L. Nika, A.I. Cocemasov, A.A. Balandin, Specific heat of twisted bilayer graphene: Engineering phonons by atomic plane rotations, *Appl. Phys. Lett.* 105 (2014) 031904. https://doi.org/10.1063/1.4890622.
231. H. Zhang, H. Lei, Y. Tang, J. Luo, K. Li, X. Deng, Thermal capacity of nanocrystalline copper at low temperatures, *Acta Phys. Sin.* 59 (2010) 471–475. https://doi.org/10.7666/d.y1820142.

3 Colloidal Nanofluids Thermal Conductivity

Nanofluids are advanced colloids made by dispersing 1–100 nm nanoparticles in a liquid. According to recent research, nanofluids have superior thermophysical characteristics in comparison to traditional fluids. Multiple factors and flow conditions influence the thermophysical properties of nanofluids. As a result, it is essential to investigate the critical factors influencing their behavior thoroughly. This section summarizes studies on the thermophysical characteristics of nanofluids. Theoretical research, model development from classical to advanced models, thermophysical property measurement methodologies, and critical factors impacting the thermophysical properties of nanofluids are all covered in-depth in this study.

3.1 THEORETICAL MODELS

The main thermophysical feature of nanoparticles is thermal conductivity (k_{nf}), influenced by the material, volume fraction, size, aspect ratio, base fluid thermophysical properties, temperature, and surfactant. Numerous studies in recent years have demonstrated that the thermal conductivity of nanofluids is greater than that of ordinary fluids due to the unpredictable motion of nanoparticles, and many models have been suggested [1–5].

The history of traditional thermal conductivity models is seen schematically in Figure 3.1. The solid–liquid thermal conductivity model was proposed by Maxwell, and the first model discussed below is built on Maxwell's theory of electrical conductivity [6].

The particles in this model are assumed to be spherical and randomly distributed in solution. The effective thermal conductivity of nanofluids is expressed as follows:

$$k_{nf} = \frac{k_p + 2k_{bf} + 2\Phi_p\left(k_p - k_{bf}\right)}{k_p + 2k_{bf} - \Phi_p\left(k_p - k_{bf}\right)} k_{bf} \tag{3.1}$$

Here k_p and k_{bf} represent the thermal conductivity of the particles and the base fluid, respectively. The volume fraction of the nanoparticles is denoted by Φ_p. It is important to note that this model is only valid in the situation where k_p resembles k_{bf}.

Hamilton [7] expanded the model to make it applicable to non-spherical particles. The model is transformed into the following form, taking into consideration the shape factor of particles:

$$k_{nf} = \frac{k_p + (n-1)k_{bf} + (n-1)\Phi_p\left(k_p - k_{bf}\right)}{k_p + (n-1)k_{bf} - \Phi_p\left(k_p - k_{bf}\right)} k_{bf} \tag{3.2}$$

DOI: 10.1201/9781032664118-3

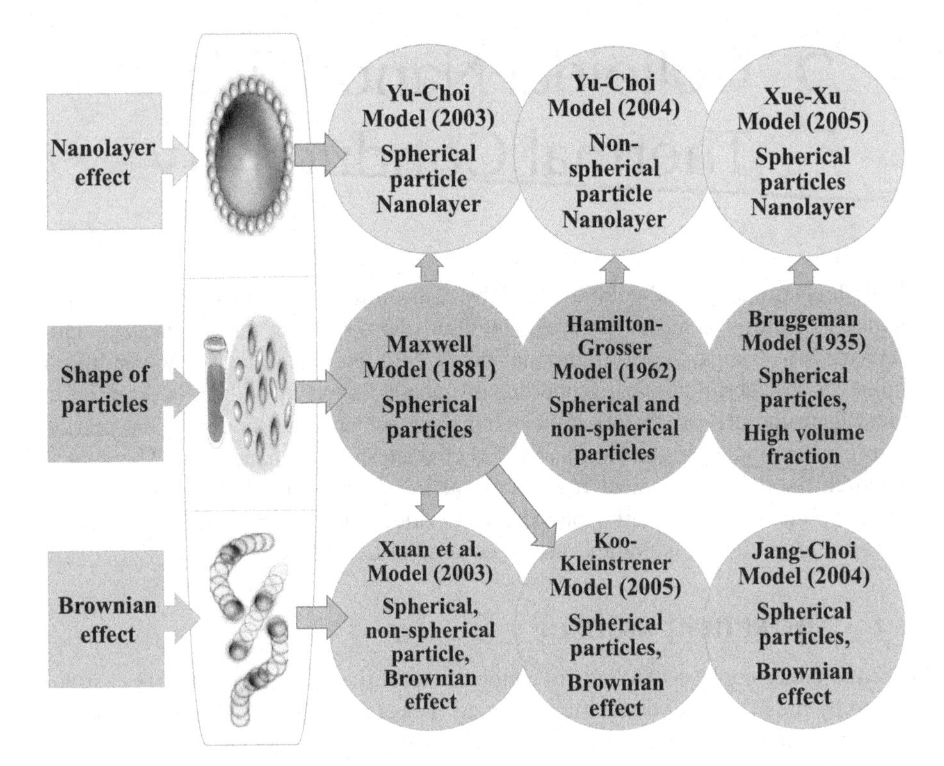

FIGURE 3.1 Schematic diagram of the development history of the classical thermal conductivity models for nanofluids and their characteristics.

Here n is the empirical shape factor given by $n=3/\Psi$, and Ψ is the surface area ratio of an arbitrarily shaped particle to that of a spherical particle of the same volume.

In terms of spherical particle interactions, the Bruggeman [8] model is more reliable for nanofluids with a high volume percentage of spherical nanoparticles.

$$\frac{k_{\mathrm{nf}}}{k_{\mathrm{bf}}}=\frac{(3\Phi_{\mathrm{p}}-1)\frac{k_{\mathrm{p}}}{k_{\mathrm{bf}}}+\left[3(1-\Phi_{\mathrm{p}})-1\right]+\sqrt{\left\{(3\Phi_{\mathrm{p}}-1)\frac{k_{\mathrm{p}}}{k_{\mathrm{bf}}}+\left[3(1-\Phi_{\mathrm{p}})-1\right]\right\}^{2}+8\frac{k_{\mathrm{p}}}{k_{\mathrm{bf}}}}}{4} \tag{3.3}$$

The nanolayer effect is described as an interface layer that forms on the surface of the nanoparticles dispersed in the fluid. Yu and Choi [9] examined the nanolayer effect, and the Maxwell model is enhanced as follows:

$$k_{\mathrm{nf}}=\frac{k_{\mathrm{p}}+2k_{\mathrm{bf}}+2\Phi_{\mathrm{p}}(k_{\mathrm{p}}-k_{\mathrm{bf}})(1+\beta)^{3}\Phi_{\mathrm{p}}}{k_{\mathrm{p}}+2k_{\mathrm{bf}}-\Phi_{\mathrm{p}}(k_{\mathrm{p}}-k_{\mathrm{bf}})(1+\beta)^{3}\Phi_{\mathrm{p}}}k_{\mathrm{bf}} \tag{3.4}$$

β is obtained by dividing the thickness of the nanolayer by particle diameter. When the nanoparticles' diameter is smaller than 10 nm, the findings of the revised model show a significant improvement in the amount of k_{nf} compared to the Maxwell model. Similarly, Xue and Xu [10] changed the Bruggeman model [8] by taking into account the nanolayer effect as shown below:

$$\left(1-\frac{\Phi_p}{\lambda_{nf}}\right)\frac{k_{nf}-k_{bf}}{2k_{nf}+k_{bf}}+\frac{\Phi_p}{\lambda_{nf}}\frac{(k_{nf}-k_l)(2k_l+k_p)-\lambda_{nf}(k_p-k_l)(2k_l+k_{nf})}{(2k_{nf}+k_l)(2k_l+k_p)+2\lambda_{nf}(k_p-k_l)(k_l-k_{nf})}=0 \tag{3.5}$$

Here k_l indicates the k of the nanolayer creating a layer on the particle's surface, and λ_{nf} shows the ratios of k_{bf} to k_p. CuO/water and CuO/ethanol nanofluids' empirical results are in good agreement with this model's predictions.

Because of the dispersion of nanoparticles, Brownian motion is present throughout the fluid. The random mobility of nanoparticles increases k_{bf} [11]. Xuan et al. [12] modified the Maxwell model by taking the Brownian effect into account because of this significant effect.

$$\frac{k_{nf}}{k_{bf}}=\frac{k_p+2k_{bf}-2\Phi_p(k_{bf}-k_P)}{k_p+2k_{bf}+\Phi_p(k_{bf}-k_P)}+\frac{\rho_p\Phi_p c_p}{2k_{bf}}\sqrt{\frac{k_B T}{3\pi r_{cl}\mu}} \tag{3.6}$$

Here k_B is the Boltzmann constant, μ is the dynamic viscosity of the liquid, r_p shows the diameter of a single nanoparticle, and r_{cl} is the apparent radius of the cluster when nanoparticles aggregate in clusters. As a result, this model could be used in nanofluids with randomly distributed or agglomerated particles. Koo and Kleinstreuer [13] presented a model for thermal conductivity that takes into account the influence of nanoparticle size, concentration, and temperature, as well as particles influenced by Brownian motion. As a result, this new thermal conductivity model incorporates both the static and Brownian components. As a result of the random motion of the nanoparticles, the conduction heat transfer is achieved as follows:

$$k_{nf}=k_{static}+k_{Brownian}$$

$$=\frac{k_p+2k_{bf}+2\Phi_p(k_p-k_{bf})}{k_p+2k_{bf}-\Phi_p(k_p-k_{bf})}k_{bf}+5\times10^4 f\Phi_p\rho_p c_p\sqrt{\frac{k_B T}{\rho_p r_p}}f_2(T,\Phi_p) \tag{3.7}$$

Here f_1 and f_2 represent empirical functions derived from empirical data and relate to particle volume percentage, shape, and liquid temperature. Brownian motion is more significant at higher temperatures; thus, the model predictions are more accurate. Chon et al. [14] included the Reynolds number in the model based on the fitting of experimental data of Al_2O_3/deionized (DI) water nanofluids, where the Reynolds number is linked to the speed of Brownian motion. The following is a description of the model:

$$\frac{k_{nf}}{k_{bf}}=1+64.7\Phi_p^{0.74}\left(\frac{r_{bf}}{r_p}\right)^{0.369}\left(\frac{k_{bf}}{k_p}\right)^{0.747}\times\mathrm{Pr}^{0.9955}\times\mathrm{Re}^{1.2321} \tag{3.8}$$

$$\mathrm{Pr} = \frac{\mu_{\mathrm{bf}}}{\rho_{\mathrm{bf}}\alpha_{\mathrm{bf}}},\ \mathrm{Re} = \frac{\rho_{\mathrm{bf}} v_{\mathrm{p}} r_{\mathrm{p}}}{\mu_{\mathrm{bf}}} = \frac{\rho_{\mathrm{bf}} \mathrm{k_B} T}{3\pi \mu_{\mathrm{bf}}^2 \Lambda_{\mathrm{bf}}} \tag{3.9}$$

Here, r_{bf} is the size of the base fluid molecule, Λ_{bf} is the mean free path of the base fluid molecules, and v_{p} demonstrates the nanoparticles' Brownian motion velocity. This model can show the influence of the Brownian motion of the nanoparticles on the k_{f}. Wang et al. [15] also investigated the impacts of adsorption and nanoparticle size, which are as follows:

$$\frac{k_{\mathrm{nf}}}{k_{\mathrm{bf}}} = \frac{\left(1-\Phi_{\mathrm{p}}\right) + 3\Phi_{\mathrm{p}} \int_0^{\infty} \frac{k_{\mathrm{cl}} n}{k_{\mathrm{cl}} + 2k_{\mathrm{bf}}} dr_{\mathrm{p}}}{\left(1-\Phi_{\mathrm{p}}\right) + 3\Phi_{\mathrm{p}} \int_0^{\infty} \frac{k_{\mathrm{bf}} n}{k_{\mathrm{cl}} + 2k_{\mathrm{bf}}} dr_{\mathrm{p}}} \tag{3.10}$$

Due to surface adsorption, it is assumed that nanoparticles would form clusters in this model. Therefore, k_{cl} could be used instead of k_{p}. The fractal theory states that any curve or geometric shape component has the same statistical character as the whole figure. When using fractals to model structures where similar patterns recur at progressively more minor scales, it is possible to fully envisage the k_{nf}, which contains various particle cluster shapes. Jang and Choi [11] also examined the collisions between molecules of the base fluid and the interaction of nanoparticles and the base fluid. Their suggested model is shown below:

$$k_{\mathrm{nf}} = k_{\mathrm{bf}}\left(1-\Phi_{\mathrm{p}}\right) + k_{\mathrm{p}}\Phi_{\mathrm{p}} + 3\mathrm{C}\frac{r_{\mathrm{bf}}}{r_{\mathrm{p}}} k_{\mathrm{bf}} \mathrm{Re}_{\mathrm{r}}^2 \mathrm{Pr} \Phi_{\mathrm{p}} \tag{3.11}$$

C denotes an empirical constant.

To summarize, the aforementioned simulations demonstrated the critical role of Brownian motion in nanofluid thermal conduction. Even though the model mentioned above fits well with fundamental heat transfer theory, there is no universal model for all nanofluids. The model should be altered for particular situations.

3.1.1 Role of Nanolayer Thickness

Nanofluids have attracted a great deal of interest with many potential uses. Many studies have been done on this subject. Particles' impact on nanofluid heat transfer was the subject of early study. The Hamilton–Crosser model [7] took particle form into account, the Cheng-Vachon model [16] took particle dispersion, and the Jeffery model [17] concentrated on particle interactions. However, since layered molecules occur between particles and liquid [18], the impact of the nanolayer on heat transmission cannot be overlooked. Yu and Choi [9] presented a modified Maxwell model that includes particles, liquid, and a nanolayer at the particle-liquid interface [6] and assumes that the nanolayer surrounding the particles is more ordered than the bulk liquid, resulting in a larger nanolayer k. In addition, when the nanoparticle diameters are smaller than 10 nm, the ordered nanolayer is effective in heat transmission. Since the above-mentioned modified Maxwell model can only be applied to nanofluids with spherical nanoparticles, Yu and Choi [19] have proposed a modified Hamilton–Crosser model, which could be applied to non-spherical nanoparticles by assuming the nanolayer is the confocal ellipsoid with the inside particles. This model's effective thermal conductivity matches nanotube-in-oil suspensions well, but it cannot estimate the nonlinear k_{nf}.

Yu and Choi [19] have developed a novel model that considers the impact of the thickness of the interfacial layer under the assumption that particles do not interact and that the temperature fields are constant throughout the fluid, particles, and nanolayer with various k values [20].

For various kinds of nanofluids, this novel model shows excellent agreement with existing experimental evidence [21]. Figure 3.2 shows the differences and connections between several models while considering the influcnce of nanolayers. Nevertheless, there remained one unsolved issue: how to measure the nanolayer's thickness? Hill et al. [21] have suggested various methods for estimating the thickness of spherical and cylindrical nanoparticles by manipulating heat conditions inside the fluid, particle, and nanolayer to address this issue. Finally, the nanolayer thickness can be calculated by solving a single equation that includes the crucial nanolayer parameter ($\delta = R/rp$). Research by Zou et al. [22] following that of Choi et al. [9] examined the impact of non-aggregating and aggregating (cluster) particles on nanofluids.

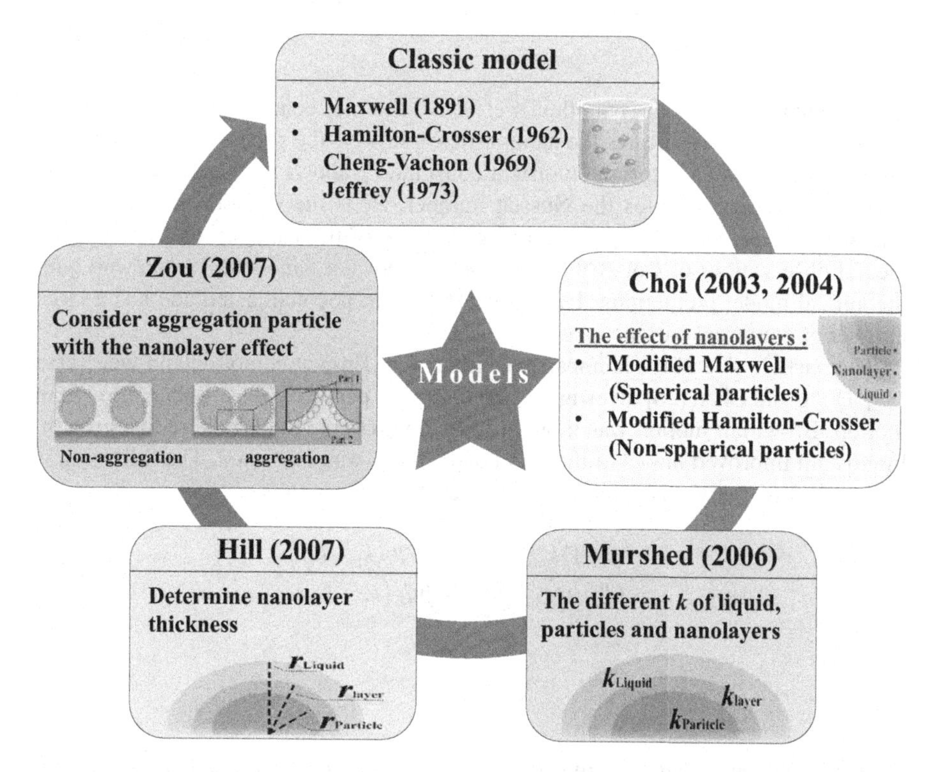

FIGURE 3.2 Schematic diagram of the differences and linkages among different models considering the effects of nanolayers.

Their research focused on the heat transfer of aggregating particles. They proposed an expression for the effective thermal conductivity, k_{eff}: $k_{eff} = (1 - y)k_m + yk_a$, where k_{eff} is the effective thermal conductivity, k_m is the predicted k from the modified Maxwell model, k_a is the thermal conductivity of the aggregation model, and y is the ratio of aggregating particles to all nanoparticles. As a result, the particle distribution in this model is more realistic. It closely matches experimental data in almost every way.

3.1.2 Nanoparticle Size Effect

Brownian motion has a significant role in raising k_{nf}. Because of their small size and intense Brownian motion, nanoparticle size has a significant impact on k_{nf} [23]. Xu et al. [24] suggest the following model compute k_{nf} by taking Brownian motion's effect on convective heat transfer into account.

$$\frac{k_{nf}}{k_{bf}} = \frac{k_p + 2k_{bf} - 2\Phi_p(k_{bf} - k_p)}{k_p + 2k_{bf} + \Phi_p(k_{bf} - k_p)} + C\frac{\text{Nu} \cdot r_{bf}}{\text{Pr}} \frac{(2 - D_{bf})}{(1 - D_{bf})^2} \frac{\left[\left(\frac{r_{p,max}}{r_{p,min}}\right)^{1-D_{bf}} - 1\right]^2}{\left(\frac{r_{p,max}}{r_{p,min}}\right)^{2-D_{bf}} - 1} \frac{1}{\bar{r}_p} \tag{3.12}$$

In this equation, k_{nf} is the nanofluid's effective thermal conductivity, k_{bf} represents the base fluid's k, k_p indicates the nanoparticle's k, Φ_p is the particle concentration in the nanofluid, C stands for an empirical constant, r_{bf} shows the base fluid molecule diameter, Nu demonstrates the Nusselt number, Pr is the Prandtl number, fractal dimension is shown by D_{bf}, $\bar{r}_p$ is the average size of the nanoparticle. The experimental observations of k of Al_2O_3/water and CuO/water nanofluids match well with theoretical model predictions. This model, however, is complicated and has a large number of empirical parameters.

The particle size of the nanoparticles influences Brownian motion and vice versa [25]. Dong and Chen [26] investigated the effects of particle size as well as Brownian motion. Brownian motion was thought to be comparable to particle radius growth. Finally, an improved model of thermal conductivity was suggested. The following is the connection between k_{eff} and particle size:

$$k_{nf} = \frac{(\beta+1)k_{bf} + \beta k_p + 2\beta\Phi_e(k_p - k_{bf})}{(\beta+1)k_{bf} + \beta k_p - \beta\Phi_e(k_p - k_{bf})} k_{bf} \tag{3.13}$$

$$\varphi_e = \varphi_s \frac{r_{eff}^3}{r_p^3} \tag{3.14}$$

Here β is the coefficient, φ_s is the volume fraction of particles in fluid, and the effective diameter is shown by r_{eff}, which depends on nanoparticles Brownian velocity.

Several studies argue that the interfacial shell produced between the nanoparticles and the base liquid influences the k_{nf}. Therefore, this interfacial impact should be taken into account. Xue and Wu [10] created a model of nanofluids that takes into account the interface shell, as shown below:

$$\left(1-\frac{\Phi_p}{z}\right)\frac{k_{nf}-k_{bf}}{2k_{nf}+k_{bf}}+\frac{\Phi_p}{z}\frac{(k_{nf}-k_{is})(2k_{is}+k_p)-z(k_p-k_{is})(2\kappa_{is}+k_{nf})}{(2k_{nf}+k_{is})(2k_{is}+k_p)+2z(k_p-k_{is})(k_{is}-k_{nf})}=0 \tag{3.15}$$

$$z=\left[\frac{r_p/2}{r_p/2+d_{is}}\right]^3 \tag{3.16}$$

where k_{is} represents the interfacial shell's k, d_{is} demonstrates the thickness of the interfacial shell, and r_p shows the diameter of the nanoparticles. The interface shell and the nanoparticles may be considered composite nanoparticles, with the volume fraction represented as $\Phi_{p/z}$. The experimental data of k of CuO/DI water nanofluids and CuO/ethylene glycol (EG) nanofluids show that k_{nf} is dependent on nanoparticle size and interface characteristics, which is in perfect accord with this model.

According to earlier research, Brownian motion-induced convection and heat transfer routes are two of the most critical variables that impact nanofluid heat transmission [27]. Ganesan et al. [28] investigated the k of nanocluster-based nanofluids with very high particle sizes (115–530 nm). The following formulas for k are based on the interfacial thermal resistance and mixed convection of nanoparticles:

$$\frac{k_{nf}}{k_{bf}}=\left(1+A\mathrm{Re}^{\gamma}\mathrm{Pr}^{0.333}\Phi_p\right)\frac{1+2\beta\Phi_p}{1-\beta\Phi_p} \tag{3.17}$$

where A is a constant and γ is the index given by the system. The root means square velocity p of the nanoparticle is used to calculate the convection velocity:

$$v_p=\sqrt{\frac{18k_BT}{\pi\rho_p d^3}} \tag{3.18}$$

k_B denotes the Boltzmann constant, T denotes the temperature, and p denotes the particle density. In general, as particle size rises, k_{nf} decreases. This model differs significantly from the conventional model, which is based on the dynamic development of nanocluster aggregates.

3.1.3 Temperature Influence

In addition to the aforementioned parameters, it has been shown that k_{nf} has a substantial temperature dependency. The thermodynamic conductivity of nanofluids is determined using mathematical models based on Brownian motion and kinetic theory, which are quickly impacted by temperature changes. The thermal conductivity of nanofluids rises with increasing temperature in most situations. The modified Maxwell model presented by Xuan et al. [12] (Eq. 3.37) shows that it not only analyzes the influence of Brownian motion on k but also the effect of temperature. Koo and Kleinstreuer [13] also took into account the impact of temperature on k in Eq. (3.38). They think that temperature directly impacts the Brownian motion of particles in nanofluids; therefore, Eq. (38) has a temperature-related component in the second term. Brownian motion is more significant at higher temperatures, and the model forecasts more accurately. However, it is only suitable for temperatures between 293 and 325 K. Based on this model, Vajjha and Das [29] gathered further data and developed a thermal conductivity model for three EG-water-based nanofluids containing Al_2O_3, ZnO, and CuO nanoparticles, respectively. The k_{nf} is a temperature function with the following expression:

$$k_{\mathrm{nf}} = \frac{k_{\mathrm{p}} + 2k_{\mathrm{bf}} + 2\Phi_{\mathrm{p}}\left(k_{\mathrm{bf}} - k_{\mathrm{p}}\right)}{k_{\mathrm{p}} + 2k_{\mathrm{bf}} - \Phi_{\mathrm{p}}\left(k_{\mathrm{bf}} - k_{\mathrm{p}}\right)} k_{\mathrm{bf}} + 5\times 10^{4} \beta \Phi_{\mathrm{p}} \rho_{\mathrm{bf}} c_{\mathrm{p,bf}} \sqrt{\frac{k_{\mathrm{B}} T}{\rho_{\mathrm{p}} r_{\mathrm{p}}}} f\left(T, \Phi_{\mathrm{p}}\right) \quad (3.19)$$

$$f\left(T,\Phi_{\mathrm{p}}\right) = \left(2.8217\times 10^{-2}\Phi_{\mathrm{p}} + 3.917\times 10^{-3}\right)\left(\frac{T}{T_0}\right) + \left(-3.0669\times 10^{-2}\Phi_{\mathrm{p}} - 3.91123\times 10^{-3}\right) \quad (3.20)$$

where k_{nf} represents the nanofluid's effective thermal conductivity, k_p is the particles' k, k_{bf} is the base fluid's k, p is the nanoparticle's concentration, is the proportion of liquid volume moving with a particle (the additional mass), bf is the base fluid's density, ρ_p is the particle's density, k_B is the Boltzmann constant, T is temperature, T_0 signifies 273 K, and this model has a more significant agreement with the experimental data than other models.

The Yu model [19] and the Moghadassi and Hosseini model [30] do not consider the influence of temperature on nanofluids.

Xuan et al. [12] and Gao et al. [31] developed a new model for estimating k of graphene nanofluids based on the Chu et al. model [32]. The influence of interfacial thermal resistance, length, thickness, the average flatness ratio of the graphene nanoparticles, and Brownian motion between the particles are all included in this model. They presented a mathematical equation based on the experimentally determined k of the graphene nanofluid in the temperature range of 253–303 K:

$$\frac{k_{\mathrm{nf}}}{k_{\mathrm{bf}}} = \frac{3 + 2\eta^2\Phi_{\mathrm{p}} \Big/ k_{\mathrm{bf}}\left(\dfrac{2R_{\mathrm{k}}}{L_{\mathrm{p}}} + 13.4\sqrt{t}\right)}{3 - \Phi_{\mathrm{p}}} + \frac{\rho_{\mathrm{p}}\Phi_{\mathrm{p}}c_{\mathrm{p,p}}}{2k_{\mathrm{bf}}}\sqrt{\frac{k_{\mathrm{B}}T}{3\pi\mu r_{\mathrm{c}}}} \quad (3.21)$$

where η is the graphene's average flatness ratio, and R_k indicates the interfacial thermal resistance, L_p is the particle length, d_p is the particle thickness, p is the particle density, $c_{p,p}$ is the particle isobaric heat capacity, is the dynamic viscosity, and r_c

is the mean radius of gyration, which is approximately half the graphene particle diameter $r_{p,eq}$. The effective k_{nf} for a static fluid is the first term on the right side of the equation. The second term is the k_{nf} produced by Brownian motion between the nanoparticles [33]. Non-spherical particles having the same volume can be considered as spherical particles, and the corresponding diameter is:

$$r_{p,eq} = \left(\frac{6V_{non-sph}}{\pi} \right)^{1/3} \tag{3.22}$$

3.1.4 Investigation of Nanoparticle Layer Thickness Change and Particle Diameter According to Yu and Choi Study [9]

In Figure 3.3 [9], a decrease in the thermal conductivity of the nanofluid is observed by changing and increasing the radius of the nanoparticle and also increasing the thickness of the nanolayer from 1 to 2 nm reducing the thermal conductivity by 5% to 10%.

3.1.5 The Results of Examining Different Calculation Formulas for the Same Conditions

In this section, the specifications of Table 3.1 for the base fluid and nanoparticles are considered, thus the nanofluid thermal conductivity is calculated using the various formulas presented above.

Considering the parameters of Table 3.1, different results at the studied temperatures have been obtained in the various articles mentioned below, and these results

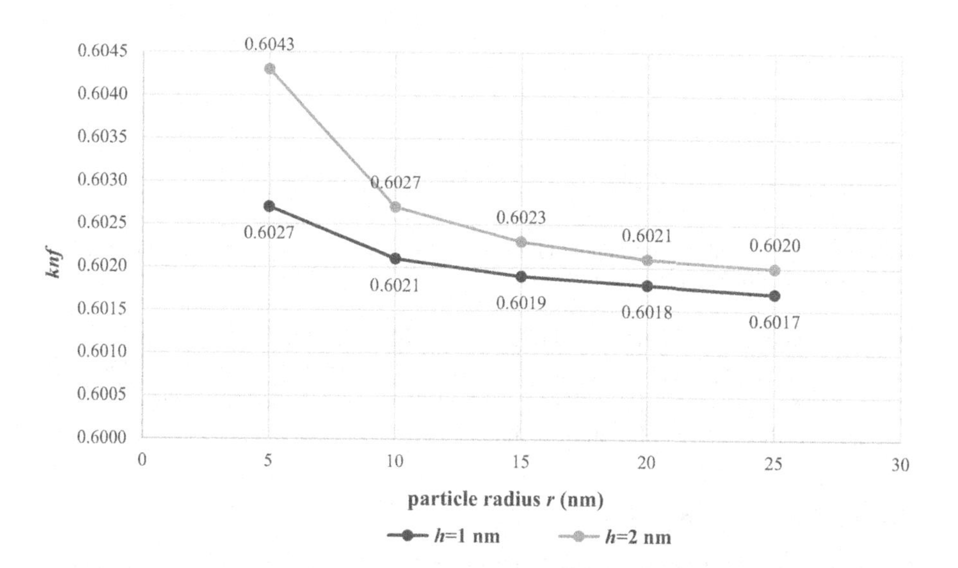

FIGURE 3.3 k_{nf} changes in particle radius r (nm).

TABLE 3.1
Parameters considered to calculate the heat transfer coefficient of AL_2O_3 / water nanofluid

K_p	thermal conductivity of particle (Al_2O_3)	40
Φ	volume concentration	0.03
k_{bf}	thermal conductivity of the base fluid (Water)	0.6
K_B	Boltzmann constant	$1.381*10^{-23}$
d_p	diameter of particle	$20*10^{-9}$

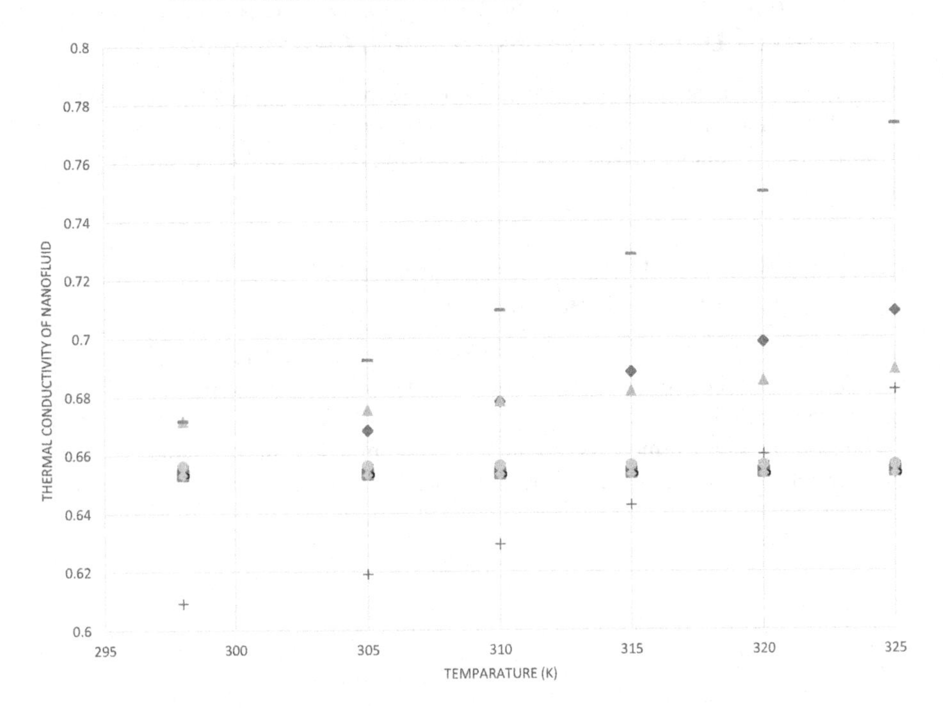

FIGURE 3.4 Calculation of nanofluid heat transfer coefficient (k_{nf}) for different temperatures and with different methods for Al_2O_3.

are also illustrated in Figure 3.4. The results of Maxwell [6], Hamilton and Crosser [7], Bruggeman [8], Xuan et al. [189], and Koo et al. [187] methods show close values, and the results of Vajjha et al. [29] method show much better and more consistent results according to the laboratory results.

3.1.6 Aggregation Effect

Many researchers have investigated the processes of heat transfer enhancement in nanofluids and have experimentally demonstrated that, in addition to Brownian motion, nanoparticle aggregation may also contribute to an increase in k_{nf} [34,35]. On the other hand, most heat transfer models assume that the nanoparticles in the base fluid are evenly disseminated, even though numerous researchers have developed novel models that take into account the aggregation effect, which will be described in this section. The aggregation impact of nanoparticles was studied by Feng et al. [22], who presented a novel model for the k_{nf} of nanofluids. The nanofluids' k_{nf} was connected to the nanoparticle size, nanoparticle volume fraction, interface nanolayer thickness, nanocluster k, and the base fluid in this model, which may be represented as:

$$\frac{k_{nf}}{k_{bf}} = \left(1-\Phi_{p,eq}\right)\frac{k_{pe}+2k_{bf}+2\left(k_{p,eq}-k_{bf}\right)\left(1+\beta\right)^3\Phi_p}{\kappa_{pe}+2k_{bf}-\left(k_{p,eq}-k_{bf}\right)\left(1+\beta\right)^3\Phi_p}$$

$$+\Phi_{p,eq}\left[\left(1-\frac{3}{2}\Phi_{p,eq}\right)+\frac{3\Phi_{p,eq}}{\iota}\left[\frac{1}{\iota}\ln\frac{r_p/2+d_{is}}{\left(r_p/2+d_{is}\right)\left(1-\iota\right)}-1\right]\right] \tag{3.23}$$

where $\iota = (1 - k_f/k_{pe})$ k_{nf} shows the nanofluid's effective thermal conductivity, k_{bf} demonstrates the base fluid's k, $k_{p,eq}$ is the equivalent k of the equivalent particles, p is the particle concentration in the nanofluid, $\Phi_{p,eq}$ is the equivalent volume fraction of equivalent particles, r_p is the particle diameter, and d_{is} is the interfacial shell thickness. This model's academic outcomes are in good accord with experimental findings. Pang et al. [36] looked at both the static and dynamic contributions to the increase in k for nanofluids, with the latter being primarily ascribed to nanoparticle aggregation. As a result, the nanofluids model may be expressed as follows:

$$k_{nf} = 1+\frac{\Phi_a\left(k_a-k_{bf}\right)}{\left(1-\Phi_a\right)/n\left(k_a-k_{bf}\right)+k_{bf}}+$$

$$\left[A_1\ln\left(A_2\Phi_a Re^m Pr^{0.333}\right)+A_3\right]\frac{1+2\Phi_a+2\left(1-\Phi_a\right)\alpha}{1-\Phi_a+\left(2+\Phi_a\right)\alpha} \tag{3.24}$$

where R_b and r_p are the heat resistance and particle diameter, respectively, and $\alpha = 2R_b.k_{bf}/r_p$. The composite theory [37] states that Φ a represents the aggregates' volume fraction in the nanofluid, and k_a indicates the aggregates' effective thermal conductivity. A_1, A_2, A_3, and m are experimentally determined constants; Nu denotes the Nusselt number, and Pr shows the Prandtl number.

It is possible to construct a theoretically effective thermal conductivity model for nanofluids using Hamilton et al. [7], Xu and Yu [38], and Wei et al. [39] models. The model is presented as a function of the fractal size and concentration by taking into account two distinct processes of heat conduction: particle aggregation and convention. The model shows that the variation in aggregation shape relates to the changes in fractal dimension. With the suggested k_{nf}, the mathematical equation is as follows:

$$\frac{k_{nf}}{k_{bf}} = \frac{m+(n-1)-(n-1)(1-m)\Phi_p}{m+(n-1)+(1-m)\Phi_p}$$

$$+C\frac{Nu\cdot r_{bf}}{Pr}\frac{(2-D)D}{(1-D)^2}\frac{\left(\Phi_p^{\zeta_1}-1\right)^2}{\Phi_p^{\zeta_2}-1}\frac{1}{\overline{L_a}} \tag{3.25}$$

$$F = 3\frac{D}{D-1}\frac{3-D}{2-D}\frac{\Phi_e^{\zeta_2}-1}{\Phi_e^{-1}-1} \tag{3.26}$$

Here $m = k_p/k_{bf}$, k_p stands for the particle's k, n shows the empirical shape factor calculated by the Φ_p and the fractal dimension D, $\zeta_1=(D-1)/(3-D)$, $\zeta_2=(D-2)/(3-D)$, C shows empirical constant, and r_{bf} is the diameter of the base fluid molecules, and the average size of aggregation is represented by L_a. The size effect and Brownian motion impact of nanoparticle aggregation are reflected in the k enhancement of the nanofluid in the two terms on the right. This prediction method is accurate.

3.1.7 Evolution of Advanced Models

As stated earlier, as research into thermal transport for nanofluids is expanded and enjoys substantial depth, numerous factors, such as nanoparticle interaction, nanolayers, aggregation, Brownian motion, and turbulent effects, find their way into research equations. The multi-parameter advanced models display the slightest difference and distance from reality, as compared to classical models. This section gives a schematic survey of advanced k models' background for nanofluids and illustrates their features (see Figure 3.5). Tillman et al. [21] propounded two models of k_{nf} with spherical and cylindrical nanoparticles considering the nanolayer's impact. For spherical nanoparticles, k_{nf} was defined as follows:

$$k_{nf} = \left[\left(k_p - k_{nl}\right)\Phi_p k_{nl}\left(\gamma_1^2 - \gamma^2 + 1\right) + \left(k_p + k_{nl}\right)\gamma_1^2\left(\Phi_p\gamma^2\left(k_{nl} - k_{bf}\right) + k_{bf}\right)\right] \cdot \left[\gamma_1^2\left(k_p + k_{nl}\right) - \left(k_p - k_{nl}\right)\Phi_p\left(\gamma_1^2 + \gamma^2 - 1\right)\right]^{-1} \tag{3.27}$$

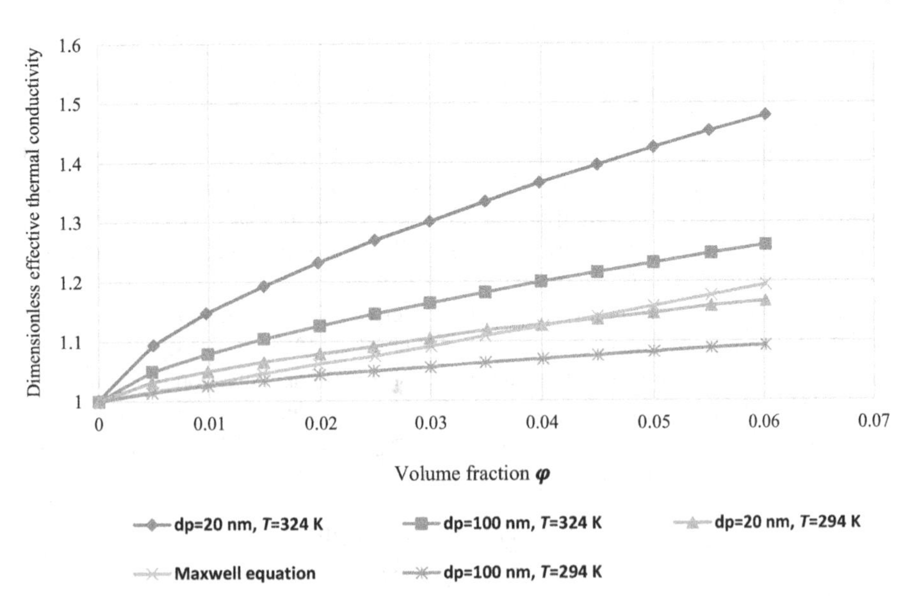

FIGURE 3.5 Comparison of k_{eff} of different d_p and T in different volume fractions with regard to the above equations for Al_2O_3/water.

As for cylindrical nanoparticles in nanofluids, k_{nf} can be reshaped into:

$$k_{nf} = \left[(k_p - k_{nl}) \Phi_p k_{nl} (2\gamma_1^3 - \gamma^3 + 1) + (k_p + 2k_{nl}) \gamma_1^3 (\Phi_p \gamma^3 (k_{nl} - k_{bf}) + k_{bf}) \right] \cdot \left[\gamma_1^3 (k_p + 2k_{nl}) - (k_p - k_{nl}) \Phi_p (\gamma_1^3 + \gamma^3 - 1) \right]^{-1} \quad (3.28)$$

where k_{nf} represents the nanofluid's effective thermal conductivity, the particles' k, base fluid, and nanolayer are, respectively, represented by k_p, k_{bf}, and k_{nl}, and Φ_p shows the particles' volume fraction $1 + \frac{2d_{nl}}{r_p} = \gamma$ and $1 + \frac{d_{nl}}{r_p} = \gamma_1$, where d_{nl} shows the nanolayer thickness, and r_p is the particle diameter. Compared with other reported models during development, these models correlated with the experimental results more effectively.

In relevant research on the particles' Brownian motion and turbulent effects, Feng and Kleinstreuer [40] reported a reputed F–K model, which, when combined with the Maxwell model [6], could be reshaped into the following:

$$k_{nf} = k_{static} + k_{mm} \quad (3.29)$$

where k_{nf} is the k of nanofluids, k_{static} represents the static part of the k_{nf}, and k_{mm} denotes the micro-mixing part of k_{nf}. Maxwell's model gives the static part:

$$k_{static} = \left[1 + \frac{3(k_p/k_{bf} - 1)}{(k_p/k_{bf} + 2) - (k_p/k_{bf} - 1)} \right] k_{bf} \quad (3.30)$$

The micro-mixing part was given by:

$$k_{mm} = 49500 \cdot 38 \cdot \frac{k_B \tau_p}{2m_p} \cdot (\rho c_p)_{nf} \cdot \Phi_p{}^2 \cdot (T \ln T - T) \cdot \frac{\exp(-\xi \omega_n \tau_p) \sinh\left(\sqrt{\frac{(3\pi\mu_{bf} r_p)^2}{4m_p^2} - \frac{Q_{p-p}}{m_p}} \frac{m_p}{3\pi\mu_{bf} r_p} \right)}{\tau_p \sqrt{\frac{(3\pi\mu_{bf} r_p)^2}{4m_p^2} - \frac{K_{p-p}}{m_p}}} \quad (3.31)$$

where m_p shows the particle mass, r_p is the nanoparticles' diameter, c_p is the isobaric heat capacity, Q_{p-p} is the intensity of the particle–particle interaction, and μ_f is the fluid's dynamic viscosity. As defined below, the damping coefficient ξ, the natural frequency ω_n, and the characteristic time interval τ_p are given as follows:

$$\xi = \frac{3\pi r_p \mu_{bf}}{2m_p \omega_n}, \quad \omega_n = \sqrt{\frac{Q_{P-P}}{m_p}}, \quad \tau_p = \frac{m_p}{3\pi\mu_{bf} r_p} \quad (3.32)$$

In an attempt to resolve the description inaccuracy association with the CNT-containing nanofluids, Das et al. [41] propounded a new model. This inaccuracy often derives from the considerable difference between the k_f and that of the particles,

$$k_{nf} = k_{bf}\left[1+\frac{k_p\Phi_p L_{mol}}{k_{bf}\left(1-\Phi_p\right)r_{cnt}}\right] \tag{3.33}$$

where L_{mol} and r_{cnt}, respectively, represent the CNTs' average diameter and molecule size. The most significant advantage of this proposed model is that it eliminated the need for empirical constants besides fitting well with the experimental data.

The model propounded by Li et al. [42] addresses the CNTs' non-isotropic issue in nanofluids. This model can be regarded as a step forward in the field, explaining the k enhancement of the CNT nanofluids by the sample function, concluding that the interfacial thermal resistance dominates the k of the CNT composites, including CNT-containing nanofluids. k_{nf} is explainable for CNT-containing nanofluids as:

$$\frac{k_{nf}}{k_{bf}} = \frac{3+\Phi_p\left(\beta_T+\beta_L\right)}{3-\Phi_p\beta_T} \tag{3.34}$$

where:

$$\beta_T = \frac{2\left(k_T^{nf}-k_{bf}\right)}{k_T^{nf}+k_{bf}},\ \beta_L = \frac{k_L^{nf}}{k_{bf}}-1 \tag{3.35}$$

where a nanofluid unit cell's transverse and longitudinal k are shown by k_T^{nf} and k_L^{nf} show, respectively, which are given by:

$$k_T^{nf} = \frac{k_{nf}}{1+\dfrac{2L_K}{r_{cnt}}\dfrac{k_{nf}}{k_{bf}}},\ k_{33}^{nf} = \frac{k_{nf}}{1+\dfrac{2L_K}{L_{cnt}}\dfrac{k_{nf}}{k_{bf}}} \tag{3.36}$$

where $L_K = R_K \cdot k_{bf}$. For nanotube composites, L_K is 16–40 nm when k_{bf} is 0.2–0.5 W/m·K, and R_K is $83\times10^{-8}\text{m}^2$ K/W, r_{cnt} is the CNT diameter is the CNT length.

Furthermore, Corcione [43] introduced a model offering quick k predictability under various scenarios (nanoparticle diameter: 10–150 nm, volume fraction: 0.002–0.09, temperature: 294–324 K), and provides a more accessible engineering application:

$$\frac{k_{nf}}{k_{bf}} = 1+4.4\text{Re}^{0.4}\text{Pr}^{0.66}\left(\frac{T}{T_{fp}}\right)^{10}\left(\frac{k_p}{k_{bf}}\right)^{0.03}\Phi_{0.66} \tag{3.37}$$

where $\text{Re} = \dfrac{2\rho_{bf}k_BT}{\pi\mu_{bf}^2 d_p}$ and T_{fp} refers to the base fluid's freezing point.

The classical models address only a limited number of parameters, focusing on the nanoparticles' state. In contrast, advanced models encompass more parameters

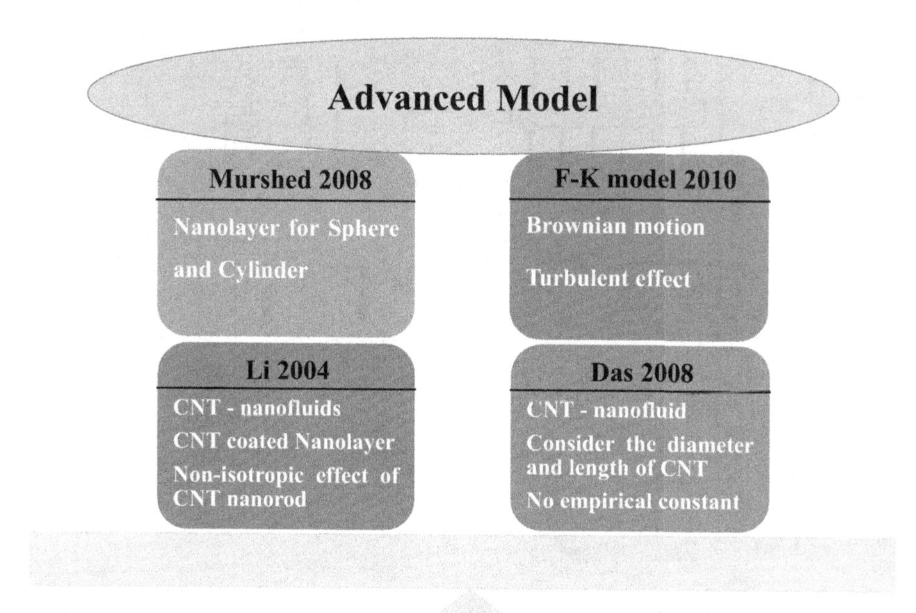

FIGURE 3.6 Schematic diagram of the development history of the advanced k models for the nanofluids and their characteristics.

specified to the actual application of nanofluids, hence the higher accuracy of the effective thermal conductivity sourced from these models. Furthermore, researchers opt to work specifically on expanding nanofluids' engineering applications to do which they have developed empirical heat transfer models for a more accurate prediction of k_{nf} (Figure 3.6).

3.2 EXPERIMENTAL MEASUREMENT TECHNIQUES

As stated in the previous section, this section gives a complete summary and presents the theoretical models of k_{nf}, beginning from the detailed development of classical models to the advanced models considering the effects of nanolayers, nanoparticle size, temperature, and aggregation on k based on the classical models, and then walks the reader to every step in the development of these models. This section deals with several classical and popular methods, such as the top classical hot-wire method, the popular 3ω method, the most standardized laser flash method, the steady-state parallel method, and the transient plane-source method (see Figure 3.7) to measure k_{nf} at this stage [44–75].

3.2.1 Transient Hot-Wire Approach

Stalhane and Pyk were the first developers of the transient hot-wire (THW) method, as seen in Figure 3.7b, introducing the model in 1931. After several decades of

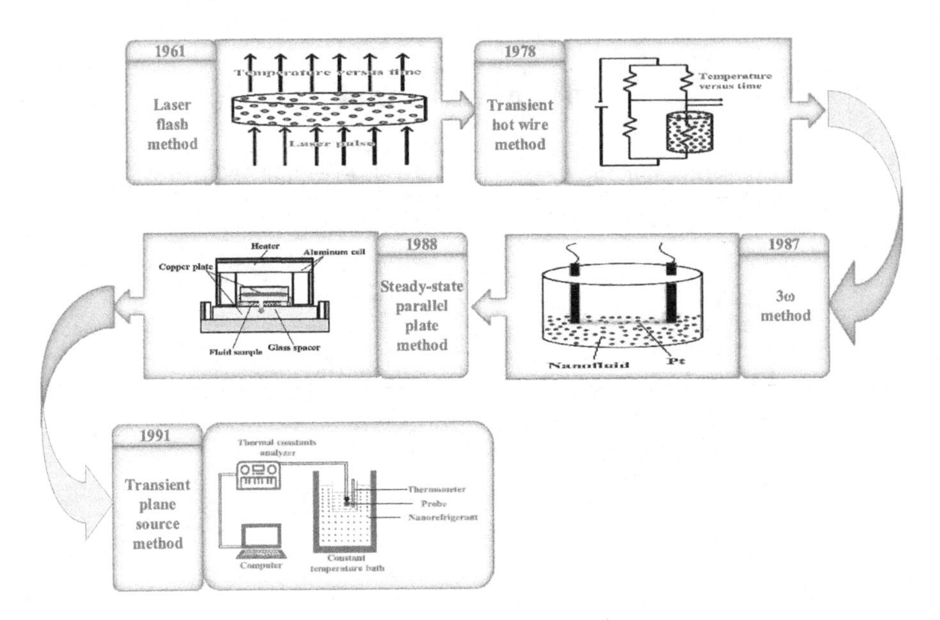

FIGURE 3.7 Schematic diagram of thermal conductivity measurement principle. (a) Laser flash method, (b) THW method, (c) 3ω method, (d) SSPP method, (e) TPS method.

development, the hot-wire method has become one of the most accurate methods for measuring k_f [47–51]. This method is based on immersing a wire (hot-wire) heated at constant power. The hot-wire rises and transfers heat to the surrounding liquid as its temperature increases, and this increase is a function of the liquid's k; therefore, so the k_f is computable by measuring the short-term temperature change using the bridge circuit indirectly [47]. Researchers can determine the hot-wire's temperature increase curve by measuring the bridge voltage difference data.

Theoretically, the hot-wire method is founded on two hypotheses: (i) the hot-wire has an infinite length and small diameter and is placed vertically in an isotropic fluid to prevent convection and the associated effects, (ii) it has an infinite k and zero heat capacity. The k can be calculated based on the relationship between the hot-wire's temperature increase rate and the k_f using this equation $k = \frac{q}{4\pi}\frac{\mathrm{d}\ln t}{\mathrm{d}\Delta T(t)}$, in which q is the hot-wire's heating power and t is time.

Antoniadis et al. [48] reported that the solid phase of nanoparticle-containing fluids often has polarity and is electrically conductive which can warp the electrical signal and lead to computational accuracy. They proposed the idea of using an electrolytic oxidized tantalum wire to build an insulated hot-wire. In a relevant study, Guo et al. [47] used this theory to investigate particle size's impact on the k_{nf}, adding SiO_2 nanoparticles to DI water and EG. They observed a 9.6% and 3.2% increase in the k values of SiO_2/EG with a volume fraction of 1.0% and SiO_2/EG with a volume fraction of 0.5%, respectively, suppressing DW-based nanofluids by 1.0% and 3.4%. Lee et al. [49] minimized the effects of capacitance and natural convection in an advanced hot-wire system to investigate the association between particle size and volume fraction and Al_2O_3/EG nanofluids' effective thermal conductivity. Their findings referred to the more significant effect of particle size on volume fraction under a below-0.25% volume fraction, insisting that the smaller the volume fraction, the higher the significance. The Brownian motion added to the particle number density's significant influence on the k_{nf} under above 0.2% volume fraction. Aparna et al. [50] conducted an experiment in which the Al_2O_3 and Ag nanoparticles had equal distribution in DI water, and the flux of Brownian motion on the nanoparticles on the thermal boundary layer was quantitatively described to compare the different measured k_{nf} by the hot-wire and the laser flash methods. The observations indicated that using the former method caused the nanoparticles to display a higher collision flux with the wall than the latter method, leading to a higher k value.

3.2.2 3ω Method

As mentioned earlier, the 3ω method (see Figure 3.7c) can effectively help characterize the k of solid nanomaterials as well as nanofluids [7,52,53]. Oh et al. [52] used the 3ω method to measure the k of an aluminum oxide/deionized water (Al_2O_3/DI water) nanofluid in an experimental design in which the homemade micro-device merely needed one nanofluid drop to complete the k measurement, which was the design's most significant advantage. The observations corroborated the increasing effect of the k of Al_2O_3/DI nanofluid on the k_{bf} by 13.3% at RT. Turgut et al. [53] used the 3ω method to measure the TiO_2/DI nanofluids with a volume fraction from 0.2% to 3.0%. Their findings correlated with the Hamilton–Crosser model [7]. Qiu et al. [54] introduced a so-called reusable freestanding sensor to measure the k in conventional fluids, such as DI water, EG, and ethanol, as well as nanofluids. Using a metallic heater/thermometer wire that is immersed into the nanofluids is the most common practice in nanofluids' test structure for this method, and accordingly, Karthik et al. [55] thought to use a pure platinum wire for their purpose. The k of 0.05 vol% CuO/DI water at RT and pH = 1 was reported as 0.666 W/m·K with a very low measurement uncertainty of less than 1.2%. Han et al. [56] even introduced a famous extended 3ω-wire method to measure the k_{nf}. This method used a small temperature oscillation to ensure that the nanofluids' thermophysical properties remained constant during the measurement. The phase-locking technology keeps the background noise to a minimum, allowing the authors to measure the k of poly-alpha-olefin (PAO) oil with dispersed composite carbon nanoparticles and the temperature-dependent k of nanorod-based nanofluids [57], confirming the method's strength for the thermophysical measurement of nanofluids containing non-spherical particles. Choi et al. [58] floated a single platinum wire into the fluid and were able to measure the k of a CNT/H_2O nanofluid. The 3ω method detects the electrical signal in the frequency domain. It should be averaged over a certain period, generally much longer than the characteristic time of the nanoparticles' random motion in the fluid, eliminating the spurious signals caused by particle motion and confirming the method's use in measuring the colloidal nanofluid's thermal conductivity.

3.2.3 Laser Flash Method

The laser flash method is another prevalent method for measuring k_{nf}, introduced by Parker et al. In cases of colloidal nanofluids; the laser method mainly functions as the heating source to increase the temperature at the nanofluids' bottom surface, while a thermometer on the top surface measures the temperature risc; so the α can be calculated based on the short-term temperature response; finally, the k_{nf} is obtained from $k=\rho c_p \alpha$ [59]. The laser flash method's principal value, as a standard short-term method, increases the temperature quickly, keeping radiation and convection from resulting in inaccuracies in measurement [60]. Since, under infrared or visible light, several nanofluids are transparent, to reduce measurement errors, using a metal film can avert the flash and the infrared light and prevent them from respectively entering the surface below and the sample on the top [61].

Harikrishnan et al. [62] used the laser flash method to measure the k of TiO_2/stearic acid nanofluids and corroborated the increasing effect of TiO_2 nanoparticles' volume fraction on the k_{nf}. Zhang et al. [63] also measured the k_{nf} with different Al_2O_3 nanoparticle sizes at high temperatures, stressing that the k_{nf} was improved to a different extent than the base fluids. In similar research, the authors used the laser flash method to compare the k of TiO_2/EG-water with TiO_2/propylene glycol/water nanofluids and calculated the theoretical value of k_{nf} using ultrasonic velocity [61]. Their results correlated with the values in the experiment.

Note that the laser flash method would compromise the accuracy in k_{nf} measurement, especially for nanofluids with low k. Aparna et al. [50] used this method to measure the k of Al_2O_3/aqueous and Ag/aqueous nanofluids, finding the laser flash method for the small liquid pool (~50 μL) limited the nanoparticles' Brownian motion compared with other methods; hence, the lowered measured k value. Using the laser flash method, Buonomo et al. [64] observed a 4.95% increase in the k of Al_2O_3/water nanofluids. Although Beck et al. [23] used the THW method to measure the Al_2O_3/water nanofluids with the same nanoparticle size and volume fraction, the k increased by 16.5%, which, compared with the findings of Buonomo et al. [64], was significantly higher. In a study investigating the different measured k values by various experimental methods, Zagabathuni et al. [44] used a collision-mediated model, reporting lower values when using the laser flash method, associated with the limitation of the Brownian motion and the nanoparticles' reduced collision frequency. Kleinstreuer et al. [65] employed three methods to measure the k of PAO-based nanofluids: the laser flash, the transient plane heat source, and the hot-wire methods. Their findings indicated the laser flash method's inaccuracy in measuring nanofluids with low k since their k approximately resembles the liquid pool, leading to significant heat flow through the liquid pool. Given the discussions mentioned above, the flash method results in the lowest measured k. However, for pure fluid, the difference is non-significant.

3.2.4 The Steady-State Parallel Plate Method

The steady-state parallel plate method (see Figure 3.7d) is an essential experimental technique for measuring the k_f during which a tiny amount of the fluid is placed between two parallel copper round plates (99.9% purity) with thermocouples arranged. The k is determined based on the relationship between the heat flowing through the test structure and the temperature difference between the upper and lower copper plates [45]. This technique is vital to make robust and integrated insulation and heat control systems to prevent heat loss from the sample. Wang et al. [45] employed this method to measure the k_{nf} Al_2O_3 and CuO dispersed in the same base fluid. The gained k values surpassed the current theoretical models considerably higher, suggesting the drawbacks of current models in describing these mixed nanofluids. Li and Peterson [66] measured the k of Al_2O_3/H_2O nanofluids with two different particle diameters, confirming a 27% improvement in k_{nf} compared with the base fluid by adding nanoparticles with 36–47 nm diameters. Sridhara et al. [67] comprehensively compared the differences in results among various experimental measurement techniques.

They stressed that the liquid's small thickness spreading on the plate led to the nanoparticle aggregation phenomenon being the least effective on the steady-state parallel plate method.

The concentric cylindrical structure is different from the steady-state method called the steady-state coaxial cylinder method is derived from the above structure. Kurt and Kayfeci [68] employed this method to measure the k of EG-water fluids, obtaining an accurate prediction of k_f, combined with artificial neural networks. Barbés et al. [69] measured the k of CuO/water nanofluids and CuO/EG nanofluids using this method joined with microcalorimetry. The results correlated with the theoretical prediction and displayed acceptable applicability to the micro-determination of fluids, thanks to the minimal temperature gradients during the measurement. Putra et al. [4] investigated the influence of natural convection in nanofluids using a horizontally placed cylinder cooled at one end and heated at the other end, concluding that for their prepared Al_2O_3/water nanofluid (concentration: 4%, mean particle diameter 131n m), k was improved by 24% compared to the base fluid. Notice that the steady-state method's apparatus is somewhat uninvolved and easy. However, the research does not suffice, and more development work is required.

3.2.5 Transient Plane-Source Method

The transient plane-source (TPS) method (see Figure 3.7e) has evolved into a commercial technique for measuring k_{nf}, which is often used by hot disk technology where the hot disk is considered both the plane source and the temperature sensor. Once immersed in the colloidal materials, the heat loss from the hot disk is monitored by its electrical resistance change versus time, resulting in the k of the colloidal materials [70–72]. The hot disk element is sandwiched between two pieces of thin polyimide films and comprises a continuous double spiral of electrically conducting nickel (Ni) metal. The double spiral nickel strip's temperature soars when energized, and the heat spreads to the samples through the thin polyimide films. The k of the colloidal materials can gradually be obtained following the real-time recorded temperature versus time curves [73].

Later, Harris et al. [74] suggested a modified transient plane source (MTPS) to measure nanofluids' thermophysical properties in motion. By using a fast response sensor, minimal sample volume, and low energy power flux, the modified method can distinguish the contribution from different heat transfer modes and eliminate the effect of heat convection. This method has been widely used for measurements of k_{nf}. Li et al. [75] measured Al_2O_3/water nanofluids in various conditions using the MTPS method to obtain the best thermal behavior. Compared with the primary fluid, k_{nf} was increased by 10.1%. Peng et al. [46] used the MTPS method to characterize the k_{nf} containing different-shaped CNTs and concluded that the aspect ratio of CNTs is positively correlated with the k of CNT-based nano refrigerants.

3.3 EXPERIMENTAL STUDIES

The findings of many research studies focusing on testing new k_{nf} measurement methods have led to the introduction of the following list of seven factors with stark influence on the obtained k values:

The list includes these features for nanoparticles:

- concentration
- temperature
- size
- shape
- agglomeration
- pH
- sonication time.

Figure 3.8 displays the intrinsic mechanisms of various factors on k_{nf}, the technical handling of which is a significant challenge in terms of measurement precision. After that, this chapter focuses on the critical advances around the impact of the seven factors mentioned above on k_{nf}. This chapter summarizes the critical k experimental data of representative nanofluids in previous studies (see Table 3.2), which, hopefully, helps readers grasp a decent overview of nanofluids' thermophysical properties. Besides, reviewing the critical measurement methods and nanoparticles reveals that researchers also use other k_{nf} measurement methods than the THW method, such as the 3ω, the steady-state parallel plate (SSPP), and the TPS methods. Moreover, the steady-state cut-bar (SSCB), the transient short hot-wire (TSHW), the flash, and the micron-scale beam deflection (MSBD) are general k_{nf} measurement methods. Many studies have also employed the Thermal-Wave Resonator Cavity (TWRC) technique.

3.3.1 Nanoparticle Concentration Effect

The findings of many researchers working on k_{nf} improvement with nanoparticles confirm nanoparticle concentration will directly influence the k_{nf} [76–80]. Li et al. [75] propounded the idea that an increase in the MgO nanoparticle concentration (from 0.5 to 2 vol%) improved Nu, highly influencing k promotion. In a different study, Esfe et al. [81] also stressed the indispensable impact of nanoparticle concentration, as at a maximum concentration of ZnO-DWCNT nanoparticles (1.95 vol% for EG nanofluid and one vol% for water-EG nanofluid), the k improved significantly:

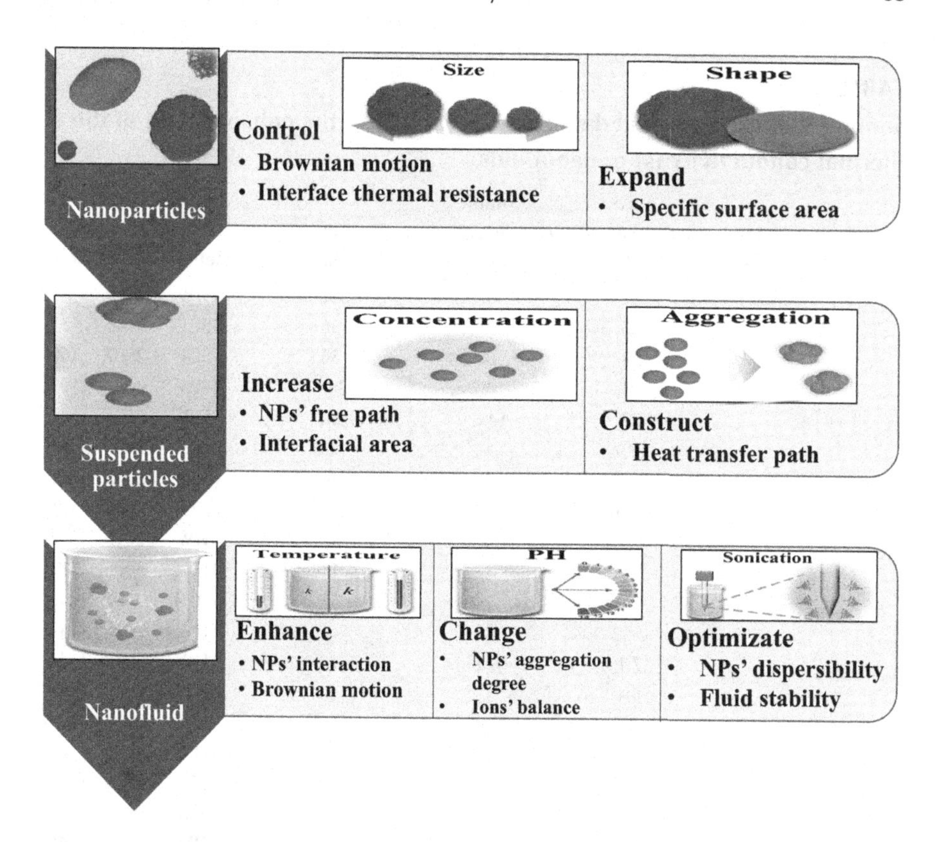

FIGURE 3.8 Intrinsic mechanism of various factors on the k_{nf}.

24.9% promotion for ZnO-DWCNT/EG nanofluid and more than 32% promotion for ZnO-DWCNT/water-EG nanofluid [77,78]. Rostamian et al. [79] reported similar observations as a 0.55 increase in CuO nanoparticle concentration from 0.02 to 0.75 vol% at least improved the k of the CuO/water-EG nanofluid by 28.9%. Higher nanoparticle concentration had more positive results in nanofluids combined with the temperature-related influences, such as improving the k of SiC/EG nanofluids by 16.21% compared with the base EG nanofluids [80].

The addition of nanoparticles with high k influences the nanofluids' general heat transfer capacity positively, which can be enhanced further by increasing the concentration and considering the study results mentioned above. Some researchers' findings attribute this positive effect of increased concentration to the reduction of the free path between nanoparticles by the percolation effect, increasing the frequency of the lattice vibration [82,83]. Other factors also contribute to the improved k value, including raising the interfacial area between nanoparticles and the base fluid [84] and the higher heat transport motion at immense amounts of particles [85].

Note that the improvement of the k value is only a function of higher nanoparticle concentration, although the dispersion and stability of the nanoparticles inside the base fluid can influence the effect of concentration [86,87], in which the dispersion is a function of sonication time. Later the following sections discuss this effect.

TABLE 3.2
Summary of experimental data from research into the enhancement of the thermal conductivity (k) of nanofluids

Particle type	Base fluid	k enhancement (%)	Volume concentration (%)	Size (nm)	Method	Year	Ref.
Al_2O_3	Water	31	18	41	THW	2009	[89]
		30	10	36	SSCB	2006	[88]
		28	6	36	SSCB	2007	[66]
		24	4	38.4	THW	2003	[93]
		22	14.6	20	TSHW	2007	[141]
		17.7	4	282	THW	2009	[23]
		14.4	4	40	Flash	2015	[121]
		10.1	4	15-50	TPS	2009	[75]
		9.7	3	43	THW	2010	[142]
		9	1	11	THW	2005	[14]
		8	3	38	SSCB	2007	[143]
		7.52	2	43	THW	2010	[142]
		7.1	0.08	10	THW	2018	[144]
		5.4	4	12	THW	2010	[145]
		4.7	0.05	10	THW	2018	[144]
		3.28	0.75	43	THW	2010	[142]
		3.1	0.04	10	THW	2018	[144]
		1.64	0.33	43	THW	2010	[142]
	EG	30	5	60.4	THW	2002	[146]
		22	0.08	10	THW	2018	[144]
		18	5	38	THW	1999	[147]
		17.3	0.06	10	THW	2018	[144]
		16.3	3	282	THW	2009	[23]
		14.3	4	12	THW	2010	[145]
		10.6	3	38	THW	2007	[143]
		9.7	4	45	3ω	2008	[52]
	EG/ Water	12.6	2	13	THW	2015	[148]
		11.3	3	10	THW	2010	[145]
		10.4	3	50	THW	2010	[145]
		8.4	2	13	THW	2015	[148]
		16.2	2	13	THW	2015	[148]
	EO	30	7.4	28	SSPP	1999	[45]
	PO	38	5	60.4	THW	2002	[146]
		20	7.1	28	SSPP	1999	[45]
	Glycerol	27	5	60.4	THW	2002	[146]
	DI	13.3	4	45	3ω	2008	[52]
		4	1	48	THW	2007	[149]

(Continued)

TABLE 3.2 (*Continued*)
Summary of experimental data from research into the enhancement of the thermal conductivity (k) of nanofluids

Particle type	Base fluid	k enhancement (%)	Volume concentration (%)	Size (nm)	Method	Year	Ref.
TiO_2	Water	11.4	3	10	THW	2007	[143]
		8.7	3	34	THW	2007	[143]
		7	2	21	THW	2009	[90]
		6.5	2.6	40	THW	2007	[141]
		6.4	3	70	THW	2007	[143]
		4.2	2	20	THW	2007	[150]
	EG	19.52	7	5	THW	2016	[151]
		18	5	15	THW	2008	[21]
		14.4	3	10	THW	2007	[143]
		12.3	3	34	THW	2007	[143]
		7.5	3	70	THW	2007	[143]
	DI	33	5	Ø10×40	THW	2005	[100]
		30	5	15	THW	2005	[100]
		14.4	1	20.5	THW	2007	[149]
		7.2	3	21	3ω	2009	[53]
CuO	Water	52	6	29	SSCB	2006	[88]
		34	9.7	23	SSPP	1999	[45]
		32.3	7.5	25	THW	2012	[152]
		24	2	55-66	THW	2016	[153]
		16.5	4.68	33	TSHW	2007	[141]
		12	3.41	24	THW	1999	[147]
		5	1	33	THW	2007	[154]
	EG	54	14.8	23	SSPP	1999	[45]
		23	4	24	THW	1999	[147]
		21	2	55-66	THW	2016	[153]
		9	1	33	THW	2006	[155]
	MEG	21.3	7.5	25	THW	2012	[152]
	EO	14	2	55-66	THW	2016	[153]
CeO_2	EG	22	2.5	10-30	THW	2018	[156]
ZnO	Water	14.2	3	10	THW	2007	[143]
		7.3	3	60	THW	2007	[143]
	EG	21	3	30	THW	2007	[143]
		13	2.4	50	THW	2014	[157]
WO_3	EG	13.8	0.3	38	THW	2007	[149]
Fe_3O_4	Kerosene	34.6	1	15	THW	2010	[158]
	Water	11.5	3	15-23	THW	2010	[159]
		2.9	4.8	15-20	3ω	2019	[160]
		1.1	1	15-20	3ω	2019	[160]
$NiFe_2O_4$	DI	17.2	2	8	THW	2015	[161]

(*Continued*)

TABLE 3.2 (*Continued*)
Summary of experimental data from research into the enhancement of the thermal conductivity (k) of nanofluids

Particle type	Base fluid	k enhancement (%)	Volume concentration (%)	Size (nm)	Method	Year	Ref.
MgO	EG/ Water	34.43	3	40	THW	2015	[162]
Cu	Oil	43	7.5	100	THW	2000	[1]
	Water	78	7.5	100	THW	2000	[1]
		23.8	0.1	75-100	THW	2006	[163]
	EG	40	0.3	<10	THW	2001	[164]
Au	Water	21	0.00026	10-20	THW	2003	[165]
	Toluene	8.8	0.011	10-20	THW	2003	[165]
		8	0.003	1.65	TSHW	2007	[141]
		1.4	0.024	2	MSBD	2006	[166]
	Ethanol	1.3	0.018	4	MSBD	2006	[166]
Ag	Water	20.8	1.7×10^{-5}	96	TWRC	2019	[167]
		16.5	0.001	10-20	THW	2003	[165]
		4	3.5×10^{-6}	96	TWRC	2019	[167]
	DI	16	0.5	5-25	THW	2015	[168]
Fe	EG	38.8	4	20	Theory	2018	[169]
		18	0.55	10	THW	2005	[115]
		16.5	0.3	10	THW	2007	[149]
		15.5	2	50	Theory	2018	[169]
Al	EG	45	5	80	THW	2008	[21]
MWCNT	Water	80	0.49	Ø40	THW	2006	[170]
		34	0.6	Ø130	THW	2005	[171]
		7	1	Ø10-30	THW	2007	[154]
		5	0.48	Ø10-30	THW	2016	[172]
	HTO	160	1	Ø25×50,000	THW	2001	[173]
		15	2	Ø5-20	THW	2012	[174]
	DI	7	1	Ø15×30,000	THW	2003	[175]
	EG	12.7	1	Ø15×30,000	THW	2003	[175]
		12.4	1	Ø20-50	THW	2005	[176]
	EO	8.5	1	Ø20-50	THW	2005	[176]
	DE	19.6	1	Ø15×30,000	THW	2003	[175]
DWCNT	Water	8	1	Ø5	THW	2005	[171]
SWCNT	Water	16.2	0.48	Ø1-2×5000-30,000	THW	2016	[172]
		8.1	0.48	Ø1-2×1000-3000	THW	2016	[172]
	EG	15.5	0.21	100-600	THW	2012	[177]
Graphene	EG	86	0.05	0.7-1.3	TSHW	2011	[178]

(*Continued*)

TABLE 3.2 (*Continued*)
Summary of experimental data from research into the enhancement of the thermal conductivity (k) of nanofluids

Particle type	Base fluid	k enhancement (%)	Volume concentration (%)	Size (nm)	Method	Year	Ref.
Graphene Oxide	EG	61	0.05	0.7-1.4	TSHW	2011	[178]
C_{60}-C_{70}	Toluene	0.816	0.378	-	MSBD	2006	[166]
fullerenes	MO	6	5	10	THW	2007	[154]
SiC (sphere)	DI	15.8	4.2	26	THW	2002	[97]
		7.2	3	100	THW	2011	[177]
	EG	13	3.5	26	THW	2002	[97]
	DO	7.36	0.8	30	THW	2016	[178]
SiC	DI	22.9	4	600	THW	2002	[99]
(cylinder)	EG	23	4	600	THW	2002	[99]
SiO_2	Water	38.2	3	40-50	THW	2019	[179]
		3.2	1	12	THW	2007	[154]
		3	1	12	THW	2006	[155]
Ag/Au	PO	33	0.006	10	THW	2006	[180]
		15	0.003	10	THW	2006	[180]
Sn/SiO_2	TH66	13	5	50-100	THW	2013	[181]
DWCNT/ ZnO	EG/ Water	33	1	Ø3/10-30	THW	2015	[79]

3.3.2 Temperature Effect

Many authors have stressed the undeniable role of temperature as a critical factor in k_{nf} investigations [14,21,85,88–92]. Li and Peterson [66] used the steady-state method to measure the k of Al_2O_3/water nanofluids at different temperatures. Temperature changes from 27.5°C to 34.5°C, improved the Al_2O_3/water nanofluid's effective thermal conductivity nearly three times compared with the base. They contended that the Brownian motion of the Al_2O_3 nanoparticles resulted in a micro-convection effect within the base fluid molecules, escalated by increased temperature, hence the nanofluid's enhanced heat transfer capacity. Similarly, Chon et al. [14] had the same observations confirming the increasing effect of higher temperatures on the Brownian velocity. Besides, Mintsa et al. [89] experimented to confirm the k of CuO/ water and Al_2O_3/water nanofluids' temperature dependence.

Duangthongsuk and Wongwises [90] used the THW method to measure the k change based on temperature for TiO_2/water nanofluids. These studies identified the Brownian motion improvement as the enhancing factor of the k under varying temperatures, an association that the findings of a relevant experiment by Godson et al. [91] further confirmed. Using the THW method for 0.9% vol

Ag/DI water nanofluids, they observed that a 40°C increase in temperature from 50°C to 90°C caused the Brownian velocity to change from about 1.5×10^{-9} to 2.1×10^{-9} m/s, increasing the k_{nf} from 0.8 to 1.6 W/m·K. The current data on nanofluid heat transfer reflects the temperature impact. Li et al. [92] stated that as the temperature increased, the nanoparticles fell short of the required surface energy for agglomeration, which improved the Brownian motion, and thus the k. Brownian motion is not only dependent on temperature, as the measured k by Harandi et al. [85] of F-MWCNTs-Fe_3O_4/EG under a changing temperature from 25°C to 50°C represented an almost linear dependence on the temperature. This dependence is associated with the enhanced interaction between nanoparticles due to increased temperature.

Das et al. [93] did a theoretical and experimental investigation of the temperature influence on the k of CuO/water and Al_2O_3/water nanofluids, reporting that at above 50°C range, their experimental data correlated with the Hamilton–Crosser thermal conductivity model. In a similar experiment, Tillman et al. [21] confirmed the palpable positive effect of temperature on k_{nf} the data which correlated with theoretical models at high temperatures, indicating the significant influence of temperature on the k, an influence intensified in response to higher temperatures.

3.3.3 Nanoparticle Size Effect

As stated, nanoparticle plays a critical role in k_{nf} measurement. The classical Maxwell model states that if nanoparticle size is excluded, k_{nf} will quite unreasonably be a function of volume fraction only [94]. Accordingly, many theoretical models consider the nanoparticle size effect's impact, and many researchers have devised experimental evidence. However, this impact is disputed, as numerous models and results indicate the negative association between the k_{nf} and nanoparticle size. In contrast, others have described the association between the two as positive.

Since the classical Maxwell model ignores the nanoparticles' Brownian motion, its predictions of k_{nf} lack accuracy. Dong and Chen [26] have described nanoparticle size and Brownian motion as two different factors influencing the k. However, the k_{nf} increased in response to decreased nanoparticle size. Nevertheless, since Brownian motion increased the nanoparticle radius, it led to an increase in the k_{nf} because of the increased nanoparticle volume fraction. Experiments have revealed the decreasing effect of larger nanoparticle sizes on the effective k of Al_2O_3/water nanofluids, which correlated with the model prediction. In their study on the k of Al_2O_3/DI water nanofluids as a function of nanoparticle size, Xu et al. [24] devised a fractal model that addressed the thermal convection and the fractal distribution of nanoparticle size, revealing that, under the effect of the Brownian motion, the effective k decreased as the nanoparticle size increased.

They reported that the Brownian motion velocity and k increased at smaller nanoparticle sizes. The findings in a relevant experiment by Jang and Choi [95] corroborated the increasing effect of nanoparticle/nanotube size on the k values of oxide nanoparticles, metallic nanoparticles, and CNT-based nanofluids. However, nanoparticles were not being suspended at nanoparticle sizes close to micrometers; therefore, the Brownian motion was stopped, and the k_{nf} would not increase.

Nevertheless, some researchers have made a different conclusion. Studying nanoparticle size in the range of 16–90 nm, Timofeeva et al. [96] asserted the improvement of the k of α-SiC/water nanofluids, which could be associated with the smaller nanoparticles' more significant surface area at the solid/liquid interface, blocking the heat flow and reducing the k. In a similar study, Chen et al. [97] investigated the association between k and nanoparticle size in SiO_2/water nanofluids in the range of 10–30 nm, finding that since the increased nanoparticle's size weakens interfacial thermal resistance, the k_{nf} increased linearly in response to increased nanoparticle size. Note that this linear relationship remained verifiable in a limited size range. Hence, more experiments were essential to explore the k_{nf}'s dependence on large nanoparticle size. Accordingly, Beck et al. [23] investigated the influence of nanoparticle size (8–282 nm) on the k of Al_2O_3/water or Al_2O_3/EG nanofluids, observing the higher k values in the Al_2O_3/water or Al_2O_3/EG nanofluids with larger nanoparticles. Also, they reported that the k remained constant when the diameter went up to 50 nm, confirming the Maxwell equations. To further explore the effect of nanoparticle size on the k_{nf}, Ganesan et al. [28] investigated the k of nanoclusters-based Fe_3O_4 nanofluids with a huge particle size (115–530 nm). Their findings contradicted the theoretical model's prediction that k of Fe_3O_4 nanofluids would increase by 5.3%–12.6% in response to declining nanoparticle size. It is assumed that nanoclusters growing kinetic into fractal-like aggregates in the suspensions are associated with this bucking the predicted trend [98].

3.3.4 The Impact of Nanoparticle's Shape

The shape of nanoparticles has been another determining factor influencing k_{nf} in the development of the k models. Using the THW method in the first experiment ever focusing on the effect of shape on k_{nf} to measure the k of SiC/water nanofluids, Xie et al. [99] revealed the significantly lower k of nanofluid with spherical SiC particles than that contained cylindrical particles under constant volume fraction. In a similar study, Murshed et al. [100] found that the k of TiO_2/DI water with spherical and rod nanoparticle shapes was lower than nanofluids containing the rod-shaped nanoparticles, which, based on the Hamilton and Crosser model (Eq. 33), can be associated with the rod-shaped nanoparticles' huger shape factor than the spherical counterpart. Jeong et al. [101] did not limit their studies to cylindrical and spherical nanoparticles and investigated nearly rectangular ZnO nanoparticles, finding rectangular nanoparticles' k was 5.9% higher than spherical nanoparticles, which was associated with the rectangular nanoparticles' slightly larger effective agglomeration radius. Ferrouillat et al. [102] experimented with the suspension of SiO_2 and ZnO nanoparticles with spherical and banana-shaped nanoparticles. For both nanofluids, the non-spherical nanoparticle-containing nanofluid displayed a higher k than the spherical nanoparticle-containing nanofluid, attributed to the former's larger surface area, thus larger contact area with the base fluid that improved heat transfer.

Timofeeva et al. [103] took a step forward and investigated alumina fluid with four shapes, that is, platelets, blades, cylinders, and bricks, which, if ordered based on their k-enhancing ability of the form high to low, would be: cylinders $>$ bricks $>$ blades $\approx$ platelets. This is because the decrease of the sphericity leads to increased particle shape and surface area, which positively influences heat transfer. Here, the sphericity represented the spheres' surface area ratio with the same volume as the particles to the particles' surface area. Notice that the surface thermal resistance also increased simultaneously. Under below-0.6 sphericity, the surface influence reduced the value of the k. In a study on four nanoparticle shapes, including long nanowires, short nanowires, nanospheres, and nanocubes, Bhanushali et al. [104] measured different k values, concluding that nanowire particles with a volume fraction of 0.25% improved k_{bf} by 40%. In contrast, nanosphere particles resulted in a mere 9.3% improvement. The significant improvement in k_{bf} by nanowire particles was attributed to the particles' large aspect ratios, which provided two advantages for k_{bf} enhancement: minimizing the negative impacts of interfacial thermal resistance on heat transfer and exciting thermal penetration. Ghosh et al. [105] also reported that nanoparticles with large aspect ratios raised the heat transfer rate and enhanced the k_{nf} during the nanoparticle collision. The collected experimental data indicate the undeniable effect of particle shape on k_{nf}. In an analytical study, Michaelides [5] discussed the significance of nanoparticles' shape and orientation in the nanofluid in influencing the k values.

3.3.5 Nanoparticles Aggregation Effect

Based on theoretical and experimental findings, the aggregation of nanoparticles can considerably improve k_{nf}. The occurrence of aggregation highly improves k_{nf} by creating an effective heat transfer path [106].

Based on Maxwell's practical medium theory [27], Hong et al. [35] investigated the influence of nanoparticle aggregation on k_{nf} in a series of experiments revealing that k of Al_2O_3/DI water nanofluids increased relative to the degree of aggregation, and the nanofluids with a volume fraction of 5% gained the highest k improvement: 22%, significantly lower than the Maxwell upper limit. The degree of aggregation was calculated by measuring μ_{nf}. Pang et al. [36] also reported a new model for predicting k_{nf} based on the practical medium theory. Notably, the current data give aggregation an indispensable positive role in improving k_{nf}, especially at low concentrations. They compared the effective k_{nf} with different shapes of SiO_2/methanol aggregates at 0.1%, concentration including fiber, ellipsoid, and spherical shapes, observing that for ellipsoid and fiber shapes, effective k_{nf} had a direct, positive relationship with the sphericity of the aggregation. Furthermore, they found that the ellipsoid shape had a less improving effect on the k values than the fiber shape.

In addition, based on the aggregation kinetics of the colloids, Prasher et al. [98] argued that k_{nf} first increases and then decreases with the increasing volume fraction of aggregation and reaches a maximum when the volume fraction is 0.35. Feng et al. [22] proposed a model for predicting the k of Al_2O_3/EG nanofluids with volume fractions of 0.03, 0.05, and 0.07, respectively, considering the nanolayer and the aggregation of nanoparticles. The results indicated that the nanoparticles' aggregation increases with decreases in nanoparticle size at the same volume fraction, and the more massive the volume fraction of the nanoparticles, the more the nanoparticles aggregate. Wei et al. [39] introduced a fractal model addressing the nanoparticles' aggregate distribution and Brownian motion to have a more substantial understanding of the influence of aggregation on the nanofluids' thermal properties. This model is remarkably in line with the experimental data for the k of SiO_2/ethanol nanofluids [15], Al_2Cu/water-EG nanofluids [107], and TiO_2/water nanofluids [100]. It was found that when the empirical shape factor $F < 6$, the shape of the aggregation is close to a circle, and the k_{nf} gradually decreases with increasing volume fraction. When the $F > 6$, the aggregation shape appears as a chain, so the k_{nf} increases with volume fraction much faster. F is the empirical shape factor related to the fractal dimension and volume fraction of the nanoparticles. Since the fractal dimension might be affected by aggregation shape, more experiments are needed for further analysis.

3.3.6 Effect of pH

Many experiments have corroborated the remarkable effect of modulating the nanofluids' pH values on changing the aggregation degree of nanoparticles and influencing k_{nf}. Under a specific pH value, the nanoparticles and the base liquid reach a balance between the positive and negative ions, called the isoelectric point (IEP). Several scholars have carried out experiments in this area. Wang et al. [108] investigated the dependence of the k on the pH value of Al_2O_3/water nanofluids and Cu/water nanofluids at a concentration of 0.1%. They observed that k_{nf} increases with the increasing pH value until it reaches the IEP and then decreases. The Al_2O_3/water nanofluids and the Cu/water nanofluids had an IEP of 7.5–8 and 9–9.5. This phenomenon is that the charge on the surface of the nanoparticles increases the IEP, increasing the electrostatic repulsion force between the nanoparticles. The subsequent decreasing aggregation of the suspension and the nanoparticles' increasing mobility led to enhanced heat transfer of the nanofluids. Li et al. [109] proposed that surface charge provides a more efficient channel for heat transfer. They reported that the k of CuO/DI water nanofluids increases with the increasing pH value in the 3–9.5 range. When the pH value is near the IEP of the CuO particles (~8.5–9.5), more surface charge is attached to the nanoparticles, which results in a more stable suspension. Finally, an optimum k is reached. At higher pH values, electrostatic repulsion between the nanoparticles is reduced, and aggregation occurs, leading to a decrease in k. Krishnakumar et al. [110] studied the change of k of Al_2O_3/ethanol nanofluids in the pH value range of 2.5 to 11. They observed the same trend in k as in Ref. [109], which suggests that the k increases to the maximum and then decreases with increasing pH value. Under the optimum pH, ≈6.0 k of the Al_2O_3/ethanol nanofluids with 0.1 and 0.5 vol% is increased by 11.2% and 15.5%, respectively. Habibzadeh et al. [111] studied the effect of pH on k of SnO_2/DI water nanofluids and pointed out that with a pH value of 8.0, the SnO_2 nanoparticles have zero repulsive force. The suspension is stable enough, and thus k_{nf} reaches the maximum. Hence, it is safe to conclude that under strongly acidic or basic conditions, k_{nf} is lower.

However, Murshed et al. [112] reported that as the pH increased from 3.4 to 9, the enhancement of k for TiO_2/DI water nanofluids at a concentration of 0.2% decreased from 5.5% to 2.55%, indicating that when the pH is close to the IEP, the suspension became unstable and the nanoparticle aggregation results in a decrease of the effective k_{nf}, which can be explained by DLVO theory [113]. Xie et al. [114] experimentally studied the effect of pH on k of Al_2O_3/DI water nanofluids. The results imply that as the difference between the suspension's pH values and that of the IEP (~9.2) increases, k_{nf} becomes huger, while the k at the IEP is the smallest. The underlying mechanism is that the repulsive force at the IEP is zero, and the resulting nanoparticle aggregation leads to a decrease in k_{nf}.

3.3.7 Sonication Time Effect

Ultrasonic dispersion is a common technique used in nanofluid preparation. Previous studies have illustrated that it is critical to perform adequate sample sonication to increase k_{nf} further [115–123]. Hong et al. [115] employed the THW method to measure k of Fe/EG nanofluids for different sonication times, reporting that when the sonication time is less than 50 min, k_{nf} increases gradually. When reaching 70 min, k_{nf} ceases to increase with increased sonication time. Hypothetically, a suitable sonication time causes the nanoparticle clusters' sizes to decrease and the fluid's stability to increase, and therefore k_{nf} is enhanced accordingly.

Other studies [116,117] investigated the influence of sonication on k of CNT/water nanofluids, reporting that enhanced dispersion uniformity increased the sonication time, leading to an improved k_{nf}. When the optimal sonication time (40 min) is exceeded, k_{nf} is reduced because the CNTs are broken, and the aspect ratio is decreased. Kole and Dey [118] asserted they observed the same trends in their study. Specifically, at the optimum sonication time, k_{nf} can be increased from approximately 0.30 to 0.35 W/m·K. Sonawane et al. [119] corroborated the improving effect of sonication on the Brownian motion of the nanoparticles in TiO_2/water, EG, and paraffin oil nanofluids based on the enhanced interaction between the nanoparticles and the fluid. Nevertheless, excessive sonication made the distance between the nanoparticles too close to aggregate, leading to a decrease in k_{nf}. Sundar et al. [120] attributed the increase of the k_{nf} to the prolonged sonication time, which improved the nanoparticle uniformity and blocked nanoparticle sedimentation caused by sonication vibration. Buonomo et al. [121] studied the effect of sonication time on the k of Al_2O_3/water nanofluids by the nanoflash method for the first time. The results indicated that k_{nf} varies incredibly for different concentrations of nanofluids relative to an increase in sonication time but remains unchanged above the optimal time. Nemade et al. [122] obtained an 18% improvement in k of CuO/water nanofluids by setting the suitable sonication time during preparation, revealing the nanoparticle size's marked dependence on the sonication time. As a result, sonication time affected k_{nf}.

Furthermore, Asadi et al. [123] investigated the influence of sonication time on surfactant-added $Mg(OH)_2$ nanofluids. In marked contrast to the above-mentioned surfactant-free nanofluids, k of surfactant-added $Mg(OH)_2$ nanofluids decreases immediately as the sonication time increases. This was because the sonication destroys the excellent stability of the surfactant in the nanofluids. In summary, the effect of sonication time on k_{nf} concentrates on its uniformity, stability, nanoparticle size, and the interaction between nanoparticles and fluid molecules.

3.4 HYBRID NANOFLUIDS

Many studies on nanofluids contain only one type of nanoparticle, and many researchers have discovered and reported various thermophysical properties and critical conclusions of such colloidal suspensions very well. Nevertheless, many studies are in progress that address nanofluids containing at least two types or more (hybrid nanofluids). This section mainly aims to give a brief description and record their progress in k_{hnf} from the theoretical and experimental perspectives. Theoretically, a classical model that is suitable for hybrid nanofluids will be introduced. In experimental studies, the effects of various critical factors on k_{hnf} are explored, such as nanoparticle type, size, shape, and temperature.

3.4.1 Theory

Hybrid nanofluids are obtained by adding two or more different nanoparticles to the base fluids to increase k_{nf}, which generally renders higher k than observed in pure fluids or single-particle nanofluids. As to the theoretical studies, Takabi and Salehi [124] reported an improved Maxwell model to predict k_{hnf}:

$$\frac{k_{hnf}}{k_{bf}} = \frac{\dfrac{\Phi_{p1}k_{p1} + \Phi_{p2}k_{p2}}{\Phi_e} + 2k_{bf} + 2\left(\Phi_{p1}k_{p1} + \Phi_{p2}k_{p2}\right) - 2\Phi_e k_{bf}}{\dfrac{\Phi_{p1}k_{p1} + \Phi_{p2}k_{p2}}{\Phi_e} + 2k_{bf} - 2\left(\Phi_{p1}k_{p1} + \Phi_{p2}k_{p2}\right) + \Phi_e k_{bf}} \tag{3.38}$$

where k_{hnf} is the effective thermal conductivity of the hybrid nanofluid, k_{bf} is the k of the base fluid, k_p denotes the k of the nanoparticles, Φ_e, and Φ_p is the volume concentration of the hybrid nanofluids and nanoparticles, respectively. The comparison of this model with the experimental data was reported by Suresh et al. [125], who believe that k of Al_2O_3–Cu/water hybrid nanofluids cannot be accurately calculated using the improved Maxwell model, especially for hybrid nanofluids with high concentration.

Charab et al. [126] devised the extended Maxwell model (EMM) to predict the k_{hnf}. The EMM can be expressed as:

$$\frac{k_{nf}}{k_{bf}} = 1 + \frac{3\left(\dfrac{k_{p1}}{k_{bf}} - 1\right) \times \Phi_{FVC}^{p1}}{\left(\dfrac{k_{p1}}{k_{bf}} + 2\right) - \left(\dfrac{k_{p1}}{k_{bf}} - 1\right) \times \Phi_{FVC}^{p1}} + \frac{3\left(\dfrac{k_{p2}}{k_{bf}} - 1\right) \times \Phi_{FVC}^{p2}}{\left(\dfrac{k_{p2}}{k_{bf}} + 2\right) - \left(\dfrac{k_{p2}}{k_{bf}} - 1\right) \times \Phi_{FVC}^{p2}} \tag{3.39}$$

$$\Phi_{FVC}^{p1} = \Phi_e\left(\frac{V_{nf1}}{V_{nf1} + V_{nf2}}\right) \tag{3.40}$$

$$\Phi_{FVC}^{p1} = \Phi_e\left(\frac{V_{nf1}}{V_{nf1} + V_{nf2}}\right) \tag{3.41}$$

where V_{nf} is the nanofluid volume and Φ_{FVC} is the fractional volume concentration. The theoretical prediction differed significantly from the experimental results, and thus the particle mapping model (PMM) was proposed. This approach assumed that the hybrid nanofluids' heat transfer was in series or parallel, and k_{hnf} was calculated based on Fourier's law using 2-D simulation. The results exhibited that the enhanced k_{nf} system was nonlinearly related to the volume fraction of the nanoparticles, probably caused by the stability of the nanoparticles and Brownian motion.

Based on the above studies, it is concluded that k_{hnf} can be significantly boosted compared with conventional fluids or nanofluids, and the boost is not linear with the concentration of different nanoparticles. Currently, more predictions about k_{hnf} are based on the correlations obtained from experimental data. Many researchers use the artificial neural network (ANN) to study k_{hnf}, which can be expressed as a function of temperature, nanoparticle size, and volume fraction [127–129] However, theoretical models for the hybrid nanofluids still fail to achieve accurate prediction of k, and thus more improved models for the hybrid nanofluids need to be developed to meet practical requirements.

3.4.2 Experimental Studies

A considerable number of experimental studies have confirmed that hybrid fluids have a significant enhancement effect compared to one-substance liquids on k_{hnf}. The nanoparticle's type has an important influence on k_{nf}. Botha et al. [130] measured k of Ag-SiO_2/oil hybrid nanofluids using the THW method and observed a 15% enhancement in k_{hnf}. The improvement of k_{hnf} is regarded as a result of Ag nanoparticles' attachment to the SiO_2, which causes shorter phonon transport distances between the nanoparticles. The study of Nine et al. [131] experimentally corroborated that Cu–Cu_2O nanofluids rendered a higher k than the Cu_2O nanofluids related to the smaller size of the Cu–Cu_2O nanoparticles that enhance the collision between nanoparticles during Brownian motion. Aravind et al. [132] observed that a 10.5% enhancement in k_{hnf} occurs for graphene–MWCNTs hybrid nanofluids. Since MWCNTs possess a high aspect ratio, they can be tightly bound to graphene to reduce the interfacial resistance. Thus, this hybrid nanofluid has a higher k than pure graphene nanofluids. Batmunkh et al. [133] proved that k of Ag–TiO_2 hybrid nanofluids is increased from 0.608 to 0.616 W/m·K at RT after adding Ag nanoparticles. They concluded that both the relatively higher k of Ag nanoparticles themselves and the improved Ag/TiO_2 interface promote phonon conduction and thus increase the fluid thermal conductivity. Ho et al. [134] measured n-eicosanoid water-based fluid with Al_2O_3 nanoparticle addition. The results show that k_{hnf} is improved because Al_2O_3 had a significantly higher k than n-eicosane particles. Madhesh et al. [135] found that when the concentration of Cu–TiO_2 nanoparticles in the fluid was approximately 2%, k_{bf} was increased by 6%. They attributed this phenomenon to the configuration of Cu nanoparticles on the surface of Al_2O_3 nanoparticles, which facilitates the formation of a thermal interfacial network between the nanoparticles and the fluid, thereby promoting heat transfer.

However, k_{hnf} can also be lower than that of the base fluid. Jana et al. [86] investigated CNT–Cu and CNT–Au hybrid nanofluids and the fluids with three components, observing that Au and Cu nanoparticles expectedly did not collaborate well with the CNTs, resulting in excessive interfacial thermal resistance and evident agglomeration, leading to k_{hnf} not receiving the potential improvement. Baghbanzadeh et al. [136] compared k of the MWCNT fluids, spherical SiO_2 fluids, and the MWCNT-SiO_2 hybrid fluids. Interestingly, their results demonstrate that k_{hnf} was between the two pure component fluids. They also stated that since the MWCNTs have a large aspect ratio and a high k, the enhancement effect on k_{hnf} is most prominent. In the hybrid fluid, the mixing of SiO_2 and MWCNTs produced additional thermal resistance. Thus the enhancement is weaker than MWCNTs but stronger than SiO_2 due to the lower k of SiO_2 than MWCNTs. Whether the hybrid nanofluids can render higher k or not is somewhat controversial. It seems that the specific nanoparticle constitution, or namely the interfacial thermal transport between the nanoparticles and the nanoparticle/fluid, dramatically affects the k of the resultant hybrid nanofluids. It is expected that future work on distinguishing the microscopic thermal transport behaviors in hybrid nanofluids would unravel this mystery.

An existing challenge in many studies is identifying the critical factors and the mechanisms to affect the k_{hnf}. The additional concentration, size, and shape of the nanoparticles and the temperature have been the recognized essential factors in the past decades. Baghbanzadeh et al. [136] contend that the increased nanoparticle concentration helps create a network to improve k. They found that the effective k_{hnf} increases from 1.05 to 1.13 W/m·K in response to a mass concentration change from 0.1% to 1%. Baby and Sundara [137] confirmed the improving effect of 28% concentration enhancement from 0.01% to 0.07% on the k of CuO-decorated graphene/DI water hybrid fluids. Similarly, they ascribed this k boost to the enhanced percolation effect due to the increase in concentration. The percolation considerably reduces the distance between nanoparticles. It promotes the contact between them, enhancing the lattice vibrations, which increases k. Yarmand et al. [80] also gave experimental evidence which supported this view. Their study demonstrated that k of Ag–graphene/water nanofluids is positively linearly dependent on the addition concentration.

Many researchers identify temperature as the key external factor influencing k_{hnf}. Sundar et al. [138] confirmed that k of MWCNT-Fe_3O_4/water nanofluids increased relative to the base fluid from 13.88% at 20°C to 28.46% at 60°C. They attribute this increase to the enhancement of Brownian motion. Harandi et al. [83] represented that the k of functionalized MWCNT-Fe_3O_4/EG hybrid nanofluids is enhanced in response to an increase in temperature, mainly attributed to the improved interaction between the nanoparticles. Esfahani et al. [139] also verified the positive influence of k by studying ZnO–Ag (2 vol%)/water hybrid nanofluids under increased temperature. They corroborated that k_{hnf} increases from 0.663 W/m·K at 25°C to 0.788 W/m·K at 50°C, which is also explained by enhanced Brownian motion.

The nanoparticles' shape and size are marginal factors compared with the two dominant factors mentioned above affecting k_{hnf}. Nine et al. [140] reported the relatively more prominent improving effect of cylindrical nanoparticles in hybrid fluids ($\approx$ 4%) on k_{bf} instead of spherical nanoparticles at an additional concentration level of 4 wt%. This is ascribed to the cylinder shape typically has a higher aspect ratio, a favorable factor for the k boost. The study of Baghbanzadeh et al. [136] experimentally verified that if the nanoparticle size is large, it is easy to form larger clusters with a reduced specific area, which weakens the fluid's k enhancement. It can be seen from the above reports that the influence of various factors on the hybrid fluid is consistent with the influence on the common single-phase fluid. Table 3.2 summarizes the experimental data from research into the enhancement of k of nanofluids, corresponding to a different type of nanoparticles and base fluids, volume concentration, particle size, and measurement methods.

3.5 THERMAL CONDUCTIVITY OF SUSPENSIONS (COMPLEMENTARY)

Just like viscosity characteristics, the first approach for thermal conduction modeling of solid–liquid suspensions is to investigate the effective thermal conduction of a homogenous mixture containing base fluid and nanoparticles. This matter is often called the theory of effective media/environment. In terminology associated with multiple-phase flow, this theory is necessarily based on a homogenous model. Since this theory is not enough per se in order to predict some of the experimental data sets, numerous researchers have modified this by taking into account other mechanisms that increase heat transfer, e.g., Brownian motion and interfacial solid layer. In the next three sections, the results of this research are presented.

3.5.1 Equations Based on the Theory of Effective Media

Several analytical studies concerning heat conduction of nanoparticles are founded on the theory of effective media, which asserts that equations ruling over homogenous media can be also generalized to non-homogenous suspensions. The values associated with transitive properties of non-homogenous suspensions are different compared to base fluids and need to be experimentally determined or analyzed.

Related to thermal conduction, an effective thermal conduction k_e can be defined for non-homogenous suspensions. All analytical and experimental studies show that effective conduction is positioned between the values of two constituents, namely k_f and k_s. Differential equations are similar for heat flux transfer and electric current (electrons). As a result, the analytical solution of electric current transfer or heat transfer-related solutions is for similar problems and geometries.

This is a chance for heat transfer problems since the results achieved for numerous analytical applications related to electric conduction can be also used for thermal conduction.

Maxwell [6,182] was the first person to calculate the electric conduction of a static mixture containing two constituents with different electric conductions. His final expression for the mixtures containing spheres with thermal conduction k_s on the ground from a static fluid with lower conduction (k_f) is illustrated as follows:

$$k_e = k_f\left[1+\frac{3(k_s+k_f)\varphi}{(k_s+2k_f)-(k_s+2k_f)\varphi}\right] \tag{3.42}$$

Bruggeman [8] conducted a similar analytical study on the effective electric conduction of different solid particles with regular forms. Bruggeman's results for effective thermal conduction are illustrated as follows:

$$k_e = k_f\left[1+\frac{k_s k_f(n-1)+(n-1)(k_s+k_f)\varphi}{k_s+k_s(n-1)-(k_s+k_f)\varphi}\right] = k_f\left[1+\frac{n(k_s+k_f)\varphi}{k_s+k_s(n-1)-(k_s+k_f)\varphi}\right] \tag{3.43}$$

Factor n equals 3 for spheres and 6 for cylinders. Hamilton and Crosser [7] conducted more comprehensive studies in order to include particles with non-regular forms. They proposed the following expression for suspension of particles with non-regular forms:

$$k_e = k_f\left[1+\frac{3(k_s-k_f)\frac{\varphi}{\Psi}}{k_s+k_s\left(\frac{3}{\Psi}-1\right)-(k_s-k_f)\varphi}\right] \tag{3.44}$$

Form factor ψ is defined as the sphericity of particles. This factor equals to 1 for spheres and Eq. (3.44) is transformed into Maxwell's expression, Eq. (3.42). It is worth noting that limiting behavior in analytical equations of Maxwell, Burgmann, and Hamilton and Crosser is independent of k_s in higher values of k_s and only depends on volume fraction φ or φ / ψ in the last equation. For $k_s \gg k_f$, the above equations can be written as follows:

$$k_e = k_f\left[1+\frac{3\varphi}{1-\varphi}\right] \approx k_f(1+3\varphi) \tag{3.45}$$

In Hamilton and Crosser's equation (Eq. 3.44), the numerator of the first expression is transformed into $3\varphi / \psi$ but the result is similar in terms of quality. Conducting numerous comparisons between the final limiting equation with experimental data has proven that the last four equations show values far less than the thermal equation observed in nanoparticles.

This matter has caused abnormal characteristics in the conduction of nanoparticles. One of the reasons for such characteristics is that the asymptotic limit of all equations in $k_s \gg k_f$ is independent of the conduction of solids k_s. They also depict that the effective conduction, which is calculated for nanoparticles, is not affected by the above values of k_s / k_f ratio. Nevertheless, it must be remembered that the function form of Eqs. (3.42)–(3.45) are achieved for spherical particles. Most nanofluids consist of irregular or stretched particles.

Bonnecaze and Brady [183] developed a theoretical framework for the heat transfer calculation of liquid suspensions and used this method in order to determine the heat conduction of numerous suspensions.

Nan et al. [37] proposed a method in their studies in order to calculate heat conduction of composite materials containing fiber, in which the impact of form, symmetry, the direction of solid particles in composite, interfacial thermal resistance, or Kapitza resistance R_i are considered. They achieved the following expression for stretched ellipsoid $\alpha_1 = \alpha_2 \prec \alpha_3$ with an aspect ratio of $E > 1$ and $E = \alpha_3 / \alpha_2$.

$$k_e = \frac{[3 + [2\beta_{11}(1 - L_{11}) + \beta_{33}(1 - L_{33})]]}{3 - \varphi(2\beta_{11}L_{11} + \beta_{33}L_{33})} \tag{3.46}$$

The two parameters in the last equation, including geometry (L_{11}, L_{33}) and conduction (β_{11}, β_{33}), are achieved as follows in terms of aspect ratio E, radiuses α_1, α_2, and α_3, and conductivity of two constituents:

$$L_{11} = \frac{E^2}{2(E^2 - 1)} - \frac{E\cosh^{-1}(E)}{2(E^2 - 1)^{\frac{3}{2}}}, L_{33} = 1 - 2L_{11}$$
$$\beta_{11} = \frac{k_{ii} - k_f}{k_f + k_{ii}(k_{ii} - k_f)}, k_{ii} = \frac{k_s}{1 + 2L_{11}(2 + E^{-1})\frac{k_s R_i}{\alpha_1}} \tag{3.47}$$

If interfacial resistance (R_i) is ignored, the expression proposed by Nan et al. [37] will be similar to the equation that is known as Fricke [184]. By reviewing isotropic spheres, this expression is transformed into Maxwell's [6,182] expression. Related to nanoparticles, it is worth noting that the analytical expression is used for the conduction Eq. (3.46) regarding fiber with extremely long aspect ratio, such as carbon nanotubes (single and a multi-walled) in which $E \gg 1$ and $k_s \gg k_f$. Considering the said conditions, Nan et al. [185] proposed the following limiting expression:

$$k_e = \frac{3k_f + \varphi k_s}{3 - 2\varphi} \approx k_f\left(1 + \frac{\varphi k_s}{3k_f}\right) \tag{3.48}$$

The thermal conductance proportion of k_s/k_f determines the rise in thermal conductivity in the final limiting formulation. Due to the model's high asymptote value of efficient thermal conductance (k_s/k_f), this is an essential consideration.

In contrast to Eq. (3.45), which is centered on spherical nanoparticles, the equation suggested by Nan et al. [185] is more suited for fibrous particles in order to compare with empirical results acquired using nanofluids incorporating carbon nanotubes. It is possible to double or triple the thermal conductivity of a suspension by incorporating a modest number of carbon nanotubes with thermal conductivities in the range of 3200–3500 W/m K into the solution. It is determined that the enhancement in thermal conductivity of nanofluids is not uncommon by comparing the experimental results of nanofluids incorporating carbon nanotubes with the formula mentioned above instead of the expression linked to spherical particles. Similarly, Yu et al. [186], who recently expressed thermal conductivity models for nanofluids incorporating carbon nanotubes, reached the same result.

3.5.2 Exchanges Based on the Brownian Movement

The effective or homogeneous environment model is the macro-analysis framework connected with these models. Studying the micro-displacement of tiny particles in a fluid flowing at various speeds yields information on the fluid's efficient thermal conductance. The Brownian equations of motion define this displacement's value. Koo and Kleinstreuer [13,187] conducted analytical research on nanofluids. They found that an increased conductivity of nanofluids may be attributable to two phenomena: (a) the static influence stated by several models, particularly Maxwell, and (b) Dynamic effects due to Brownian particle movement and micro-dimension dislocation in a fluid near particles. As a result, they developed the following formulas to increase the thermal conductance of nanofluids.

$$k_e = k_f\left[1 + \frac{3(k_s - k_f)\varphi}{(k_s + 2k_f) - (k_s + k_f)\varphi}\right] + 50000\varphi\rho_s c_s\sqrt{\frac{k_B T}{2\alpha\rho_s}}\beta_k f(T,\varphi) \quad (3.49)$$

Afterward, they indicated that $f(T, \varphi)$ is described as a quasi-experimental function and that β_k is described as another quasi-experimental function connected to the influence of a fluid transported as a result of the Brownian movement. The kind of particles utilized affects this parameter. Experimental information is used to derive $f(T, \varphi)$ and β_k functions. A Reynolds number linked with Brownian movement, Re_{Br}, was utilized by Prasher et al. [188] to describe the microscopic displacements induced by the movement of particles. They developed the following equation in the situation of particles with high thermal conductance, $k_s >> k_f$, and they did so while neglecting the interface resistance, R_i:

$$k_e = k_f[1 + 40000\,Re_{Br}^{2.5}\,Pr^{0.333}\,\varphi]\left[\frac{1+2\varphi}{1-\varphi}\right] \quad (3.50)$$

According to the Brownian movement, the Reynolds number is described as follows:

$$Re_{Br} = \frac{1}{v_f}\sqrt{\frac{9k_B T}{\pi\alpha\rho_s}} \quad (3.51)$$

For nanoparticle suspensions, Xuan et al. [189] used the Langevin equation and the idea of particle random Brownian movement to characterize velocity fluctuations and heat transmission. The Brownian movement had significant and exceptional impacts on the Maxwell-predicted effective thermal conductance, which the investigators tested through a thermal chain exchange between nanoparticles and fluids [6, 182]. Thus, the effective thermal conductivity of the suspension was calculated using the following formula:

$$k_e = k_f\left[1 + \frac{3(k_s - k_f)\varphi}{(k_s + 2k_f) - (k_s + k_f)\varphi}\right] + \frac{9\varphi h_c k_B T}{8\pi\rho_s\alpha^4} \quad (3.52)$$

h_c is the thermal transmission coefficient between fluid and particles, incorporating the thermal impedance at the interface. This is a variable that can't be predicted for any given particle or form. Employing empirical information comprising volume ratio, Brownian motion-dependent Reynolds number, fluid Prandtl number, and particle characterization, Chon et al. [14] developed a more straightforward technique. As a result, a data-processing-derived equation for the nanofluid's relative thermal conductance was presented.

In order to establish a quasi-experimental expression for relative conductivity, Jang and Choi [11] used an efficient environment hypothesis in conjunction with Brownian movement and the "boundary layer" of fluid surrounding the nanoparticles. The Reynolds number of the particle, Re_s, the proportion of the nanoparticle to the liquid molecule diameter, the fluid Prandtl number, and an experimental constant are all included in this equation.

These papers assert that the formulas derived from Brownian movement research accurately anticipate empirical results. Additionally, straightforward estimations demonstrate no uniformity for the information set employed to generate the model's empirical constants. There is still a considerable disparity between the information set that was not utilized to derive the empirical constants. At this point in the study, it is not apparent how the particles' Brownian movement influences should be incorporated into the thermal conductance formulas of the nanofluids, given the current state of the knowledge.

3.5.3 Formulas Based on Interface

According to researchers, Choi et al. [173], a probable mechanism for nanofluids' high thermal conductance is a solid layer of liquid molecules (liquid layering) development at the particle-liquid boundary. Keblinski et al. [190], Yu and Choi [19], and Xue [191] quickly followed with three novel models for the thermal conductance of nanofluids. In all scenarios, the development of a solid coating at the interface is taken into consideration and the use of Maxwell's hypothesis of electrical conductivity as the foundation for estimating the thermal conductance of a combination. The Hamilton–Crosser model for non-spherical particles was utilized by Yu and Choi [19] during a comparative analysis using the similar idea of a solid layer at the interface.

There are numerous molecules from the baseline fluid that adhere to the nanoparticle surface, generating a layer with the characteristics of a solid because of the strong intermolecular interactions that exist at this solid-liquid contact. Solid-phase characteristics of the basic fluid are present in the layer created by the molecules of the basic fluid. The solid layer may, for instance, have the qualities of ice whenever the basic liquid is water. As a result, to determine the thermal conductance of nanoparticles, two factors must be taken into account:

a. Nanoparticles exhibiting k_s conductance.
b. A solid coating of basic fluid molecules surrounding the nanoparticle with k_{sf} conductance.

Nanoparticles are shown by large gray circles, while solid basic liquid molecules are represented by black circles linked to the nanoparticles. Hence, a compound containing nanoparticles and a solid layer constitutes the solid component of a nanofluid suspension. A sphere and a solid shell combine to form the nanoparticle.

It is easy to infer a conduction equal to k_{eq} for the combined sphere [192]:

$$k_{eq} = \frac{\frac{k_{sf}}{k_s}\left[2\left(1-\frac{k_{sf}}{k_s}\right)+\left(1+\frac{\delta\alpha}{\alpha}\right)\left(1+2\frac{k_{sf}}{k_s}\right)\right]}{\frac{k_{sf}}{k_s}-1+\left(1+\frac{\delta\alpha}{\alpha}\right)^3\left(1+2\frac{k_{sf}}{k_s}\right)}k_s \tag{3.53}$$

In the solid layered model, the fraction of volume occupied by the solid in the suspension increased from φ to $\varphi\left(1+\frac{\delta\alpha}{\alpha}\right)^3$. Thus, the modified equivalent thermal conduction for solids and the increased volume fraction of the solids can improve the efficient conduction of heterogeneous nanofluid environments. If k_{eq} and the new fraction of volume are inserted in Maxwell's equation for the conduction of a heterogeneous thermodynamic system, the following improved term will be obtained for the conduction of nanofluids [19].

$$k_{eq} = k_s\left[1+\frac{3(k_{eq}-k_f)\left(1+\frac{\delta\alpha}{\alpha}\right)^3\varphi}{(k_{eq}+2k_f)-(k_{eq}+k_f)\left(1+\frac{\delta\alpha}{\alpha}\right)^3\varphi}\right] \tag{3.54}$$

3.5.4 Two Constraining Items Related to the Structure of Nanoparticles

While the analytical frameworks of the thermal conduction of fluids are considered, two useful items will be investigated below according to the configuration or structure of nanoparticles in the field of fluids. In both cases, the distance between the two planes equals L, and the solid volume fractions are equal to φ. The nanoparticles get concentrated and form layers placed as series or parallel on two conduction layers that maintain a constant difference in terms of temperature ΔT.

In the first case where the layer of particles is placed as a series alongside the liquid layer, the thermal flux is transferred from one plane to the next and causes thermal differences ΔT_1 and ΔT_2 between the planes. As the volume fraction of solids is equal to φ, the dimensionless thickness values of the layer of particles and liquid equal φ and $\varphi - 1$.

The following term for the thermal difference can be obtained by a preliminary investigation of the thermal conduction between the two planes:

$$\Delta T_1 = \frac{qL(1-\varphi)}{k_f}, \Delta T_2 = \frac{qL\varphi}{k_s} \tag{3.55}$$

The sum of the thermal difference is equal to the total difference in temperature. Thus, the following equation can be obtained:

$$\Delta T = q\left[\frac{qL(1-\varphi)}{k_f}+\frac{qL\varphi}{k_s}\right] \tag{3.56}$$

If $q=k_e\Delta T$ is considered, the last equation states a term for the efficient conduction in the spaces between conductor planes:

$$\frac{1}{k_e}=\frac{1-\varphi}{k_s}+\frac{\varphi}{k_s} \tag{3.57}$$

Nanofluids usually have significantly lower volume fractions than solids $\varphi \ll 1$, and the conduction of nanoparticles far exceeds that of fluids $k_s \gg k_f$. Thus, the asymptote limit of the above term can be expressed in the following manner:

$$k_{e(\varphi\ll 1)}=k_f(1+\varphi) \tag{3.58}$$

In the second case (B) where the fluid and solid layers are placed parallel to each other, the total thermal flux is equal to the sum of the thermal flux values in the two layers:

$$q=[k_f(1-\varphi)+k_s\varphi]\frac{\Delta T}{L} \tag{3.59}$$

The above term can be turned into the following equation for thermal conduction:

$$k_e=k_f(1-\varphi)+k_s\varphi \tag{3.60}$$

In the case of nanofluids, when $k_s \gg k_f$ and $\varphi \ll 1$, the constraining thermal conduction of the second case will be shown as follows:

$$k_{e(\varphi\ll 1)}=k_f+\varphi k_s \tag{3.61}$$

It is evident that the efficient conduction in the second case far exceeds the conduction of the first case. When the above two constraining cases are considered as boundary and ultimate cases of the point where the particles are concentrated in the layers, it would be interesting to note that all of the above analytical results concerning thermal conduction (including Eqs. 3.42–3.48) will be placed between them.

3.6 MD SIMULATIONS

MD is defined as a computer simulation technique that predicts the time evolution of a particular interacting system, including the generation of atomic trajectories of a system using numerical integration of Newton's equation of motion for a specific interatomic potential defined by an initial and boundary condition.

Although most of the studies on k_{nf} are based on the experimental and theoretical results discussed so far, several researchers have employed MD simulation. Despite the time-consuming computations and intricate processes, MD simulation can be a decent tool for investigating small-scale thermal systems' behavior. Lu and Fan [193] suggested that simplified MD simulations can be utilized to effectively analyze k of Al_2O_3/water and Al_2O_3/EG nanofluids. The simplified MD simulation processes mainly focus on selecting the potential energy models between nanoparticles, such as the two-particle model and the multi-particle model [194,195], between which the latter, the potential multi-particle model, is not used for MD simulation of nanofluids since it can contain some parameters that often have indeterminacy and computational complexity. Typically, the simulation results correlate with experimental data, meaning the results are valid and reliable. Besides, simulation results reported that the k of Al_2O_3/water and Al_2O_3/EG nanofluids increase significantly with an increase in the volume fraction of nanoparticles and decrease with increasing nanoparticle size. Since the heat transfer process is mainly present on the surface of the nanoparticles, which possess a much larger surface area-to-volume ratios, k_{nf} increases. By using MD simulation, Rudyak et al. [196] investigated the influence of different dispersed nanoparticles' ratios of mass (m) and diameter (r) on nanofluids' k_{eff} using molecular models in base fluids and nanoparticles, which are established by hard-sphere systems of different masses and diameters. It was demonstrated that k_{eff} of nanofluids consistently exceeds k_{bf} due to the presence of the nanoparticles, and the enhancement in the k values depends on the volume concentration, size, and masses of the nanoparticles. In particular, the effect of mass will be more significant than the size of the nanoparticles, which reveals that the density of the nanoparticles is also a key factor for k_{nf}.

Using the EMD simulation and the Green-Kubo function, Achhal and Jabraoui et al. [197] investigated the temperature dependence and particle size dependence of the k for Cu/Ar nanofluids. The interactions between Cu and Ar atoms were explained using the reputed embedded atom method (EAM) potential and Lennard–Jones (L–J) potential. They arranged a box containing Ar atoms in a face-centered cubic (fcc) configuration as a model for the nanofluids and used Cu atoms to replace the Ar atoms in the box's center. The boundary conditions were adjusted to be periodic boundaries, enabling the sample to relax under $T=300$ K using the NVT ensemble for a period to ensure the fluid phase's presence. The observations indicated that the k improvement displayed a positive trend from 0.19% to 7.66% under increased volume fraction since more nanoparticles blocked the fluid's condensation around the nanoparticles. Similarly, Javanmardi and Jafarpur [198] proved a nonlinear volume fraction dependence of k_{eff} in SWCNT/water nanofluids. For their study, the SWCNTs were armchair with chiral vector (12, 0), and the interactions between the liquid water, SWCNTs, water, and carbon were explained by a TIP4P potential Brenner's potential and L–J potential, respectively.

Most researchers have agreed upon the influence of many nanoparticle characteristics, such as volume fraction, on k_{nf}: An increasing nanoparticle volume fraction considerably improves the k. Temperature is another critical element. Generally, as the temperature increases, the volume fraction of nanoparticles and k_{nf} increases significantly. Since the converging of heat current autocorrelation function is time-consuming, Mohebbi [199] developed a new EDM-NEDM joined simulation method to predict c_p and α of nanofluids, setting the boundary condition to a non-periodic boundary. The k_{nf} was further gained by the so-called relation: $k=\alpha\rho c_p$. They learned that the k enhancements were 15% and 50% at 140 K and 107 K, respectively. It is seen that k_{nf} increases much more at low temperatures than at high temperatures. Sankar et al. [200] also investigated the temperature dependence of k for Pt/water nanofluids using the EMD method, in which some water molecules are replaced by an equal amount of Pt nanoparticles, and the simulation was carried out under the NVT ensemble. The nanofluid k enhancement was calculated at three different temperatures. It was found that the effective k_{nf} increases proportionally with the temperature.

The advanced model mentioned above ignores the Brownian motion influence on the heat transfer enhancement of nanofluids. To investigate whether Brownian motion affects k_{nf}, Keblinski et al. [190] first investigated the Brownian motion on heat transfer using the MD simulation in 2002. They used a heat flux autocorrelation function obtained by two different simulations in the same system. In the first simulation, the movements of all atoms were required to follow Newton's equation of motion, whereas, in another simulation, the center of mass of all nanoparticles was fixed so that the position of the nanoparticles was kept fixed at all times. It turned out that there is no difference in heat flux autocorrelation function using the two different simulation approaches. Since k is obtained by integrating the heat flux autocorrelation function, it was concluded that the Brownian motion does not affect k_{nf}.

Meanwhile, Sun et al. [201] also demonstrated that the Brownian motion of nanoparticles barely influences the improvement of k_{nf}. Besides, Cui et al. [202] studied the effect of chaotic movements on heat transfer by applying different shear velocities (50 and 0 m/s) to the nanoparticles and exploring the differences. They found that the chaotic movements of the nanoparticles affected the heat transfer process. The underlying mechanism of chaotic movements was mainly due to Brownian motion.

Another critical factor with increasing influence on k_{nf} is the aggregation of nanoparticles. Kang et al. [203] employed the EMD method to measure the k of Cu/Ar nanofluids following a Green-Kubo function. Unlike other MD simulations, there were multiple nanoparticles in each simulated box with periodic boundary conditions to pave the ground for the aggregation of nanoparticles. The results clearly show that aggregation of nanoparticles can effectively improve k_{nf}. Using the same method, Lee et al. [204] investigated nanoparticle aggregation influence and drew the same conclusion, where k_{nf} was obtained in two separate states, that is, aggregated state and non-aggregated state. When the nanofluid interior is in a state of aggregation, k_{nf} is more pronounced, principally due to the collision between more nanoparticles and the more effective heat transfer in the aggregated state.

Besides, the influence of size [205], shape [205,206], and surrounding nanolayers [207] of nanoparticles is undeniable. Furthermore, many studies have drawn significant conclusions concerning nanoparticles' size and the various increasing or decreasing influences on improving k values for different nanofluids. Cui et al. [202] represented that the larger the surface area-to-volume ratio for different nanoparticle shapes, the more significant the k enhancement. Hence, the higher the nanoparticles' surface area-to-volume ratio, the higher the k_{nf}. Liang and Tsai [207] believe that the increase in the density of nanolayers will have a positive impact on k_{nf}. Despite the large body of studies concerning MD simulation application to improve the k values in nanofluids and the remarkable breakthroughs in recognizing the influential factors on k, the current understanding seems incomplete. Experiments have confirmed the significance of the nanofluid's pH value as another determining factor of k improvement. However, studies using MD simulation tools have not addressed the pH's role in this field yet, hence taking MD simulation applications more seriously to enable future breakthroughs.

Figure 3.9 illustrates the studies' discussions concerning the k improvement of nanofluid suspensions formed by standard nanoparticles and base fluids. The nanoparticles include metal oxides: Al_2O_3 [23,52,142,145,149,208], TiO_2 [21,53,143,149,150],

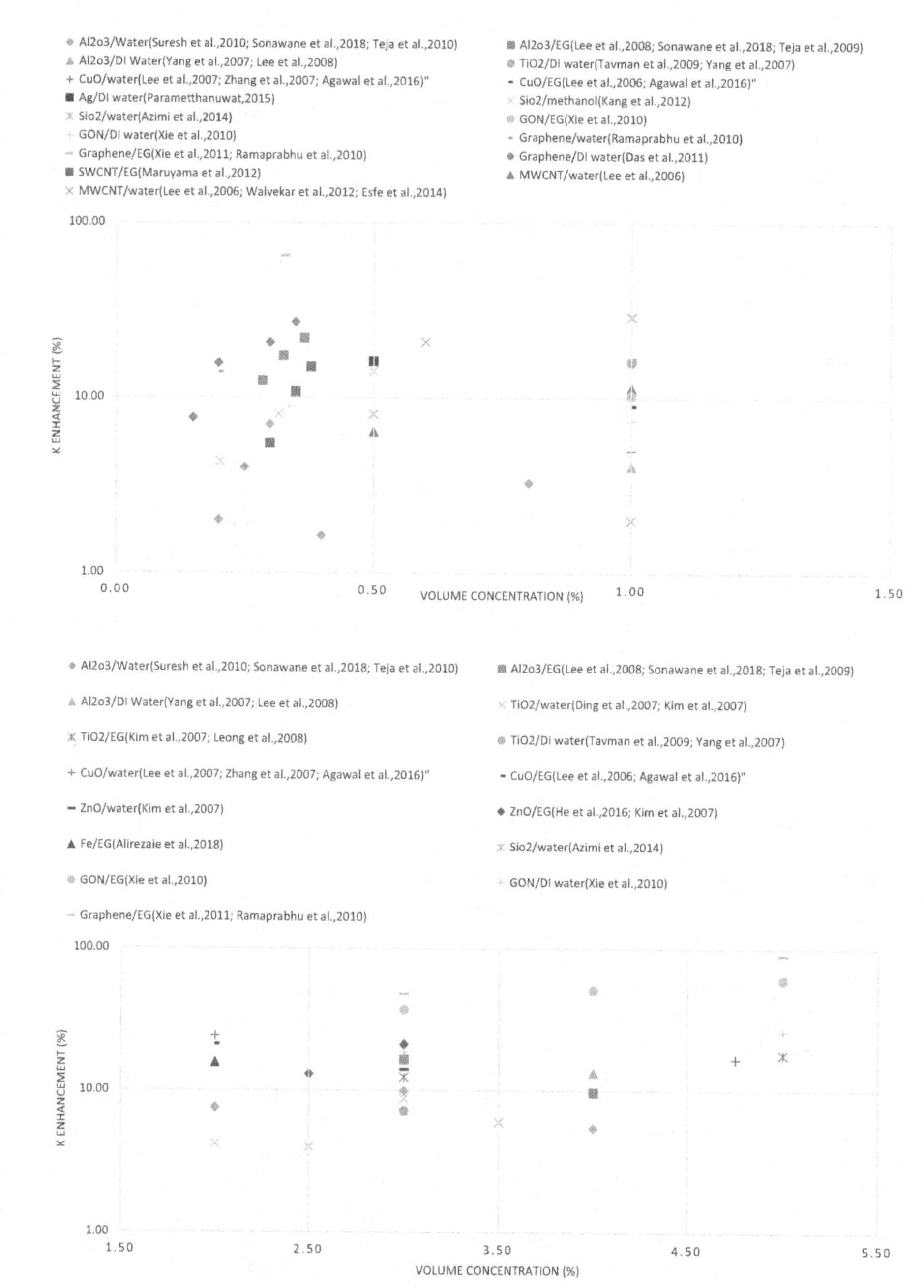

FIGURE 3.9 Summary and comparison of the enhancement of k_{nf} for various studies.

CuO [141,153,154,155], and ZnO [143,157], metals: Ag [168] and Fe [169,209], a non-metal oxide: SiO_2 [210,211], and advanced carbon nanomaterials: graphene oxide (GON) [212,213], graphene [178,214,215], CNT [155,177,216], and functionalized CNTs [217,218]. It is seen from the figure that k of some carbon nanomaterials (CNT and graphene)-based nanofluids display ultrahigh improvement compared to other metal- and metal oxide-based nanofluids. Furthermore, nanomaterials have substantial thermophysical properties, but they are also excellent fillers to improve k_{bf}. This finding can pave the way for future research to find practical applications for nanomaterials.

REFERENCES

1. Y. Xuan, Q. Li, Heat transfer enhancement of nanofuids, *Int. J. Heat Fluid Flow.* 21 (2000) 58–64. https://doi.org/10.1016/S0142-727X(99)00067-3.
2. O. Mahian, L. Kolsi, M. Amani, P. Estellé, G. Ahmadi, C. Kleinstreuer, J.S. Marshall, M. Siavashi, R.A. Taylor, H. Niazmand, S. Wongwises, T. Hayat, A. Kolanjiyil, A. Kasaeian, I. Pop, Recent advances in modeling and simulation of nanofluid flows-Part I: Fundamentals and theory, *Phys. Rep.* 790 (2019) 1–48. https://doi.org/10.1016/j.physrep.2018.11.004.
3. K. Khanafer, K. Vafai, A critical synthesis of thermophysical characteristics of nanofluids, *Int. J. Heat Mass Transf.* 54 (2011) 4410–4428. https://dx.doi.org/10.1016/j.ijheatmasstransfer.2011.04.048.
4. N. Putra, W. Roetzel, S.K. Das, Natural convection of nano-fluids, *Heat Mass Transf.* 39 (2003) 775–784. https://doi.org/10.1007/s00231-002-0382-z.
5. E.E. Michaelides, *Nanofluidics: Thermodynamic and Transport Properties*, Springer, New York, 2014.
6. J.C. Maxwell, *A Treatise on Electricity and Magnetism*, Clarendon Press, Oxford, 1873.
7. R.L. Hamilton, Thermal conductivity of heterogeneous two-component systems, *Ind. Eng. Chem. Fundam.* 1 (1962) 187–191. https://doi.org/10.1021/i160003a005.
8. D.A.G. Bruggeman, Berechnung verschiedener physikalischer Konstanten von heterogenen substanzen. I. Dielektrizitätskonstanten und leitfähigkeiten der mischkörper aus isotropen substanzen, *Ann. Phys.* 416 (1935) 636–664. https://doi.org/10.1002/andp.19354160705.
9. W. Yu, S.U.S. Choi, The role of interfacial layers in the enhanced thermal conductivity of nanofluids: A renovated Maxwell model, *J. Nanoparticle Res.* 5 (2003) 167–171. https://doi.org/10.1023/A:1024438603801.
10. Q. Xue, W. Xu, A model of thermal conductivity of nanofluids with interfacial shells, *Mater. Chem. Phys.* 90 (2005) 298–301. https://doi.org/10.1016/j.matchemphys.2004.05.029.
11. S.P. Jang, S.U.S. Choi, Role of Brownian motion in the enhanced thermal conductivity of nanofluids, *Appl. Phys. Lett.* 84 (2004) 4316–4318. https://doi.org/10.1063/1.1756684.
12. Y. Xuan, Q. Li, W. Hu, Aggregation structure and thermal conductivity of nanofluids, *AIChE J.* 49 (2003) 1038–1043. https://dx.doi.org/10.1002/aic.690490420.
13. J. Koo, C. Kleinstreuer, A new thermal conductivity model for nanofluids, *J. Nanoparticle Res.* 6 (2004) 577–588. https://doi.org/10.1007/s11051-004-3170-5.
14. C.H. Chon, K.D. Kihm, S.P. Lee, S.U.S. Choi, Empirical correlation finding the role of temperature and particle size for nanofluid (Al_2O_3) thermal conductivity enhancement, *Appl. Phys. Lett.* 87 (2005) 153107. https://doi.org/10.1063/1.2093936.
15. B. Wang, L. Zhou, X. Peng, A fractal model for predicting the effective thermal conductivity of liquid with suspension of nanoparticles, *Int. J. Heat Mass Transf.* 46 (2003) 2665–2672. https://doi.org/10.1016/S0017-9310(03)00016-4.

16. S.C. Cheng, R.I. Vachon, The prediction of the thermal conductivity of two and three phase solid heterogeneous mixtures, *Int. J. Heat Mass Transf.* 12 (1969) 249–264. https://doi.org/10.1016/0017-9310(69)90009-X.
17. D.J. Jeffrey, Conduction through a random suspension of spheres, *Proc. R. Soc. A* 335 (1973) 355–367. https://doi.org/10.1098/rspa.1973.0130.
18. C.J. Yu, A.G. Richter, A. Datta, M.K. Durbin, P. Dutta, Molecular layering in a liquid on a solid substrate: An X-ray reflectivity study, *Phys. B* 283 (2000) 27–31. https://doi.org/10.1016/S0921-4526(99)01885-2.
19. W. Yu, S.U.S. Choi, The role of interfacial layers in the enhanced thermal conductivity of nanofluids: A renovated Hamilton-Crosser model, *J. Nanoparticle Res.* 6 (2004) 355–361. https://doi.org/10.1007/s11051-004-2601-7.
20. K.C. Leong, C. Yang, S.M.S. Murshed, A model for the thermal conductivity of nanofluids-the effect of interfacial layer, *J. Nanoparticle Res.* 8 (2006) 245–254. https://doi.org/10.1007/s11051-005-9018-9.
21. P. Tillman, J.M. Hill, Determination of nanolayer thickness for a nanofluid, *Int. Commun. Heat Mass Transf.* 34.4 (2007): 399–407.
22. Y. Feng, B. Yu, P. Xu, M. Zou, The effective thermal conductivity of nanofluids based on the nanolayer and the aggregation of nanoparticles, *J. Phys. D. Appl. Phys.* 40 (2007) 3164–3171. https://doi.org/10.1088/0022-3727/40/10/020.
23. M.P. Beck, Y. Yuan, P. Warrier, A.S. Teja, The effect of particle size on the thermal conductivity of alumina nanofluids, *J. Nanoparticle Res.* 11 (2009) 1129–1136. https://doi.org/10.1007/s11051-008-9500-2.
24. J. Xu, B. Yu, M. Zou, P. Xu, A new model for heat conduction of nanofluids based on fractal distributions of nanoparticles, *J. Phys. D. Appl. Phys.* 39 (2006) 4486–4490. https://doi.org/10.1088/0022-3727/41/13/139801.
25. K.D. Kihm, C.H. Chon, J.S. Lee, S.U.S. Choi, A new heat propagation velocity prevails over Brownian particle velocities in determining the thermal conductivities of nanofluids, *Nanoscale Res. Lett.* 6 (2011) 361. https://doi.org/10.1186/1556-276X-6-361.
26. S. Dong, X. Chen, An improved model for thermal conductivity of nanofluids with effects of particle size and Brownian motion, *J. Therm. Anal. Calorim.* 129 (2017) 1255–1263. https://doi.org/10.1007/s10973-017-6256-x.
27. J. Eapen, R. Rusconi, R. Piazza, S. Yip, The classical nature of thermal conduction in nanofluids, *J. Heat Transf-T ASME* 132 (2010) 102402. https://doi.org/10.1115/1.4001304.
28. V. Ganesan, C. Louis, S.P. Damodaran, Novel nanofluids based on magnetite nanoclusters and investigation on their cluster size-dependent thermal conductivity, *J. Phys. Chem. C* 122 (2018) 6918–6929. https://doi.org/10.1021/acs.jpcc.7b12043.
29. R.S. Vajjha, D.K. Das, Experimental determination of thermal conductivity of three nanofluids and development of new correlations, *Int. J. Heat Mass Transf.* 52 (2009) 4675–4682. https://dx.doi.org/10.1016/j.ijheatmasstransfer.2009.06.027.
30. A.R. Moghadassi, S.M. Hosseini, D. Henneke, A. Elkamel, A model of nanofluids effective thermal conductivity based on dimensionless groups, *J. Therm. Anal. Calorim.* 96 (2009) 81–84. https://doi.org/10.1007/s10973-008-9843-z.
31. Y. Gao, H. Wang, A.P. Sasmito, A.S. Mujumdar, Measurement and modeling of thermal conductivity of graphene nanoplatelet water and ethylene glycol base nanofluids, *Int. J. Heat Mass Transf.* 123 (2018) 97–109. https://doi.org/10.1016/j.ijheatmasstransfer.2018.02.089.
32. K. Chu, W. Li, F. Tang, Flatness-dependent thermal conductivity of graphene-based composites, *Phys. Lett. A* 377 (2013) 910–914. https://dx.doi.org/10.1016/j.physleta.2013.02.009.
33. X.Q. Wang, A.S. Mujumdar, Heat transfer characteristics of nanofluids: A review, *Int. J. Therm. Sci.* 46 (2007) 1–19. https://doi.org/10.1016/j.ijthermalsci.2006.06.010.

34. J. Fan, L. Wang, Review of heat conduction in nanofluids, *J. Heat Transf-T ASME*. 133 (2011) 040801. https://doi.org/10.1115/1.4002633.
35. J. Hong, D. Kim, Effects of aggregation on the thermal conductivity of alumina/water nanofluids, *Thermochim. Acta*. 542 (2012) 28–32. https://dx.doi.org/10.1016/j.tca.2011.12.019.
36. C. Pang, J.Y. Jung, Y.T. Kang, Aggregation based model for heat conduction mechanism in nanofluids, *Int. J. Heat Mass Transf.* 72 (2014) 392–399. https://dx.doi.org/10.1016/j.ijheatmasstransfer.2013.12.055.
37. C.W. Nan, R. Birringer, D.R. Clarke, H. Gleiter, Effective thermal conductivity of particulate composites with interfacial thermal resistance, *J. Appl. Phys.* 81 (1997) 6692–6699. https://doi.org/10.1063/1.365209.
38. J. Xu, B. Yu, A new model for heat conduction of nanofluids based on fractal distributions of nanoparticles, *J. Phys. D Appl. Phys.* 39 (2006) 4486–4490. https://doi.org/10.1088/0022-3727/39/20/028.
39. W. Wei, J. Cai, X. Hu, Q. Han, S. Liu, Y. Zhou, Fractal analysis of the effect of particle aggregation distribution on thermal conductivity of nanofluids, *Phys. Lett. A* 380 (2016) 2953–2956. https://dx.doi.org/10.1016/j.physleta.2016.07.005.
40. Y. Feng, C. Kleinstreuer, Nanofluid convective heat transfer in a parallel-disk system, *Int. J. Heat Mass Transf.* 53 (2010) 4619–4628. https://doi.org/10.1016/j.ijheatmasstransfer.2010.06.031.
41. H.E. Patel, K.B. Anoop, T. Sundararajan, S.K. Das, Model for thermal conductivity of CNT-nanofluids, *Bull. Mater. Sci.* 31 (2008) 387–390. https://doi.org/10.1007/s12034-008-0060-y.
42. C. Nan, G. Liu, Y. Lin, M. Li, Interface effect on thermal conductivity of carbon nanotube composites, *Appl. Phys. Lett.* 85 (2004) 3549–3551. https://doi.org/10.1063/1.1808874.
43. M. Corcione, Empirical correlating equations for predicting the effective thermal conductivity and dynamic viscosity of nanofluids, *Energy Convers. Manag.* 52 (2011) 789–793. https://doi.org/10.1016/j.enconman.2010.06.072.
44. A. Zagabathuni, S. Ghosh, S.K. Pabi, The difference in the thermal conductivity of nanofluids measured by different methods and its rationalization, *Beilstein J. Nanotechnol.* 7 (2016) 2037–2044. https://doi.org/10.3762/BJNANO.7.194.
45. X. Wang, X. Xu, S.S.U. Choi, Thermal conductivity of nanoparticle-fluid mixture, *J. Thermophys. Heat Transf.* 13 (1999) 474–480. https://doi.org/10.2514/2.6486.
46. W. Jiang, G. Ding, H. Peng, Measurement and model on thermal conductivities of carbon nanotube nanorefrigerants, *Int. J. Therm. Sci.* 48 (2009) 1108–1115. https://doi.org/10.1016/j.ijthermalsci.2008.11.012.
47. W. Guo, G. Li, Y. Zheng, C. Dong, Measurement of the thermal conductivity of SiO_2 nanofluids with an optimized transient hot wire method, *Thermochim. Acta*. 661 (2018) 84–97. https://doi.org/10.1016/j.tca.2018.01.008.
48. K.D. Antoniadis, G.J. Tertsinidou, M.J. Assael, W.A. Wakeham, Necessary conditions for accurate, transient hot-wire measurements of the apparent thermal conductivity of nanofluids are seldom satisfied, *Int. J. Thermophys.* 37 (2016) 78. https://doi.org/10.1007/s10765-016-2083-8.
49. J. Lee, H. Lee, Y.J. Baik, J. Koo, Quantitative analyses of factors affecting thermal conductivity of nanofluids using an improved transient hot-wire method apparatus, *Int. J. Heat Mass Transf.* 89 (2015) 116–123. https://dx.doi.org/10.1016/j.ijheatmasstransfer.2015.05.064.
50. Z. Aparna, M.M. Michael, S.K. Pabi, S. Ghosh, Diversity in thermal conductivity of aqueous Al_2O_3- and Ag-nanofluids measured by transient hot-wire and laser flash methods, *Exp. Therm. Fluid Sci.* 94 (2018) 231–245. https://doi.org/10.1016/j.expthermflusci.2018.02.005.

51. M.H. Esfe, S. Saedodin, O. Mahian, S. Wongwises, Thermal conductivity of Al_2O_3/water nanofluids: Measurement, correlation, sensitivity analysis, and comparisons with literature reports, *J. Therm. Anal. Calorim.* 117 (2014) 675–681. https://doi.org/10.1016/10.1007/s10973-014-3771-x.
52. D.W. Oh, A. Jain, J.K. Eaton, K.E. Goodson, J.S. Lee, Thermal conductivity measurement and sedimentation detection of aluminum oxide nanofluids by using the 3ω method, *Int. J. Heat Fluid Flow.* 29 (2008) 1456–1461. https://doi.org/10.1016/j.ijheatfluidflow.2008.04.007.
53. A. Turgut, I. Tavman, M. Chirtoc, H.P. Schuchmann, C. Sauter, S. Tavman, Thermal conductivity and viscosity measurements of water-based TiO_2 nanofluids, *Int. J. Thermophys.* 30 (2009) 1213–1226. https://doi.org/10.1007/s10765-009-0594-2.
54. L. Qiu, X. Zheng, G. Su, D. Tang, Design and application of a freestanding sensor based on 3ω technique for thermal-conductivity measurement of solids, liquids, and nanopowders, *Int. J. Thermophys.* 34 (2013) 2261–2275. https://doi.org/10.1007/s10765-011-1075-y.
55. R. Karthik, R. Harish Nagarajan, B. Raja, P. Damodharan, Thermal conductivity of CuO-DI water nanofluids using 3-ω measurement technique in a suspended micro-wire, *Exp. Therm. Fluid Sci.* 40 (2012) 1–9. https://dx.doi.org/10.1016/j.expthermflusci.2012.01.006.
56. Z. Han, B. Yang, S.H. Kim, M.R. Zachariah, Application of hybrid sphere/carbon nanotube particles in nanofluids, Nanotechnology 18 (2007) 105701. https://doi.org/10.1088/0957-4484/18/10/105701.
57. B. Yang, Z.H. Han, Temperature-dependent thermal conductivity of nanorod-based nanofluids, *Appl. Phys. Lett.* 89 (2006) 2004–2007. https://doi.org/10.1063/1.2338424.
58. T.Y. Choi, M.H. Maneshian, B. Kang, W.S. Chang, C.S. Han, D. Poulikakos, Measurement of the thermal conductivity of a water-based single-wall carbon nanotube colloidal suspension with a modified 3-ω method, *Nanotechnology* 20 (2009). https://doi.org/10.1088/0957-4484/20/31/315706.
59. X. Yang, C. Liang, T. Ma, Y. Guo, J. Kong, J. Gu, M. Chen, J. Zhu, A review on thermally conductive polymeric composites: Classification, measurement, model and equations, mechanism and fabrication methods, *Adv. Compos. Hybrid Mater.* 1 (2018) 207–230. https://doi.org/10.1007/s42114-018-0031-8.
60. M.J. Assael, M. Dix, K. Gialou, L. Vozar, W.A. Wakeham, Application of the transient hot-wire technique to the measurement of the thermal conductivity of solids, *Int. J. Thermophys.* 23 (2002) 615–633. https://doi.org/10.1023/A:1015494802462.
61. M. Leena, S. Srinivasan, A comparative study on thermal conductivity of TiO_2/ethylene glycol-water and TiO_2/propylene glycol-water nanofluids, *J. Therm. Anal. Calorim.* 131 (2018) 1987–1998. https://doi.org/10.1007/s10973-017-6616-6.
62. S. Harikrishnan, S. Magesh, S. Kalaiselvam, Preparation and thermal energy storage behaviour of stearic acid-TiO_2 nanofluids as a phase change material for solar heating systems, *Thermochim. Acta.* 565 (2013) 137–145. https://dx.doi.org/10.1016/j.tca.2013.05.001.
63. Z. Zhang, Y. Yuan, L. Ouyang, Q. Sun, X. Cao, S. Alelyani, Enhanced thermal properties of Li_2CO_3-Na_2CO_3-K_2CO_3 nanofluids with nanoalumina for heat transfer in high-temperature CSP systems, *J. Therm. Anal. Calorim.* 128 (2017) 1783–1792. https://doi.org/10.1007/s10973-016-6050-1.
64. B. Buonomo, L. Colla, L. Fedele, O. Manca, L. Marinelli, A comparison of nanofluid thermal conductivity measurements by flash and hot disk techniques, *J. Phys. Conf. Ser.* 547 (2014) 012046. https://doi.org/10.1088/1742-6596/547/1/012046.
65. C. Kleinstreuer, Y. Feng, Experimental and theoretical studies of nanofluid thermal conductivity enhancement: A review, *Nanoscale Res. Lett.* 6 (2011) 229. https://doi.org/10.1186/1556-276X-6-229.

66. C.H. Li, G.P. Peterson, The effect of particle size on the effective thermal conductivity of Al_2O_3-water nanofluids, *J. Appl. Phys.* 101 (2007) 044312. https://doi.org/10.1063/1.2436472.
67. V. Sridhara, L.N. Satapathy, Al_2O_3-based nanofluids: A review, *Nanoscale Res. Lett.* 6 (2011) 456. https://doi.org/10.1186/1556-276X-6-456.
68. H. Kurt, M. Kayfeci, Prediction of thermal conductivity of ethylene glycol-water solutions by using artificial neural networks, *Appl. Energy.* 86 (2009) 2244–2248. https://dx.doi.org/10.1016/j.apenergy.2008.12.020.
69. B. Barbés, R. Páramo, E. Blanco, C. Casanova, Thermal conductivity and specific heat capacity measurements of CuO nanofluids, *J. Therm. Anal. Calorim.* 115 (2014) 1883–1891. https://doi.org/10.1007/s10973-013-3518-0.
70. S.E. Gustafsson, E. Karawacki, M.N. Khan, Transient hot-strip method for simultaneously measuring thermal conductivity and thermal diffusivity of solids and fluids, *J. Phys. D. Appl. Phys.* 12 (1979) 1411–1421. https://doi.org/10.1088/0022-3727/12/9/003.
71. S.E. Gustafsson, Transient plane source techniques for thermal conductivity and thermal diffusivity measurements of solid materials, *Rev. Sci. Instrum.* 62 (1991) 797–804. https://doi.org/10.1063/1.1142087.
72. M. Gustavsson, S.E. Gustafsson, Thermal conductivity as an indicator of fat content in milk, *Thermochim. Acta* 442 (2006) 1–5. https://doi.org/10.1016/j.tca.2005.11.037.
73. V. Bohac, M.K. Gustavsson, L. Kubicar, S.E. Gustafsson, Parameter estimations for measurements of thermal transport properties with the hot disk thermal constants analyzer, *Rev. Sci. Instrum.* 71 (2000) 2452–2455. https://doi.org/10.1063/1.1150635.
74. A. Harris, S. Kazachenko, R. Bateman, J. Nickerson, M. Emanuel, Measuring the thermal conductivity of heat transfer fluids via the modified transient plane source (MTPS), *J. Therm. Anal. Calorim.* 116 (2014) 1309–1314. https://doi.org/10.1007/s10973-014-3811-6.
75. D. Zhu, X. Li, N. Wang, X. Wang, J. Gao, H. Li, Dispersion behavior and thermal conductivity characteristics of Al_2O_3-H_2O nanofluids, *Curr. Appl. Phys.* 9 (2009) 131–139. https://doi.org/10.1016/j.cap.2007.12.008.
76. M.H. Esfe, S. Saedodin, Turbulent forced convection heat transfer and thermophysical properties of MgO-water nanofluid with consideration of different nanoparticles diameter, an empirical study, *J. Therm. Anal. Calorim.* 119 (2015) 1205–1213. https://doi.org/10.1007/s10973-014-4197-1.
77. M.H. Esfe, S. Esfandeh, M. Afrand, M. Rejvani, S.H. Rostamian, Experimental evaluation, new correlation proposing and ANN modeling of thermal properties of EG based hybrid nanofluid containing ZnO-DWCNT nanoparticles for internal combustion engines applications, *Appl. Therm. Eng.* 133 (2018) 452–463. https://doi.org/10.1016/j.applthermaleng.2017.11.131.
78. M.H. Esfe, W.M. Yan, M. Akbari, A. Karimipour, M. Hassani, Experimental study on thermal conductivity of DWCNT-ZnO/water-EG nanofluids, *Int. Commun. Heat Mass Transf.* 68 (2015) 248–251. https://dx.doi.org/10.1016/j.icheatmasstransfer.2015.09.001.
79. S.H. Rostamian, M. Biglari, S. Saedodin, M.H. Esfe, An inspection of thermal conductivity of CuO-SWCNTs hybrid nanofluid versus temperature and concentration using experimental data, ANN modeling and new correlation, *J. Mol. Liq.* 231 (2017) 364–369. https://dx.doi.org/10.1016/j.molliq.2017.02.015.
80. X. Li, C. Zou, X. Lei, W. Li, Stability and enhanced thermal conductivity of ethylene glycol-based SiC nanofluids, *Int. J. Heat Mass Transf.* 89 (2015) 613–619. https://dx.doi.org/10.1016/j.ijheatmasstransfer.2015.05.096.
81. M. Esfe, W.-M. Yan, M. Akbari, A. Karimipour, M. Hassani, Experimental study on thermal conductivity of DWCNT-ZnO/water-EG nanofluids, *Int. Commun. Heat Mass Transf.* 68 (2015). https://dx.doi.org/10.1016/j.icheatmasstransfer.2015.09.001.

82. H. Yarmand, S. Gharehkhani, G. Ahmadi, S.F.S. Shirazi, S. Baradaran, E. Montazer, M.N.M. Zubir, M.S. Alehashem, S.N. Kazi, M. Dahari, Graphene nanoplatelets-silver hybrid nanofluids for enhanced heat transfer, *Energy Convers. Manag.* 100 (2015) 419–428. https://dx.doi.org/10.1016/j.enconman.2015.05.023.
83. H. Yarmand, S. Gharehkhani, S.F.S. Shirazi, M. Goodarzi, A. Amiri, W.S. Sarsam, M.S. Alehashem, M. Dahari, S.N. Kazi, Study of synthesis, stability and thermo-physical properties of graphene nanoplatelet/platinum hybrid nanofluid, *Int. Commun. Heat Mass Transf.* 77 (2016) 15–21. https://dx.doi.org/10.1016/j.icheatmasstransfer.2016.07.010.
84. L.S. Sundar, E. Venkata Ramana, M.P.F. Graça, M.K. Singh, A.C.M. Sousa, Nanodiamond-Fe_3O_4 nanofluids: Preparation and measurement of viscosity, electrical and thermal conductivities, *Int. Commun. Heat Mass Transf.* 73 (2016) 62–74. https://dx.doi.org/10.1016/j.icheatmasstransfer.2016.02.013.
85. S.S. Harandi, A. Karimipour, M. Afrand, M. Akbari, A. D'Orazio, An experimental study on thermal conductivity of F-MWCNTs-Fe_3O_4/EG hybrid nanofluid: Effects of temperature and concentration, *Int. Commun. Heat Mass Transf.* 76 (2016) 171–177. https://dx.doi.org/10.1016/j.icheatmasstransfer.2016.05.029.
86. S. Jana, A. Salehi-Khojin, W.H. Zhong, Enhancement of fluid thermal conductivity by the addition of single and hybrid nano-additives, *Thermochim. Acta* 462 (2007) 45–55. https://doi.org/10.1016/j.tca.2007.06.009.
87. S.M. Abbasi, A. Rashidi, A. Nemati, K. Arzani, The effect of functionalisation method on the stability and the thermal conductivity of nanofluid hybrids of carbon nanotubes/gamma alumina, *Ceram. Int.* 39 (2013) 3885–3891. https://dx.doi.org/10.1016/j.ceramint.2012.10.232.
88. C.H. Li, G.P. Peterson, Experimental investigation of temperature and volume fraction variations on the effective thermal conductivity of nanoparticle suspensions (nanofluids), *J. Appl. Phys.* 99 (2006) 084314. https://doi.org/10.1063/1.2191571.
89. H.A. Mintsa, G. Roy, C.T. Nguyen, D. Doucet, New temperature dependent thermal conductivity data for water-based nanofluids, *Int. J. Therm. Sci.* 48 (2009) 363–371. https://dx.doi.org/10.1016/j.ijthermalsci.2008.03.009.
90. W. Duangthongsuk, S. Wongwises, Measurement of temperature-dependent thermal conductivity and viscosity of TiO_2-water nanofluids, *Exp. Therm. Fluid Sci.* 33 (2009) 706–714. https://dx.doi.org/10.1016/j.expthermflusci.2009.01.005.
91. L. Godson, B. Raja, D.M. Lal, S. Wongwises, Experimental investigation on the thermal conductivity and viscosity of silver-deionized water nanofluid, *Exp. Heat Transf.* 23 (2010) 317–332. https://doi.org/10.1080/08916150903564796.
92. Y. Li, W. Qu, J. Feng, Temperature dependence of thermal conductivity of nanofluids, *Chin. Phys. Lett.* 25 (2008) 3319–3322. https://doi.org/10.1088/0256-307X/25/9/060.
93. S.K. Das, N. Putra, P. Thiesen, W. Roetzel, Temperature dependence of thermal conductivity enhancement for nanofluids, *J. Heat Transf-T ASME* 125 (2003) 567–574. https://doi.org/10.1115/1.1571080.
94. S. Kakaç, A. Pramuanjaroenkij, Review of convective heat transfer enhancement with nanofluids, *Int. J. Heat Mass Transf.* 52 (2009) 3187–3196. https://dx.doi.org/10.1016/j.ijheatmasstransfer.2009.02.006.
95. S.P. Jang, S.U.S. Choi, Effects of various parameters on nanofluid thermal conductivity, *J. Heat Transf-T ASME* 129 (2007) 617–623. https://doi.org/10.1115/1.2712475.
96. E.V. Timofeeva, D.S. Smith, W. Yu, D.M. France, D. Singh, J.L. Routbort, Particle size and interfacial effects on thermo-physical and heat transfer characteristics of water-based α-SiC nanofluids, *Nanotechnology* 21 (2010) 215703. https://doi.org/10.1088/0957-4484/21/21/215703.
97. G. Chen, W. Yu, D. Singh, D. Cookson, J. Routbort, Application of SAXS to the study of particle-size-dependent thermal conductivity in silica nanofluids, *J. Nanoparticle Res.* 10 (2008) 1109–1114. https://doi.org/10.1007/s11051-007-9347-y.

98. R. Prasher, P.E. Phelan, P. Bhattacharya, Effect of aggregation kinetics on the thermal conductivity of nanoscale colloidal solutions (nanofluid), *Nano Lett.* 6 (2006) 1529–1534. https://doi.org/10.1021/nl060992s.
99. H. Xie, J. Wang, T. Xi, Y. Liu, Thermal conductivity of suspensions containing nanosized SiC particles, *Int. J. Thermophys.* 23 (2002) 571–580. https://doi.org/10.1023/A:1015121805842.
100. S.M.S. Murshed, K.C. Leong, C. Yang, Enhanced thermal conductivity of TiO_2-Water based nanofluids, *Int. J. Therm. Sci.* 44 (2005) 367–373. https://doi.org/10.1016/j.ijthermalsci.2004.12.005.
101. J. Jeong, C. Li, Y. Kwon, J. Lee, S.H. Kim, R. Yun, Particle shape effect on the viscosity and thermal conductivity of ZnO nanofluids, *Int. J. Refrig.* 36 (2013) 2233–2241. https://dx.doi.org/10.1016/j.ijrefrig.2013.07.024.
102. S. Ferrouillat, A. Bontemps, O. Poncelet, O. Soriano, J.A. Gruss, Influence of nanoparticle shape factor on convective heat transfer and energetic performance of water-based SiO_2 and ZnO nanofluids, *Appl. Therm. Eng.* 51 (2013) 839–851. https://doi.org/10.1016/j.applthermaleng.2012.10.020.
103. E.V. Timofeeva, J.L. Routbort, D. Singh, Particle shape effects on thermophysical properties of alumina nanofluids, *J. Appl. Phys.* 106 (2009) 014304. https://doi.org/10.1063/1.3155999.
104. S. Bhanushali, N.N. Jason, P. Ghosh, A. Ganesh, G.P. Simon, W. Cheng, Enhanced thermal conductivity of copper nanofluids: The effect of filler geometry, *ACS Appl. Mater. Interfaces* 9 (2017) 18925–18935. https://doi.org/10.1021/acsami.7b03339.
105. M.M. Ghosh, S. Ghosh, S.K. Pabi, Effects of particle shape and fluid temperature on heat-transfer characteristics of nanofluids, *J. Mater. Eng. Perform.* 22 (2013) 1525–1529. https://doi.org/10.1007/s11665-012-0441-7.
106. D. Song, D. Jing, W. Ma, X. Zhang, Effect of particle aggregation on thermal conductivity of nanofluids: Enhancement of phonon MFP, *J. Appl. Phys.* 125 (2019) 015103. https://doi.org/10.1063/1.5062600.
107. M. Chopkar, S. Sudarshan, P.K. Das, I. Manna, Effect of particle size on thermal conductivity of nanofluid, *Metall. Mater. Trans. A.* 39 (2008) 1535–1542. https://doi.org/10.1007/s11661-007-9444-7.
108. X. Wang, X. Li, S. Yang, Influence of pH and SDBS on the stability and thermal conductivity of nanofluids, *Energy Fuels.* 23 (2009) 2684–2689. https://doi.org/10.1021/ef800865a.
109. X. Li, D. Zhu, X. Wang, N. Wang, J. Gao, H. Li, Thermal conductivity enhancement dependent pH and chemical surfactant for Cu-H_2O nanofluids, *Thermochim. Acta* 469 (2008) 98–103. https://doi.org/10.1016/j.tca.2008.01.008.
110. T.S. Krishnakumar, S.P. Viswanath, S.M. Varghese, J. Prakash M, Experimental studies on thermal and rheological properties of Al_2O_3-ethylene glycol nanofluid, *Int. J. Refrig.* 89 (2018) 122–130. https://doi.org/10.1016/j.ijrefrig.2018.03.008.
111. S. Habibzadeh, A. Kazemi-Beydokhti, A.A. Khodadadi, Y. Mortazavi, S. Omanovic, M. Shariat-Niassar, Stability and thermal conductivity of nanofluids of tin dioxide synthesized via microwave-induced combustion route, *Chem. Eng. J.* 156 (2010) 471–478. https://doi.org/10.1016/j.cej.2009.11.007.
112. S.M.S. Murshed, K.C. Leong, C. Yang, Characterization of electrokinetic properties of nanofluids, *J. Nanosci. Nanotechnol.* 8 (2008) 5966–5971. https://doi.org/10.1166/jnn.2008.329.
113. H.J. Kim, S.H. Lee, J.H. Lee, S.P. Jang, Effect of particle shape on suspension stability and thermal conductivities of water-based bohemite alumina nanofluids, *Energy* 90 (2015) 1290–1297. https://doi.org/10.1016/j.energy.2015.06.084.
114. H. Xie, W. Yu, Y. Li, L. Chen, Discussion on the thermal conductivity enhancement of nanofluids, *Nanoscale Res. Lett.* 6 (2011) 124. https://doi.org/10.1186/1556-276X-6-124.
115. T.K. Hong, H.S. Yang, C.J. Choi, Study of the enhanced thermal conductivity of Fe nanofluids, *J. Appl. Phys.* 97 (2005) 064311. https://doi.org/10.1063/1.1861145.

116. P. Garg, J.L. Alvarado, C. Marsh, T.A. Carlson, D.A. Kessler, K. Annamalai, An experimental study on the effect of ultrasonication on viscosity and heat transfer performance of multi-wall carbon nanotube-based aqueous nanofluids, *Int. J. Heat Mass Transf.* 52 (2009) 5090–5101. https://dx.doi.org/10.1016/j.ijheatmasstransfer.2009.04.029.
117. R. Sadri, G. Ahmadi, H. Togun, M. Dahari, S.N. Kazi, E. Sadeghinezhad, N. Zubir, An experimental study on thermal conductivity and viscosity of nanofluids containing carbon nanotubes, *Nanoscale Res. Lett.* 9 (2014) 151. https://doi.org/10.1186/1556-276X-9-151.
118. M. Kole, T.K. Dey, Thermophysical and pool boiling characteristics of ZnO-ethylene glycol nanofluids, *Int. J. Therm. Sci.* 62 (2012) 61–70. https://dx.doi.org/10.1016/j.ijthermalsci.2012.02.002.
119. S.S. Sonawane, R.S. Khedkar, K.L. Wasewar, Effect of sonication time on enhancement of effective thermal conductivity of nano TiO_2-water, ethylene glycol, and paraffin oil nanofluids and models comparisons, *J. Exp. Nanosci.* 10 (2015) 310–322. https://dx.doi.org/10.1080/17458080.2013.832421.
120. L.S. Sundar, E. Venkata Ramana, M.K. Singh, A.C.M. Sousa, Thermal conductivity and viscosity of stabilized ethylene glycol and water mixture Al_2O_3 nanofluids for heat transfer applications: An experimental study, *Int. Commun. Heat Mass Transf.* 56 (2014) 86–95. https://dx.doi.org/10.1016/j.icheatmasstransfer.2014.06.009.
121. B. Buonomo, O. Manca, L. Marinelli, S. Nardini, Effect of temperature and sonication time on nanofluid thermal conductivity measurements by nano-flash method, *Appl. Therm. Eng.* 91 (2015) 181–190. https://dx.doi.org/10.1016/j.applthermaleng.2015.07.077.
122. K. Nemade, S. Waghuley, A novel approach for enhancement of thermal conductivity of CuO/H_2O based nanofluids, *Appl. Therm. Eng.* 95 (2016) 271–274. https://dx.doi.org/10.1016/j.applthermaleng.2015.11.053.
123. A. Asadi, M. Asadi, M. Siahmargoi, T. Asadi, M. Gholami Andarati, The effect of surfactant and sonication time on the stability and thermal conductivity of water-based nanofluid containing $Mg(OH)_2$ nanoparticles: An experimental investigation, *Int. J. Heat Mass Transf.* 108 (2017) 191–198. https://dx.doi.org/10.1016/j.ijheatmasstransfer.2016.12.022.
124. B. Takabi, S. Salehi, Augmentation of the heat transfer performance of a sinusoidal corrugated enclosure by employing hybrid nanofluid, *Adv. Mech. Eng.* 6 (2014) 147059. https://doi.org/10.1155/2014/147059.
125. S. Suresh, K.P. Venkitaraj, P. Selvakumar, M. Chandrasekar, Synthesis of Al_2O_3-Cu/water hybrid nanofluids using two step method and its thermo physical properties, *Colloids Surf. A Physicochem. Eng. Asp.* 388 (2011) 41–48. https://dx.doi.org/10.1016/j.colsurfa.2011.08.005.
126. A.A. Charab, S. Movahedirad, R. Norouzbeigi, Thermal conductivity of Al_2O_3+TiO_2/water nanofluid: Model development and experimental validation, *Appl. Therm. Eng.* 119 (2017) 42–51. https://dx.doi.org/10.1016/j.applthermaleng.2017.03.059.
127. M.H. Esfe, S. Wongwises, A. Naderi, A. Asadi, M.R. Safaei, H. Rostamian, M. Dahari, A. Karimipour, Thermal conductivity of Cu/TiO_2-water/EG hybrid nanofluid: Experimental data and modeling using artificial neural network and correlation, *Int. Commun. Heat Mass Transf.* 66 (2015) 100–104. https://dx.doi.org/10.1016/j.icheatmasstransfer.2015.05.014.
128. M. Vafaei, M. Afrand, N. Sina, R. Kalbasi, F. Sourani, H. Teimouri, Evaluation of thermal conductivity of MgO-MWCNTs/EG hybrid nanofluids based on experimental data by selecting optimal artificial neural networks, *Physica E* 85 (2017) 90–96. https://dx.doi.org/10.1016/j.physe.2016.08.020.
129. M.H. Esfe, S. Esfandeh, S. Saedodin, H. Rostamian, Experimental evaluation, sensitivity analyzation and ANN modeling of thermal conductivity of ZnO-MWCNT/EG-water hybrid nanofluid for engineering applications, *Appl. Therm. Eng.* 125 (2017) 673–685. https://dx.doi.org/10.1016/j.applthermaleng.2017.06.077.

130. S.S. Botha, P. Ndungu, B.J. Bladergroen, Physicochemical properties of oil-based nanofluids containing hybrid structures of silver nanoparticles supported on silica, *Ind. Eng. Chem. Res.* 50 (2011) 3071–3077. https://doi.org/10.1021/ie101088x.
131. M.J. Nine, B. Munkhbayar, M.S. Rahman, H. Chung, H. Jeong, Highly productive synthesis process of well dispersed Cu_2O and Cu/Cu_2O nanoparticles and its thermal characterization, *Mater. Chem. Phys.* 141 (2013) 636–642. https://dx.doi.org/10.1016/j.matchemphys.2013.05.032.
132. S.S.J. Aravind, S. Ramaprabhu, Graphene-multiwalled carbon nanotube-based nanofluids for improved heat dissipation, *RSC Adv.* 3 (2013) 4199–4206. https://doi.org/10.1039/C3RA22653K.
133. M. Batmunkh, M.R. Tanshen, M.J. Nine, M. Myekhlai, H. Choi, H. Chung, H. Jeong, Thermal conductivity of TiO_2 nanoparticles based aqueous nanofluids with an addition of a modified silver particle, Ind. *Eng. Chem. Res.* 53 (2014) 8445–8451. https://doi.org/10.1021/ie403712f.
134. C.J. Ho, J.B. Huang, P.S. Tsai, Y.M. Yang, Preparation and properties of hybrid water-based suspension of Al_2O_3 nanoparticles and MEPCM particles as functional forced convection fluid, *Int. Commun. Heat Mass Transf.* 37 (2010) 490–494. https://dx.doi.org/10.1016/j.icheatmasstransfer.2009.12.007.
135. D. Madhesh, R. Parameshwaran, S. Kalaiselvam, Experimental investigation on convective heat transfer and rheological characteristics of Cu-TiO_2 hybrid nanofluids, *Exp. Therm. Fluid Sci.* 52 (2014) 104–115. https://dx.doi.org/10.1016/j.expthermflusci.2013.08.026.
136. M. Baghbanzadeh, A. Rashidi, D. Rashtchian, R. Lotfi, A. Amrollahi, Synthesis of spherical silica/multiwall carbon nanotubes hybrid nanostructures and investigation of thermal conductivity of related nanofluids, *Thermochim. Acta* 549 (2012) 87–94. https://dx.doi.org/10.1016/j.tca.2012.09.006.
137. T.T. Baby, R. Sundara, Synthesis and transport properties of metal oxide decorated graphene dispersed nanofluids, *J. Phys. Chem. C* 115 (2011) 8527–8533. https://doi.org/10.1021/jp200273g.
138. L.S. Sundar, M.K. Singh, A.C.M. Sousa, Enhanced heat transfer and friction factor of MWCNT-Fe_3O_4/water hybrid nanofluids, *Int. Commun. Heat Mass Transf.* 52 (2014) 73–83. https://dx.doi.org/10.1016/j.icheatmasstransfer.2014.01.012.
139. N.N. Esfahani, D. Toghraie, M. Afrand, A new correlation for predicting the thermal conductivity of ZnO-Ag (50%-50%)/water hybrid nanofluid: An experimental study, *Powder Technol.* 323 (2018) 367–373. https://doi.org/10.1016/j.powtec.2017.10.025.
140. M.J. Nine, M. Batmunkh, J.H. Kim, H.S. Chung, H.M. Jeong, Investigation of Al_2O_3-MWCNTs hybrid dispersion in water and their thermal characterization, *J. Nanosci. Nanotechnol.* 12 (2012) 4553–4559. https://doi.org/10.1166/jnn.2012.6193.
141. X. Zhang, H. Gu, M. Fujii, Effective thermal conductivity and thermal diffusivity of nanofluids containing spherical and cylindrical nanoparticles, *Exp. Therm. Fluid Sci.* 31 (2007) 593–599. https://doi.org/10.1016/j.expthermflusci.2006.06.009.
142. M. Chandrasekar, S. Suresh, A. Chandra Bose, Experimental investigations and theoretical determination of thermal conductivity and viscosity of Al_2O_3/water nanofluid, *Exp. Therm. Fluid Sci.* 34 (2010) 210–216. https://dx.doi.org/10.1016/j.expthermflusci.2009.10.022.
143. S.H. Kim, S.R. Choi, D. Kim, Thermal conductivity of metal-oxide nanofluids: Particle size dependence and effect of laser irradiation, *J. Heat Transfer.* 129 (2007) 298. https://doi.org/10.1115/1.2427071.
144. N. Kumar, S.S. Sonawane, S.H. Sonawane, Experimental study of thermal conductivity, heat transfer and friction factor of Al_2O_3 based nanofluid, *Int. Commun. Heat Mass Transf.* 90 (2018) 1–10. https://doi.org/10.1016/j.icheatmasstransfer.2017.10.001.

145. M.P. Beck, Y. Yuan, P. Warrier, A.S. Teja, The thermal conductivity of alumina nanofluids in water, ethylene glycol, and ethylene glycol+water mixtures, *J. Nanoparticle Res.* 12 (2010) 1469–1477. https://doi.org/10.1007/s11051-009-9716-9.
146. H. Xie, J. Wang, T. Xi, Y. Liu, F. Ai, Q. Wu, Thermal conductivity enhancement of suspensions containing nanosized alumina particles, *J. Appl. Phys.* 91 (2002) 4568–4572. https://doi.org/10.1063/1.1454184.
147. S. Lee, S.U.S. Choi, S. Li, J.A. Eastman, Measuring thermal conductivity of fluids containing oxide nanoparticles, *J. Heat Transfer.* 121 (1999) 280–289. https://doi.org/10.1115/1.2825978.
148. N.A. Usri, W.H. Azmi, R. Mamat, K.A. Hamid, G. Najafi, Thermal conductivity enhancement of Al_2O_3 nanofluid in ethylene glycol and water mixture, *Energy Procedia* 79 (2015) 397–402. https://dx.doi.org/10.1016/j.egypro.2015.11.509.
149. D.H. Yoo, K.S. Hong, H.S. Yang, Study of thermal conductivity of nanofluids for the application of heat transfer fluids, *Thermochim. Acta* 455 (2007) 66–69. https://doi.org/10.1016/j.tca.2006.12.006.
150. Y. He, Y. Jin, H. Chen, Y. Ding, D. Cang, H. Lu, Heat transfer and flow behaviour of aqueous suspensions of TiO_2 nanoparticles (nanofluids) flowing upward through a vertical pipe, *Int. J. Heat Mass Transf.* 50 (2007) 2272–2281. https://doi.org/10.1016/j.ijheatmasstransfer.2006.10.024.
151. R.S. Khedkar, N. Shrivastava, S.S. Sonawane, K.L. Wasewar, Experimental investigations and theoretical determination of thermal conductivity and viscosity of TiO_2-ethylene glycol nanofluid, *Int. Commun. Heat Mass Transf.* 73 (2016) 54–61. https://dx.doi.org/10.1016/j.icheatmasstransfer.2016.02.004.
152. R.S. Khedkar, S.S. Sonawane, K.L. Wasewar, Influence of CuO nanoparticles in enhancing the thermal conductivity of water and monoethylene glycol based nanofluids, *Int. Commun. Heat Mass Transf.* 39 (2012) 665–669. https://doi.org/10.1016/j.icheatmasstransfer.2012.03.012.
153. R. Agarwal, K. Verma, N.K. Agrawal, R.K. Duchaniya, R. Singh, Synthesis, characterization, thermal conductivity and sensitivity of CuO nanofluids, *Appl. Therm. Eng.* 102 (2016) 1024–1036. https://dx.doi.org/10.1016/j.applthermaleng.2016.04.051.
154. Y. Hwang, J.K. Lee, C.H. Lee, Y.M. Jung, S.I. Cheong, C.G. Lee, B.C. Ku, S.P. Jang, Stability and thermal conductivity characteristics of nanofluids, *Thermochim. Acta* 455 (2007) 70–74. https://doi.org/10.1016/j.tca.2006.11.036.
155. Y.J. Hwang, Y.C. Ahn, H.S. Shin, C.G. Lee, G.T. Kim, H.S. Park, J.K. Lee, Investigation on characteristics of thermal conductivity enhancement of nanofluids, *Curr. Appl. Phys.* 6 (2006) 1068–1071. https://doi.org/10.1016/j.cap.2005.07.021.
156. M. Keyvani, M. Afrand, D. Toghraie, M. Reiszadeh, An experimental study on the thermal conductivity of cerium oxide/ethylene glycol nanofluid: Developing a new correlation, *J. Mol. Liq.* 266 (2018) 211–217. https://doi.org/10.1016/j.molliq.2018.06.010.
157. H. Li, L. Wang, Y. He, Y. Hu, J. Zhu, B. Jiang, Experimental investigation of thermal conductivity and viscosity of ethylene glycol based ZnO nanofluids, *Appl. Therm. Eng.* 88 (2014) 363–368. https://dx.doi.org/10.1016/j.applthermaleng.2014.10.071.
158. W. Yu, H. Xie, L. Chen, Y. Li, Enhancement of thermal conductivity of kerosene-based Fe_3O_4 nanofluids prepared via phase-transfer method, *Colloids Surf. A Physicochem. Eng. Asp.* 355 (2010) 109–113. https://doi.org/10.1016/j.colsurfa.2009.11.044.
159. M. Abareshi, E.K. Goharshadi, S.M. Zebarjad, H.K. Fadafan, A. Youssefi, Fabrication, characterization and measurement of thermal conductivity of Fe_3O_4 nanofluids, *J. Magn. Magn. Mater.* 322 (2010) 3895–3901. https://dx.doi.org/10.1016/j.jmmm.2010.08.016.
160. S. Doganay, A. Turgut, L. Cetin, Magnetic field dependent thermal conductivity measurements of magnetic nanofluids by 3ω method, *J. Magn. Magn. Mater.* 474 (2019) 199–206. https://doi.org/10.1016/j.jmmm.2018.10.142.

161. A. Karimi, M.A.A. Sadatlu, B. Saberi, H. Shariatmadar, M. Ashjaee, Experimental investigation on thermal conductivity of water based nickel ferrite nanofluids, *Adv. Powder Technol.* 26 (2015) 1529–1536. https://dx.doi.org/10.1016/j.apt.2015.08.015.
162. M.H. Esfe, M. Afrand, A. Karimipour, W.M. Yan, N. Sina, An experimental study on thermal conductivity of MgO nanoparticles suspended in a binary mixture of water and ethylene glycol, *Int. Commun. Heat Mass Transf.* 67 (2015) 173–175. https://dx.doi.org/10.1016/j.icheatmasstransfer.2015.07.009.
163. M.S. Liu, M.C.C. Lin, C.Y. Tsai, C.C. Wang, Enhancement of thermal conductivity with Cu for nanofluids using chemical reduction method, *Int. J. Heat Mass Transf.* 49 (2006) 3028–3033. https://doi.org/10.1016/j.ijheatmasstransfer.2006.02.012.
164. J.A. Eastman, S.U.S. Choi, S. Li, W. Yu, L.J. Thompson, Anomalously increased effective thermal conductivities of ethylene glycol-based nanofluids containing copper nanoparticles, *Appl. Phys. Lett.* 78 (2001) 718–720. https://dx.doi.org/10.1063/1.1341218.
165. H.E. Patel, S.K. Das, T. Sundararajan, A. Sreekumaran Nair, B. George, T. Pradeep, Thermal conductivities of naked and monolayer protected metal nanoparticle based nanofluids: Manifestation of anomalous enhancement and chemical effects, *Appl. Phys. Lett.* 83 (2003) 2931–2933. https://dx.doi.org/10.1063/1.1602578.
166. S.A. Putnam, D.G. Cahill, P.V. Braun, Thermal conductivity of nanoparticle suspensions, *J. Appl. Phys.* 99 (2006) 084308. https://doi.org/10.1063/1.2189933.
167. R. Carbajal-Valdéz, A. Rodríguez-Juárez, J.L. Jiménez-Pérez, J.F. Sánchez-Ramírez, A. Cruz-Orea, Z.N. Correa-Pacheco, M. Macias, J.L. Luna-Sánchez, Experimental investigation on thermal properties of Ag nanowire nanofluids at low concentrations, *Thermochim. Acta* 671 (2019) 83–88. https://doi.org/10.1016/j.tca.2018.11.015.
168. T. Parametthanuwat, N. Bhuwakietkumjohn, S. Rittidech, Y. Ding, Experimental investigation on thermal properties of silver nanofluids, *Int. J. Heat Fluid Flow.* 56 (2015) 80–90. https://dx.doi.org/10.1016/j.ijheatfluidflow.2015.07.0.
169. A. Alirezaie, M.H. Hajmohammad, M.R. Hassani Ahangar, M. Hemmat Esfe, Price-performance evaluation of thermal conductivity enhancement of nanofluids with different particle sizes, *Appl. Therm. Eng.* 128 (2018) 373–380. https://doi.org/10.1016/j.applthermaleng.2017.08.143.
170. Y. Ding, H. Alias, D. Wen, R.A. Williams, Heat transfer of aqueous suspensions of carbon nanotubes (CNT nanofluids), *Int. J. Heat Mass Transf.* 49 (2006) 240–250. https://doi.org/10.1016/j.ijheatmasstransfer.2005.07.009.
171. M.J. Assael, I.N. Metaxa, J. Arvanitidis, D. Christofilos, C. Lioutas, Thermal conductivity enhancement in aqueous suspensions of carbon multi-walled and double-walled nanotubes in the presence of two different dispersants, *Int. J. Thermophys.* 26 (2005) 647–664. https://doi.org/10.1007/s10765-005-5569-3.
172. M. Xing, J. Yu, R. Wang, Experimental investigation and modelling on the thermal conductivity of CNTs based nanofluids, *Int. J. Therm. Sci.* 104 (2016) 404–411. https://dx.doi.org/10.1016/j.ijthermalsci.2016.01.024.
173. S.U.S. Choi, Z.G. Zhang, W. Yu, F.E. Lockwood, E.A. Grulke, Anomalous thermal conductivity enhancement in nanotube suspensions, *Appl. Phys. Lett.* 79 (2001) 2252–2254. https://doi.org/10.1063/1.1408272.
174. M.F. Pakdaman, M.A. Akhavan-Behabadi, P. Razi, An experimental investigation on thermo-physical properties and overall performance of MWCNT/heat transfer oil nanofluid flow inside vertical helically coiled tubes, *Exp. Therm. Fluid Sci.* 40 (2012) 103–111. https://dx.doi.org/10.1016/j.expthermflusci.2012.02.005.
175. H. Xie, H. Lee, W. Youn, M. Choi, Nanofluids containing multiwalled carbon nanotubes and their enhanced thermal conductivities, *J. Appl. Phys.* 94 (2003) 4967–4971. https://doi.org/10.1063/1.1613374.

176. M.S. Liu, M.C.C. Lin, I.T. Huang, C.C. Wang, Enhancement of thermal conductivity with carbon nanotube for nanofluids, *Int. Commun. Heat Mass Transf.* 32 (2005) 1202–1210. https://doi.org/10.1016/j.icheatmasstransfer.2005.05.005.
177. S. Harish, K. Ishikawa, E. Einarsson, S. Aikawa, S. Chiashi, J. Shiomi, S. Maruyama, Enhanced thermal conductivity of ethylene glycol with single-walled carbon nanotube inclusions, *Int. J. Heat Mass Transf.* 55 (2012) 3885–3890. https://dx.doi.org/10.1016/j.ijheatmasstransfer.2012.03.001.
178. W. Yu, H. Xie, X. Wang, X. Wang, Significant thermal conductivity enhancement for nanofluids containing graphene nanosheets, *Phys. Lett. A* 375 (2011) 1323–1328. https://dx.doi.org/10.1016/j.physleta.2011.01.040.
179. S.W. Lee, S.D. Park, S. Kang, I.C. Bang, J.H. Kim, Investigation of viscosity and thermal conductivity of SiC nanofluids for heat transfer applications, *Int. J. Heat Mass Transf.* 54 (2011) 433–438. https://dx.doi.org/10.1016/j.ijheatmasstransfer.2010.09.026.
180. X. Li, C. Zou, L. Zhou, A. Qi, Experimental study on the thermo-physical properties of diathermic oil based SiC nanofluids for high temperature applications, *Int. J. Heat Mass Transf.* 97 (2016) 631–637. https://dx.doi.org/10.1016/j.ijheatmasstransfer.2016.02.056.
181. R. Ranjbarzadeh, A. Moradikazerouni, R. Bakhtiari, A. Asadi, M. Afrand, An experimental study on stability and thermal conductivity of water/silica nanofluid: Eco-friendly production of nanoparticles, *J. Clean. Prod.* 206 (2019) 1089–1100. https://doi.org/10.1016/j.jclepro.2018.09.205.
182. J.C. Maxwell, *The Scientific Papers of James Clerk Maxwell*. Vol. 2. Cambridge University Press, New York, 1890.
183. R.T. Bonnecaze, J.F. Brady, A method for determining the effective conductivity of dispersions of particles, *Proc. R. Soc. Lond. A* 430 (1990) 285–313 https://doi.org/10.1098/rspa.1990.0092
184. H. Fricke, A mathematical treatment of the electric conductivity and capacity of disperse systems I. The electric conductivity of a suspension of homogeneous spheroids, *Phys. Rev.* 24 (1924) 575–587.
185. C.-W. Nan, Z. Shi, Y. Lin, A simple model for thermal conductivity of carbon nanotube-based composites, *Chem. Phys. Lett.* 375 (2003) 666–669.
186. W. Yu, D. France, E. Timofeeva, D. Singh, Effective thermal conductivity models for carbon nanotube-based nanofluids, J. Nanofluids 2 (2013) 69–73. 10.1166/jon.2013.1036.
187. J. Koo, C. Kleinstreuer, Impact analysis of nanoparticle motion mechanisms on the thermal conductivity of nanofluids, *Int. Commun. Heat Mass Transf.* 32 (2005) 1111–1118.
188. R. Prasher, W. Evans, P. Meakin, J. Fish, P. Phelan, P. Keblinski, Effect of aggregation on thermal conduction in colloidal nanofluids, *Appl. Phys. Lett.* 89 (2006) 143119.
189. Y. Xuan, Q. Li, X. Zhang, M. Fujii, Stochastic thermal transport of nanoparticle suspensions, *J. Appl. Phys.* 100 (2006) 043507.
190. P. Keblinski, S.R. Phillpot, S.U.S. Choi, J.A. Eastman, Mechanisms of heat flow in suspensions of nano-sized particles (nanofluids), *Int. J. Heat Mass Transf.* 45 (2002) 855–863. https://doi.org/10.1016/S0017-9310(01)00175-2.
191. Q.-Z. Xue, Model for effective thermal conductivity of nanofluids, *Phys. Lett. A* 307 (2003) 313–317.
192. M.N. Ozisik, *Heat Conduction*, J. Wiley and Sons, New York, 1980.
193. W. Lu, Q. Fan, Study for the particle's scale effect on some thermophysical properties of nanofluids by a simplified molecular dynamics method, *Eng. Anal. Bound. Elem.* 32 (2008) 282–289. https://doi.org/10.1016/j.enganabound.2007.10.006.
194. R.J. Sadus, *Molecular Simulation of Fluids (Theory, Algorithms and Object-Orientation)*, Elsevier, New York, 1999.

195. D.K. Dysthe, A.H. Fuchs, B. Rousseau, Prediction of fluid mixture transport properties by molecular dynamics, *Int. J. Thermophys.* 19 (1998) 437–448. https://doi.org/10.1023/A:1022513411043.
196. V.Y. Rudyak, A.A. Belkin, E.A. Tomilina, On the thermal conductivity of nanofluids, *Tech. Phys. Lett.* 36 (2010) 660–662. https://doi.org/10.1134/S1063785010070229.
197. E.M. Achhal, H. Jabraoui, S. Zeroual, H. Loulijat, A. Hasnaoui, S. Ouaskit, Modeling and simulations of nanofluids using classical molecular dynamics: Particle size and temperature effects on thermal conductivity, *Adv. Powder Technol.* 29 (2018) 2434–2439. https://doi.org/10.1016/j.apt.2018.06.023.
198. M.J. Javanmardi, K. Jafarpur, A molecular dynamics simulation for thermal conductivity evaluation of carbon nanotube-water nanofluids, *J. Heat Transfer.* 135 (2013) 042401. https://doi.org/10.1115/1.4022997.
199. A. Mohebbi, Prediction of specific heat and thermal conductivity of nanofluids by a combined equilibrium and non-equilibrium molecular dynamics simulation, *J. Mol. Liq.* 175 (2012) 51–58. https://dx.doi.org/10.1016/j.molliq.2012.08.010.
200. N. Sankar, N. Mathew, C.B. Sobhan, Molecular dynamics modeling of thermal conductivity enhancement in metal nanoparticle suspensions, *Int. Commun. Heat Mass Transf.* 35 (2008) 867–872. https://doi.org/10.1016/j.icheatmasstransfer.2008.03.006.
201. C. Sun, W. Lu, B. Bai, J. Liu, Anomalous enhancement in thermal conductivity of nanofluid induced by solid walls in a nanochannel, *Appl. Therm. Eng.* 31 (2011) 3799–3805. https://dx.doi.org/10.1016/j.applthermaleng.2011.07.021.
202. W. Cui, Z. Shen, J. Yang, S. Wu, Effect of chaotic movements of nanoparticles for nanofluid heat transfer augmentation by molecular dynamics simulation, *Appl. Therm. Eng.* 76 (2015) 261–271. https://dx.doi.org/10.1016/j.applthermaleng.2014.11.030.
203. H. Kang, Y. Zhang, M. Yang, L. Li, Molecular dynamics simulation on effect of nanoparticle aggregation on transport properties of a nanofluid, *J. Nanotechnol. Eng. Med.* 3 (2012) 021001. https://doi.org/10.1115/1.4007044.
204. S.L. Lee, R. Saidur, M.F.M. Sabri, T.K. Min, Molecular dynamic simulation on the thermal conductivity of nanofluids in aggregated and non-aggregated states, *Numer. Heat Transf. Part A Appl.* 68 (2015) 432–453. https://dx.doi.org/10.1080/10407782.2014.986366.
205. W. Cui, Z. Shen, J. Yang, S. Wu, M. Bai, Influence of nanoparticle properties on the thermal conductivity of nanofluids by molecular dynamics simulation, *RSC Adv.* 4 (2014) 55580–55589. https://doi.org/10.1039/c4ra07736a.
206. W. Cui, M. Bai, J. Lv, G. Li, X. Li, On the influencing factors and strengthening mechanism for thermal conductivity of nanofluids by molecular dynamics simulation, *Ind. Eng. Chem. Res.* 50 (2011) 13568–13575. https://doi.org/10.1021/ie201307w.
207. Z. Liang, H.L. Tsai, Thermal conductivity of interfacial layers in nanofluids, *Phys. Rev. E* 83 (2011) 041602. https://doi.org/10.1103/PhysRevE.83.041602.
208. T. Armaghani, A. Kasaeipoor, M. Izadi, I. Pop., MHD natural convection and entropy analysis of a nanofluid inside T-shaped baffled enclosure, *Int. J. Numer. Meth. Heat Fluid Flow* 28.12 (2018) 2916–2941.
209. A. Abedini, T. Armaghani, A.J. Chamkha, MHD free convection heat transfer of a water-Fe_3O_4 nanofluid in a baffled C-shaped enclosure, *J. Therm. Anal. Calorim.* 135.1 (2019) 685–695.
210. C. Pang, J.Y. Jung, J.W. Lee, Y.T. Kang, Thermal conductivity measurement of methanol-based nanofluids with Al_2O_3 and SiO_2 nanoparticles, *Int. J. Heat Mass Transf.* 55 (2012) 5597–5602. https://dx.doi.org/10.1016/j.ijheatmasstransfer.2012.05.048.
211. W.H. Azmi, K. V. Sharma, P.K. Sarma, R. Mamat, G. Najafi, Heat transfer and friction factor of water based TiO_2 and SiO_2 nanofluids under turbulent flow in a tube, *Int. Commun. Heat Mass Transf.* 59 (2014) 30–38. https://dx.doi.org/10.1016/j.icheatmasstransfer.2014.10.007.

212. W. Yu, H. Xie, D. Bao, Enhanced thermal conductivities of nanofluids containing graphene oxide nanosheets, *Nanotechnology* 21 (2010). https://doi.org/10.1088/0957-4484/21/5/055705.
213. W. Yu, H. Xie, W. Chen, Experimental investigation on thermal conductivity of nanofluids containing graphene oxide nanosheets, *J. Appl. Phys.* 107 (2010) 094317. https://doi.org/10.1063/1.3372733.
214. T.T. Baby, S. Ramaprabhu, Investigation of thermal and electrical conductivity of graphene based nanofluids, *J. Appl. Phys.* 108 (2010) 124308. https://doi.org/10.1063/1.3516289.
215. S.S. Gupta, V.M. Siva, S. Krishnan, T.S. Sreeprasad, P.K. Singh, T. Pradeep, S.K. Das, Thermal conductivity enhancement of nanofluids containing graphene nanosheets, *J. Appl. Phys.* 110 (2011) 084302. https://doi.org/10.1063/1.3650456.
216. R. Walvekar, I.A. Faris, M. Khalid, Thermal conductivity of carbon nanotube nanofluid-Experimental and theoretical study, *Heat Transf. Res.* 41 (2012) 145–163. https://doi.org/10.1002/htj.20405.
217. M.H. Esfe, S. Saedodin, O. Mahian, S. Wongwises, Thermophysical properties, heat transfer and pressure drop of COOH-functionalized multi walled carbon nanotubes/water nanofluids, *Int. Commun. Heat Mass Transf.* 58 (2014) 176–183. https://dx.doi.org/10.1016/j.icheatmasstransfer.2014.08.037.
218. M.H. Esfe, S. Saedodin, O. Mahian, S. Wongwises, Heat transfer characteristics and pressure drop of of COOH-functionalized DWCNTs/water nanofluid in turbulent flow at low concentrations, *Int. J. Heat Mass Transf.* 73 (2014) 186–194. https://dx.doi.org/10.1016/j.ijheatmasstransfer.2014.01.069.

APPENDIX

TABLE 3.3
Thermal Conductivity Models for Nanofluids and Their Characteristics

Model	Year	Equation	Features	Refs.
Maxwell	1873	$k_{nf} = \frac{k_p + 2k_{bf} + 2\Phi_p(k_p - k_{bf})}{k_p + 2k_{bf} - \Phi_p(k_p - k_{bf})} k_{bf}$	Spherical and randomly scattered particles	[6]
Hamilton	1962	$k_{nf} = \frac{k_p + (n-1)k_{bf} + (n-1)\Phi_p(k_p - k_{bf})}{k_p + (n-1)k_{bf} - \Phi_p(k_p - k_{bf})} k_{bf}$	Non-spherical particles	[7]
Bruggeman	1935	$\frac{k_{nf}}{k_{bf}} = \frac{(3\Phi_p - 1)\frac{k_p}{k_{bf}} + [3(1-\Phi_p) - 1] + \sqrt{\left\{(3\Phi_p - 1)\frac{k_p}{k_{bf}} + [3(1-\Phi_p) - 1]\right\}^2 + 8\frac{k_p}{k_{bf}}}}{4}$	Large volume fraction of a spherical nanoparticles	[8]
Yu and Choi	2003	$k_{nf} = \frac{k_p + 2k_{bf} + 2\Phi_p(k_p - k_{bf})(1+\beta)^3 \Phi_p}{k_p + 2k_{bf} - \Phi_p(k_p - k_{bf})(1+\beta)^3 \Phi_p} k_{bf}$	Maxwell model with the consideration of nanolayer effect	[9]
Xue and Xu	2005	$\left(1 - \frac{\Phi_p}{\lambda_{nf}}\right)\frac{k_{nf} - k_{bf}}{2k_{nf} + k_{bf}} + \frac{\Phi_p}{\lambda_{nf}} \frac{(k_{nf} - k_l)(2k_l + k_p) - \lambda_{nf}(k_p - k_l)(2k_l + k_{nf})}{(2k_{nf} + k_l)(2k_l + k_p) + 2\lambda_{nf}(k_p - k_l)(k_l - k_{nf})} = 0$	Bruggeman model with the consideration of nanolayer effect	[10]
Xuan et al.	2003	$\frac{k_{nf}}{k_{bf}} = \frac{k_p + 2k_{bf} - 2\Phi_p(k_{bf} - k_P)}{k_p + 2k_{bf} + \Phi_p(k_{bf} - k_P)} + \frac{\rho_p \Phi_p c_p}{2k_{bf}} \sqrt{\frac{k_B T}{3\pi r_{cl} \mu}}$	The Maxwell model with considering the Brownian effect	[12]
Koo and Kleinstreuer	2004	$k_{nf} = k_{static} + k_{Brownian} = \frac{k_p + 2k_{bf} + 2\Phi_p(k_p - k_{bf})}{k_p + 2k_{bf} - \Phi_p(k_p - k_{bf})} k_{bf} + 5 \times 10^4 f_1 \Phi_p \rho_p c_p \sqrt{\frac{k_B T}{\rho_p r_p}} f_2(T, \Phi_p)$	The effect of size, concentration and temperature of nanoparticles as well as particles affected by Brownian motion	[13]
Chon et al.	2005	$\frac{k_{nf}}{k_{bf}} = 1 + 64.7\Phi_p^{0.74}\left(\frac{r_{bf}}{r_p}\right)^{0.369}\left(\frac{k_{bf}}{k_p}\right)^{0.747} \times Pr^{0.9955} \times Re^{1.2321}$	Reynolds number is related to the velocity of Brownian motion in Al_2O_3/deionized (DI) water nanofluids	[14]

(Continued)

TABLE 3.3 (*Continued*)
Thermal Conductivity Models for Nanofluids and Their Characteristics

Model	Year	Equation	Features	Refs.
Wang et al.	2003	$\frac{k_{nf}}{k_{bf}} = \frac{(1-\Phi_p)+3\Phi_p\int_0^\infty \frac{k_{cl}n}{k_{cl}+2k_{bf}}dr_p}{(1-\Phi_p)+3\Phi_p\int_0^\infty \frac{k_{bf}n}{k_{cl}+2k_{bf}}dr_p}$	A fractal model of the effects of adsorption and nanoparticle size	[15]
Jang and Choi	2004	$k_{nf} = k_{bf}(1-\Phi_p)+k_p\Phi_p+3C\frac{r_{bf}}{r_p}k_{bf}\mathrm{Re}_r^2\mathrm{Pr}\Phi_p$	Collisions between base fluid molecules and the interaction between nanoparticles and base fluid	[11]
Xu et al.	2006	$\frac{k_{nf}}{k_{bf}} = \frac{k_p+2k_{bf}-2\Phi_p(k_{bf}-k_p)}{k_p+2k_{bf}+\Phi_p(k_{bf}-k_p)}+C\frac{\mathrm{Nu}\cdot r_{bf}}{\mathrm{Pr}}\frac{(2-D_{bf})}{(1-D_{bf})^2}\frac{\left[\left(\frac{r_{p,max}}{r_{p,min}}\right)^{1-D_{bf}}-1\right]^2}{\left(\frac{r_{p,max}}{r_{p,min}}\right)^{2-D_{bf}}-1}\frac{1}{r_p}$	Considering the convection heat transfer produced by Brownian motion.	[24]
Dong and Chen	2017	$k_{nf} = \frac{(\beta+1)k_{bf}+\beta k_p+2\beta\Phi_e(k_p-k_{bf})}{(\beta+1)k_{bf}+\beta k_p-\beta\Phi_e(k_p-k_{bf})}k_{bf}$ Where $\Phi_e = \rho_s\frac{r_{eff}^3}{r_p^3}$	Contributions of both particle size and Brownian motion	[26]
Xue and Wu	2005	$\left(1-\frac{\Phi_p}{z}\right)\frac{k_{nf}-k_{bf}}{2k_{nf}+k_{bf}}+\frac{\Phi_p}{z}\frac{(k_{nf}-k_{is})(2k_{is}+k_p)-z(k_p-k_{is})(2k_{is}+k_{nf})}{(2k_{nf}+k_{is})(2k_{is}+k_p)+2z(k_p-k_{is})(k_{is}-k_{nf})}=0$ where $z=\left[\frac{r_p/2}{r_p/2+d_{is}}\right]^3$	Interfacial effect of interface shell	[10]
Ganesan et al.	2018	$\frac{k_{nf}}{k_{bf}} = (1+A\mathrm{Re}^\gamma\mathrm{Pr}^{0.333}\Phi_p)\frac{1+2\beta\Phi_p}{1-\beta\Phi_p}$	Nanoclusters-based nanofluids with a very large particle size, considering the interfacial thermal resistance and the mixed convection of nanoparticles	[28]

(Continued)

TABLE 3.3 (*Continued*)
Thermal Conductivity Models for Nanofluids and Their Characteristics

Model	Year	Equation	Features	Refs.
Vajjha and Das	2009	$k_{nf}=\frac{k_p+2k_{bf}-2\Phi_p(k_{bf}-k_p)}{k_p+2k_{bf}+\Phi_p(k_{bf}-k_p)}k_{bf}+5\times10^4\beta\Phi_p\rho_{bf}c_{p,bf}\sqrt{\frac{k_BT}{\rho_p r_p}}f(T,\Phi_p)$, where $f(T,\Phi_p)=(2.8217\times10^{-2}\Phi_p+3.917\times10^{-3})\left(\frac{T}{T_0}\right)+(-3.0669\times10^{-2}\Phi_p-3.91123\times10^{-3})$	Conductivity model for three EG-water-based nanofluids containing Al_2O_3, ZnO, and CuO nanoparticles	[29]
Xuan et al. and Gao et al.	2003	$\frac{k_{nf}}{k_{bf}}=\frac{3+2\eta^2\Phi_p/k_{bf}\left(\frac{2R_k}{L_p}+13.4\sqrt{t}\right)}{3-\Phi_p}+\frac{\rho_p\Phi_p c_{p,p}}{2k_{bf}}\sqrt{\frac{k_BT}{3\pi\mu r_c}}$	New model for estimating k of graphene nanofluids based on Chu et al. model [33]	[12,31]
Feng et al.	2018	$\frac{k_{nf}}{k_{bf}}=(1-\Phi_{p,eq})\frac{k_{pe}+2k_{bf}+2(k_{p,eq}-k_{bf})(1+\beta)^3\Phi_p}{\kappa_{pe}+2k_{bf}-(k_{p,eq}-k_{bf})(1+\beta)^3\Phi_p}+\Phi_{p,eq}\left[\left(1-\frac{3}{2}\Phi_{p,eq}\right)+\frac{3\Phi_{p,eq}}{2}\left[\frac{1}{\iota}\ln\frac{r_p/2+d_{is}}{(r_p/2+d_{is})(1-\beta)}-1\right]\right]$	Aggregation impact of nanoparticles	[22]
Pang et al.	2007	$k_{nf}=1+\frac{\Phi_a(k_a-k_{bf})}{(1-\Phi_a)/n(k_a-k_{bf})+k_{bf}}+\left[A_1\ln(A_2\Phi_a Re^m Pr^{0.333})+A_3\right]\frac{1+2\Phi_a+2(1-\Phi_a)\alpha}{1-\Phi_a+(2+\Phi_a)\alpha}$	The static and dynamic contributions to the increase in k for nanofluids, with the latter being primarily ascribed to nanoparticle aggregation	[36]
Hamilton et al., Xu and Yu, and Wei et al.	2014	$\frac{k_{nf}}{k_{bf}}=\frac{m+(n-1)-(n-1)(1-m)\Phi_p}{m+(n-1)+(1-m)\Phi_p}+C\frac{Nu\cdot r_{bf}}{Pr}\frac{(2-D)D}{(1-D)^2}\frac{(\Phi_p^{\zeta_1}-1)^2}{\Phi_p^{\zeta_2}-1}\frac{1}{\overline{L_a}}$	Function of the fractal size and concentration by taking into account two distinct processes of heat conduction: particle aggregation and convention	[7,38,39]

(*Continued*)

TABLE 3.3 (*Continued*)
Thermal Conductivity Models for Nanofluids and Their Characteristics

Model	Year	Equation	Features	Refs.
Murshed et al.	1962	$k_{nf}=\left[(k_p-k_{nl})\Phi_p k_{nl}(\gamma_1^2-\gamma^2+1)+(k_p+k_{nl})\gamma_1^2(\Phi_p\gamma^2(k_{nl}-k_{bf})+k_{bf})\right]\cdot\left[\gamma_1^2(k_p+k_{nl})-(k_p-k_{nl})\Phi_p(\gamma_1^2+\gamma^2-1)\right]^{-1}$, $k_{nf}=\left[(k_p-k_{nl})\Phi_p k_{nl}(2\gamma_1^3-\gamma^3+1)+(k_p+2k_{nl})\gamma_1^3(\Phi_p\gamma^3(k_{nl}-k_{bf})+k_{bf})\right]\cdot\left[\gamma_1^3(k_p+2k_{nl})-(k_p-k_{nl})\Phi_p(\gamma_1^3+\gamma^3-1)\right]^{-1}$	Spherical and cylindrical nanoparticles considering the nanolayer's impact	[21]
Feng and Kleinstreuer et al.	2006	$k_{nf}=k_{static}+k_{mm}$, $k_{static}=\left[1+\frac{3(k_p/k_{bf}-1)}{(k_p/k_{bf}+2)-(k_p/k_{bf}-1)}\right]k_{bf}$, $k_{mm}=49500\cdot 38\cdot\frac{k_B\tau_p}{2m_p}\cdot(\rho c_p)_{nf}\cdot\Phi_p{}^2\cdot(T\ln T-T)\cdot\frac{\exp(-\xi\omega_n\tau_p)\sinh\left(\sqrt{\frac{(3\pi\mu_{bf}r_p)^2}{4m_p^2}-\frac{Q_{p-p}}{m_p}}\frac{m_p}{3\pi\mu_{bf}r_p}\right)}{\tau_p\sqrt{\frac{(3\pi\mu_{bf}r_p)^2}{4m_p^2}-\frac{K_{p-p}}{m_p}}}$	Particles' Brownian motion and turbulent effects	[40]
Das et al.	2016	$k_{nf}=k_{bf}\left[1+\frac{k_p\Phi_p L_{mol}}{k_{bf}(1-\Phi_p)r_{cnt}}\right]$	The description inaccuracy association with the CNT-containing nanofluids resolvation	[41]

(*Continued*)

TABLE 3.3 (*Continued*)
Thermal Conductivity Models for Nanofluids and Their Characteristics

Model	Year	Equation	Features	Refs.
Li et al.	2008	$\frac{k_{nf}}{k_{bf}} = \frac{3+\Phi_p(\beta_T+\beta_L)}{3-\Phi_p\beta_T}$ where $\beta_T = \frac{2(k_T^{nf}-k_{bf})}{k_T^{nf}+k_{bf}}, \beta_L = \frac{k_L^{nf}}{k_{bf}}-1$	The k enhancement of the CNT nanofluids by the sample function	[42]
Corcione	2010	$\frac{k_{nf}}{k_{bf}} = 1+4.4\mathrm{Re}^{0.4}\mathrm{Pr}^{0.66}\left(\frac{T}{T_{fp}}\right)^{10}\left(\frac{k_p}{k_{bf}}\right)^{0.03}\Phi_{0.66}$	Offering quick k predictability under various scenarios	[43]
Charab et al.	2008	$\frac{k_{nf}}{k_{bf}} = 1+\frac{3\left(\frac{k_{p1}}{k_{bf}}-1\right)\times\Phi_{FVC}^{p1}}{\left(\frac{k_{p1}}{k_{bf}}+2\right)-\left(\frac{k_{p1}}{k_{bf}}-1\right)\times\Phi_{FVC}^{p1}}+\frac{3\left(\frac{k_{p2}}{k_{bf}}-1\right)\times\Phi_{FVC}^{p2}}{\left(\frac{k_{p2}}{k_{bf}}+2\right)-\left(\frac{k_{p2}}{k_{bf}}-1\right)\times\Phi_{FVC}^{p2}},$ where $\Phi_{FVC}^{p1} = \Phi_e\left(\frac{V_{nf1}}{V_{nf1}+V_{nf2}}\right),$ $\Phi_{FVC}^{p1} = \Phi_e\left(\frac{V_{nf1}}{V_{nf1}+V_{nf2}}\right)$	EMM to predict the k_{hnf}	[126]
Maxwell	1873	$k_e = k_f\left[1+\frac{3(k_s+k_f)\varphi}{(k_s+2k_f)-(k_s+2k_f)\varphi}\right]$	Electric conduction of a static mixture containing two constituents with different electric conduction	[6,182]
Bruggeman	1935	$k_e = k_f\left[1+\frac{k_sk_f(n-1)+(n-1)(k_s+k_f)\varphi}{k_s+k_s(n-1)-(k_s+k_f)\varphi}\right] = k_f\left[1+\frac{n(k_s+k_f)\varphi}{k_s+k_s(n-1)-(k_s+k_f)\varphi}\right]$	Effective electric conduction of different solid particles with regular forms	[8]

(*Continued*)

TABLE 3.3 (*Continued*)
Thermal Conductivity Models for Nanofluids and Their Characteristics

Model	Year	Equation	Features	Refs.
Hamilton and Crosser	1962	$k_e = k_f\left[1+\frac{3\left(k_S-k_f\right)\frac{\varphi}{\Psi}}{k_S+k_S\left(\frac{3}{\Psi}-1\right)-\left(k_S-k_f\right)\varphi}\right],$ For $k_s \gg k_f$ $k_e = k_f\left[1+\frac{3\varphi}{1-\varphi}\right] \approx k_f(1+3\varphi)$	Suspension of particles with non-regular forms	[7]
Nan et al.	1997	$k_e = \frac{[3+[2\beta_{11}(1-L_{11})+\beta_{33}(1-L_{33})]]}{3-\varphi(2\beta_{11}L_{11}+\beta_{33}L_{33})},$ $L_{11} = \frac{E^2}{2(E^2-1)} - \frac{E\cosh^{-1}(E)}{2(E^2-1)^{\frac{3}{2}}}, L_{33} = 1-2L_{11}$ $\beta_{11} = \frac{k_{ii}-k_f}{k_f+k_{ii}(k_{ii}-k_f)}, k_{ii} = \frac{k_s}{1+2L_{11}(2+E^{-1})\frac{k_s R_i}{\alpha_1}}$	Heat conduction of composite materials containing fiber, with consideration of the impact of form, symmetry, the direction of solid particles in composite, interfacial thermal resistance	[37]
Nan et al.	2003	$k_e = \frac{3k_f+\varphi k_s}{3-2\varphi} \approx k_f\left(1+\frac{\varphi k_s}{3k_f}\right)$	A simple model for thermal conductivity of carbon nanotube-based composites	[185]
Koo and Kleinstreuer	2004, 2005	$k_e = k_f\left[1+\frac{3(k_s-k_f)\varphi}{(k_s+2k_f)-(k_s+k_f)\varphi}\right]+50000\varphi\rho_s c_s\sqrt{\frac{k_B T}{2\alpha\rho_s}}\beta_k f(T,\varphi)$	Impact analysis of nanoparticle motion mechanisms on the thermal conductivity of nanofluids	[13,187]

(*Continued*)

TABLE 3.3 (*Continued*)
Thermal Conductivity Models for Nanofluids and Their Characteristics

Model	Year	Equation	Features	Refs.
Prasher et al.	2006	$k_e = k_f[1+40000\,\mathrm{Re_{Br}}^{2.5}\,\mathrm{Pr}^{0.333}\,\varphi]\left[\frac{1+2\varphi}{1-\varphi}\right]$ $\mathrm{Re_{Br}} = \frac{1}{v_f}\sqrt{\frac{9k_BT}{\pi\alpha\rho_s}}$	Situation of particles with high thermal conductance, $k_s >> k_f$, and they did so while neglecting the interface resistance, R_i	[188]
Xuan et al.	2006	$k_e = k_f\left[1+\frac{3(k_s-k_f)\varphi}{(k_s+2k_f)-(k_s+k_f)\varphi}\right]+\frac{9\varphi h_c k_B T}{8\pi\rho_s\alpha^4}$	Langevin equation and the idea of particle random Brownian movement to characterize velocity fluctuations and heat transmission	[189]
Ozisik	1980	$k_{eq} = \frac{\frac{k_{sf}}{k_s}\left[2\left(1-\frac{k_{sf}}{k_s}\right)+\left(1+\frac{\delta\alpha}{\alpha}\right)\left(1+2\frac{k_{sf}}{k_s}\right)\right]}{\frac{k_{sf}}{k_s}-1+\left(1+\frac{\delta\alpha}{\alpha}\right)^3\left(1+2\frac{k_{sf}}{k_s}\right)}k_s$	A sphere and a solid shell combine to form the nanoparticle	[192]
Yu et al.	2004	$k_{eq} = k_s\left[1+\frac{3(k_{eq}-k_f)\left(1+\frac{\delta\alpha}{\alpha}\right)^3\varphi}{(k_{eq}+2k_f)-(k_{eq}+k_f)\left(1+\frac{\delta\alpha}{\alpha}\right)^3\varphi}\right]$	k_{eq} and the new fraction of volume are inserted in the Maxwell's equation for the conduction of heterogeneous thermodynamic system	[19]

4 New Look at Nanofluid Thermal Conductivity Correlations

Maysam Molana

This section takes a skeptical look at the mathematical modeling of the effective thermal conductivity of nanofluids. Most related published experimental-based mathematical models have been analyzed statistically. The sensitivity analysis showed that the role of nanofluids bulk temperature in most published models is negligible. Then, much data from the models in the valid ranges of variables have been extracted. The next step was performing statistical analysis of the variances and means of different datasets (data populations). The results showed that there is no significant statistical difference between different datasets. It can be concluded that changing the nanofluid temperature does not considerably affect the nanofluid's thermal conductivity. Then a more straightforward correlation instead of each published model with acceptable accuracy is proposed. Finally, a comprehensive, more straightforward, and more accurate correlation (MAG correlation) neglecting the temperature of the nanofluid to predict the effective thermal conductivity of nanofluids is introduced. Results indicated that the predicted values using the proposed correlation are in good agreement with experimental data.

4.1 INTRODUCTION

The total energy consumption of the world has increased sharply. During the past decades, Energy will be one of the most critical challenges in the following decades, especially for developing countries. Heat transfer scientists and engineers have concentrated on different methods to reduce the energy consumption of heat transfer phenomena. Their methods included but were not limited to increasing the heat transfer fluid's velocity (forced convection) [1,2], enhancing the heat transfer area (like fins, heat sinks, plate heat exchangers) [3–5], the fast transition from laminar flow regime to the turbulent (increasing friction factor) [6,7], conjugate heat transfer and mixed convection [8–10], boiling and other phase change processes because of the high amount of latent heat (like heat pipes and phase change materials) [11,12], and so on.

Meanwhile, some scientists believe that this challenge can be solved by increasing conventional heat transfer fluids' relatively low thermal conductivity. Maxwell [13] thought considerable enhancement in thermophysical properties of a mixture of fluid and dispersed metal particles compared to the base fluid can be experienced.

DOI: 10.1201/9781032664118-4

He never could set up such an experiment because of the high mass value (and size) of the particles that had been manufactured at the time. His idea remained only a theoretical one until the last decade of the twentieth century. Choi and Eastman [14], based on the recent development of nanotechnology, suggested a new class of heat transfer fluid containing nanoparticles (particles generally smaller than 100 nm in dimensions). They proposed to disperse a low-volume concentration of metal or oxide nanoparticles in a conventional heat transfer fluid like water, oil, and ethylene glycol. They reported an anomalous enhancement in the heat transfer coefficient of nanofluids compared to the base fluid [15].

Great attention has been devoted to this new scientific horizon throughout the world. Therefore, an unbelievable number of scientific papers have been published during the past two decades.

Today, nanofluid applications in different fields and industries are under serious investigation, and some experimental successes have been achieved. The nanofluids applications include, but are not limited to heat exchangers [16–18], impingement jets [19,20], renewable energies [21,22], heating and tempering processes [23–26], automotive industries [27,28], electronic cooling [29–32], lubrication [33–35], medicine [36–38], combustion [39–41], etc.

There are two conventional approaches to simulate the nanofluids flow and heat transfer characteristics named single phase [42,43] and two-phase approaches [44,45]. The single-phase approach assumes that a nanofluid is a homogenous mixture (suspension) of nanoparticles and base fluid. This approach ignores the different interactions between nanoparticles and liquid molecules. It tends to estimate the thermophysical properties using predictive models for thermal conductivity, density, viscosity, and specific heat. Thus, the single-phase approach deals with the nanofluids as a whole.

Prediction of thermophysical properties of a mixture has a history senior to the nanofluids. In 1962, Hamilton and Crosser [46] proposed one of the oldest correlations of a mixture's thermal conductivity based on a few experiments on the spheres, disks, cylinders, and cubes of Aluminum and Balsa in Silastic ($n=3$ for spherical particles):

$$\frac{k_{\text{eff}}}{k_{\text{b}}}=k_{\text{b}}\left[\frac{k_{\text{p}}+(n-1)k_{\text{b}}-(n-1)\left(k_{\text{b}}-k_{\text{p}}\right)\varphi}{k_{\text{p}}+(n-1)k_{\text{b}}+\left(k_{\text{b}}-k_{\text{p}}\right)\varphi}\right] \tag{4.1}$$

At the same time, Hashin and Shtrikman [47] studied the multiphase thermophysical properties theoretically using a variational approach and proposed one of the first correlations on the thermal conductivity of a mixture, where k_{eff}, k_{p}, k_{b}, and φ are the thermal conductivity of mixture, the thermal conductivity of particles, the thermal conductivity of fluid and volume concentration of particles, respectively.

$$k_{\text{f}}\left(1+\frac{3\varphi(k_{\text{p}}-k_{\text{f}})}{3k_{\text{f}}+(1-\varphi)(k_{\text{p}}-k_{\text{f}})}\right)\leq k_{\text{eff}}\leq\left(1-\frac{3(1-\varphi)(k_{\text{p}}-k_{\text{f}})}{3k_{\text{p}}-\varphi(k_{\text{p}}-k_{\text{f}})}\right)k_{\text{p}} \tag{4.2}$$

Jeffery [48] investigated the heat conduction of a stationary random suspension of spherical particles with a low-volume concentration. He extended the work of

Maxwell to calculate the heat flux with consideration of interactions between pairs of particles. Eventually, he proposed a functional correlation for the thermal conductivity of a mixture:

$$k_{\text{eff}} = \left(1 + 3\beta + \left(3\beta^2 + \frac{3\beta^2}{4} + \frac{9\beta^3}{16} \frac{\alpha+2}{2\alpha+3} + \frac{3\beta^4}{64} + \cdots\right)\varphi^2\right)k_{\text{b}}$$

$$\beta = \frac{\alpha - 1}{\alpha + 2}, \quad \alpha = \frac{k_{\text{p}}}{k_{\text{b}}} \tag{4.3}$$

Turian et al. [49] conducted an experimental study on the suspension of coal particles in water, oil, and other liquids. They developed a simple correlation for the thermal conductivity of a mixture. The results of their correlation are in good agreement with the experimental data, especially for the low-volume concentration of particles.

$$k_{\text{eff}} = k_{\text{p}}^{\varphi} k_{\text{b}}^{1-\varphi} \tag{4.4}$$

There are more than 70 models for predicting the thermal conductivity of nanofluids. More than 37 models are developed using experimental data and results. Finally, 14 models predicting the thermal conductivity of nanofluids, including nanofluids temperature, using experimental data are found. Table 4.1 shows all published experimental models for nanofluids' thermal conductivity, taking into account the temperature of the nanofluid.

4.2 METHODOLOGY

All of the mentioned models in Table 4.1 consider the thermal conductivity of nanofluids as a function of the thermal conductivity of the base fluid (k_b), the thermal conductivity of nanoparticle (k_p), nanoparticle volume concentration (φ), and temperature (T). Two models [51,53] additionally consider the nanoparticle mean diameter.

Generally, it seems normal that the thermal conductivity of nanofluids is enhanced with an increase in the thermal conductivity of the base fluid, thermal conductivity of nanoparticles, nanofluids temperature, and any decrease in the mean diameter of nanoparticles. However, some models predict diverse and sometimes bizarre behavior of the thermal conductivity of nanofluids according to different variables. For example, the thermal conductivity of nanofluids is usually enhanced with an increase in temperature. Conversely, some models predict the decreasing trend of thermal conductivity of nanofluids with an increase in temperature. This situation is frequent, especially when the base fluid is ethylene glycol.

There are also some contradictions in the role of the thermal conductivity of nanoparticles in the estimation of effective thermal conductivity. For example, the model developed by Aberoumand et al. [50] shows that effective thermal conductivity will reduce with an increase in the thermal conductivity of nanoparticles.

A decision is made to monitor the behavior of the effective thermal conductivity against different variables that appear in published models. Then, it is tried to evaluate the role of each variable in effective thermal conductivity based on published models.

TABLE 4.1
All Published Experimental Models for the Thermal Conductivity of Nanofluids, Including Nanofluids Temperature

Ref.		Model	Base Fluid	Nanoparticle	Limitations
1	Aberoumand et al. [50]	$k_{eff} = (3.9\times10^{-5}T - 0.0305)\varphi^2 + (0.086 - 1.6\times10^{-4}T)\varphi$ $+3.1\times10^{-4}T + 0.129 - 5.77\times10^{-6}k_p - 40\times10^{-4}$	Heat Transfer Oil	Ag Cu MWCNT	Volume concentration: 0%–2%
2	Fakoor Pakdaman et al. [51]	$\frac{k_{eff}}{k_b} = 1 + 304.47(1+\varphi)^{136.35}\exp(-0.021T)\left(\frac{1}{d_p}\right)^{0.369}\left(\frac{T^{1.2321}}{10^{610.6287/(T-140)}}\right)$	Heat Transfer Oil	MWCNT	Weight concentration 0%–0.4% Temperature ranging from 313 K to 343 K. dp: 5–20 nm
3	Ahammed et al. [52]	$\frac{k_{eff}}{k_b} = 2.5T_r^{1.032}\varphi^{0.108}$	Water	Graphene	T_r is the absolute temperature ratio between nanofluids temperature and ambient temperature Temperature range between 10°C and 50°C volume concentration range is 0.05%–0.15%
4	Patel et al. [53]	$k_{eff} = k_b\left(1 + 0.135\times\left(\frac{k_p}{k_b}\right)^{0.273}\varphi^{0.467}\left(\frac{T}{20}\right)^{0.547}\left(\frac{100}{d_p}\right)^{0.234}\right)$	Ethylene Glycol	Al_2O_3 CuO Cu Al	dp: 10–50 nm Kb: 0.1–0.7 Volume concentration: 0.1%–3% Temperature: 20°C–50°C
5	Li and Peterson [54]	$\frac{k_{eff}}{k_b} = 0.7644815\varphi + 0.018689T + 0.537853$, for Al_2O_3 $\frac{k_{eff}}{k_b} = 3.761088\varphi + 0.017924T + 0.69266$, for CuO	Water	Al_2O_3 CuO	Volume concentration: 1%–10% Temperature: 20°C–50°C

(Continued)

TABLE 4.1 (*Continued*)
All Published Experimental Models for the Thermal Conductivity of Nanofluids, Including Nanofluids Temperature

Ref.		Model	Base Fluid	Nanoparticle	Limitations
6	Hemmat Esfe et al. [55]	$k_{eff} = 0.4 + 0.0332\varphi + 0.00101T + 0.000619\varphi T + 0.0687\varphi^3 + 0.0148\varphi^5 - 0.00218\varphi^6 - 0.0419\varphi^4 - 0.0604\varphi^2$	Water–EG (60:40)	MgO	Temperature: 20°C–50°C Volume concentration: 0%–3%
7	Hemmat Esfe et al. [56]	$\frac{k_{eff}}{k_b} = 1.04 + 5.91\times10^{-5}T + 0.00154T\varphi + 0.0195\varphi^2 - 0.014\varphi - 0.00253\varphi^3 - 0.000104T\varphi^2 - 0.0357 \times \sin(1.72 + 0.407\varphi^2 - 1.67\varphi)$ $\frac{k_{eff}}{k_b} = 0.999 + 9.581\times10^{-5}T + 0.00142T\varphi + 0.0519\varphi^2 + 0.00208\varphi^2 + 0.00208\varphi^4 - 0.00719\varphi - 0.0193\varphi^3 - 8.21\times10^{-5}T\varphi^2$	Ethylene Glycol	Al_2O_3	Volume concentration: 0.2%–5% Temperature: 25°C–55°C
8	Hemmat Esfe et al. [57]	$\frac{k_{eff}}{k_b} = 1.04 + 0.000589T + \frac{-0.000184}{T\varphi} \times \cos(6.11 + 0.00673T + 4.41T\varphi - 0.0414\sin(T)) - 32.5\varphi$	Water–EG	Cu/TiO_2	Volume concentration: 0.1%–2% Temperature: 30°C–60°C
9	Hemmat Esfe et al. [58]	$\frac{k_{eff}}{k_b} = \frac{-214.83 + T}{346.58 - 106.98\varphi} + \frac{227.69}{T}$	Water	$CNTs/Al_2O_3$	Temperature: 303–332 K Volume concentration: 0%–1%
10	Harandi et al. [59]	$\frac{k_{eff}}{k_b} = 1 + 0.0162\varphi^{0.7038}T^{0.6009}$	Ethylene Glycol	F-MWCNTs–Fe_3O_4	Temperature: 25°C–55°C Volume concentration: 0.1%–2.3%

(*Continued*)

TABLE 4.1 (*Continued*)
All Published Experimental Models for the Thermal Conductivity of Nanofluids, Including Nanofluids Temperature

Ref.	Model	Base Fluid	Nanoparticle	Limitations
11 Zadkhast et al. [60]	$\frac{k_{eff}}{k_b} = 0.907\exp(0.36\varphi^{0.3111} + 0.000956T)$	Water	MWCNT–CuO	Temperature: 25°C–50°C Volume concentration: 0.05%–0.6%
12 Kakavandi and Akbari [61]	$\frac{k_{eff}}{k_b} = 0.0017 \times \varphi^{0.698} \times T^{1.386} + 0.981$	Water–EG	SiC–MWCNTs	Temperature: 25°C–55°C Volume concentration: 0%–0.75%
13 Karimi et al. [62]	$k_{eff} = (1 + 359.28\varphi\left(\frac{T}{T_{max}}\right))^{0.06418} \times (-6.58 \times 10^{-6}T^2 + 0.0018T + 0.5694)$ for Fe_3O_4 $k_{eff} = \left(1 + 418.81\varphi\left(\frac{T}{T_{max}}\right)\right)^{0.06748} \times (-6.58 \times 10^{-6}T^2 + 0.0018T + 0.5694)$ for $CoFe_2O_4$	Water	Fe_3O_4 $CoFe_2O_4$	Temperature: 20°C–60°C Volume concentration: 0%–4.8%
14 Karimipour et al. [63]	$\frac{k_{eff}}{k_b} = 0.792194 + 0.0547913\varphi + 0.00998805T + 0.000730423\varphi T - 0.00421237\varphi^2 - 0.0000643292T^2$	Liquid paraffin	CuO	Temperature: 25°C–100°C Weight concentration: 0.25%–6%
15 Ranjbarzadeh et al. [64]	$\frac{k_{eff}}{k_b} = 1 + 0.4281\left(\frac{T}{100}\right)^{1.707}\varphi^{0.8449}$	Water	SiO_2	Temperature: 25°C–55°C Volume concentration: 0.1%–3%
16 Keyvani et al. [65]	$\frac{k_{eff}}{k_b} = 0.9320 + 0.0673\varphi + 0.0021T$	Ethylene Glycol	CeO_2	Temperature: 25°C–50°C Volume concentration: 0.25%–2.5%
17 Afrand et al. [66]	$\frac{k_{eff}}{k_b} = 0.9320 + 0.0673\varphi^{0.323}T^{0.245}$	Water	Fe_3O_4	Temperature: 20°C–55°C Volume concentration: 0.1%–3%

It is predicted that maybe the effect of one variable versus the others can be neglected. Representing the contradictions in understanding the role of different variables will help attract researchers' attention to this challenging issue since it is highly believed that the single-phase approach of nanofluids investigation is severely dependent on the predictive models. The next step is devoted to extracting some more straightforward and more accurate correlations. Figure 4.1 shows the algorithm that has been employed to conduct this research.

Some methods should be put in the front before running the algorithm mentioned above. First of all, a sensitivity analysis of each proposed model is performed to find out the most sensitive variables in the model. Sensitivity analysis allows us to identify the inputs whose variation has the most impact on the key outputs.

Then, the published models in their validation range of variables mentioned by the authors (Table 4.1) are run to extract a couple of predicted data. Then, different datasets will be provided with different temperatures. For example, when a model is valid in the temperature range of 20°C–40°C, five datasets (statistical populations) for using the temperature of 20°C, 25°C, 30°C, 35°C, and 40°C are provided. It should be noted that some datasets statistically identical is assumed when two parameters of all datasets are statistically equal: mean and variance.

The null hypothesis is that all population means (or variances) are equal. In contrast, the alternative hypothesis states that at least one is different. So when the P-value is more than 0.05 (for the 0.05 level of significance), the test cannot reject the null hypothesis. This means that despite temperature changes in different datasets, the means (or variances) of them are statistically equal.

With this intention, statistical analysis for each model is conducted separately. If this analysis is sufficient for the hypothesis, then a different model considering different variables will be developed. At the final step of the algorithm, the model's goodness-of-fit will be evaluated and reported.

4.3 RESULTS AND DISCUSSION

In this section, all published models (Table 4.1) will be investigated and analyzed separately. The algorithm mentioned above will be run for each model. An effort is made to develop a simple mathematical model to predict all available experimental data. The goodness-of-fit of the final comprehensive model will be analyzed statistically.

4.4 MODEL BY ABEROUMAND ET AL.

Aberoumand et al. [50] conducted an experimental study on the rheological behavior of heat transfer oil-based nanofluids. They provided two correlations for the effective thermal conductivity and effective viscosity of nanofluids. They dispersed Ag nanoparticles with a mean diameter of 20 nm in oil heat transfer with a range of 0%–2% volume concentration. Then, they added some other experimental data about Cu and MCWNT nanoparticles to develop a predictive correlation for the effective thermal conductivity of nanofluids (Eq. 4.5).

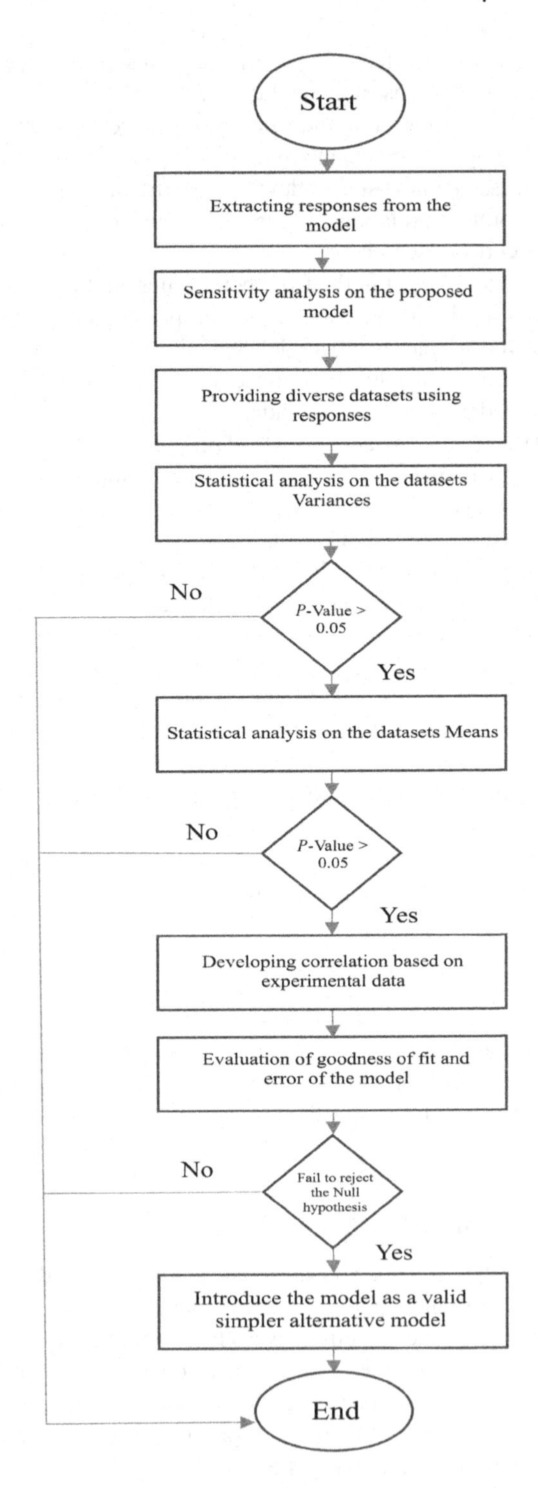

FIGURE 4.1 The employed algorithm for statistical analysis.

$$k_{\mathrm{eff}} = (3.9\times10^{-5}T - 0.0305)\varphi^2 + (0.086 - 1.6\times10^{-4}T)\varphi$$
$$+3.1\times10^{-4}T + 0.129 - 5.77\times10^{-6}k_{\mathrm{p}} - 40\times10^{-4} \tag{4.5}$$

First of all, according to their model, the effective thermal conductivity decreases with an increase in the thermal conductivity of nanoparticles. It is not clear why the authors have suggested such a mathematical model. They also did not provide any experimental data or graphs to show the behavior of the effective thermal conductivity versus the nanoparticle thermal conductivity. Their model gives 0.13023 W/m K for silver nanoparticles ($k_{\mathrm{p}}=413$), 0.13017 W/m K for copper nanoparticles ($k_{\mathrm{p}}=403$), and 0.11236 W/m K for MCWNT nanoparticles ($k_{\mathrm{p}}=3500$).

Their model predicts effective thermal conductivity of about 0.12937 W/m K for nanoparticle volume concentration of zero. It is unusual because the thermal conductivity of base fluid is 0.13 W/m K at this condition. It means that their model estimates effective thermal conductivity as more minor than the thermal conductivity of the base fluid. On the other hand, if one assumes the nanoparticle volume concentration as zero, their model converts to Eq. (4.6). This gives a different value from the thermal conductivity of the base fluid.

$$k_{\mathrm{eff}} = 3.1\times10^{-4}T + 0.129 - 5.77\times10^{-6}k_{\mathrm{p}} - 40\times10^{-4} \tag{4.6}$$

Let's check the dimensional consistency of their model. The model includes nine terms which only five terms have dimensions. The dimensional form ($[M][L][T]^{-3}[\theta]^{-1} = [\theta]+[\theta]+[\theta]-[M][L][T]^{-3}[\theta]^{-1}$) shows that this model is dimensionally inconsistent. Two sides of the equation do not give consistent dimensions.

Figure 4.2 shows the result of the sensitivity analysis of their proposed model. Generally, to identify inputs whose variation has little or no effect on the response, look for inputs with a flat line. For inputs with a flat line, the requirements (tolerances) can be eased without adversely affecting the performance, saving time and cost. The graph shows that changes in the variation of the k_{p} and T effects have little influence on the effective thermal conductivity. However, nanoparticle volume concentration plays a vital role in the prediction of the response.

Six thousand data from their model in the validation ranges of variables were extracted. The responses using different temperatures were categorized. The statistical analysis gives a mean and variance analysis *P*-values of 0.963 and 0.332, respectively (at the 0.95 level of significance). Therefore, there is no statistically significant difference between different datasets at different temperatures. Statistically speaking, the responses are the same, despite the considerable temperature change. It can be concluded that the temperature of the nanofluid does not play a vital role in the effective thermal conductivity prediction in their model with more than 95% of probability. So, the role of temperature can be easily neglected. Figure 4.3 shows the result of the variance analysis of the model of Aberoumand et al. [50].

The next step is providing a correlation for effective thermal conductivity without nanofluids temperature using experimental data based on the nonlinear regression method (see Appendix). Equation (4.7) is our proposed correlation. It is shown that

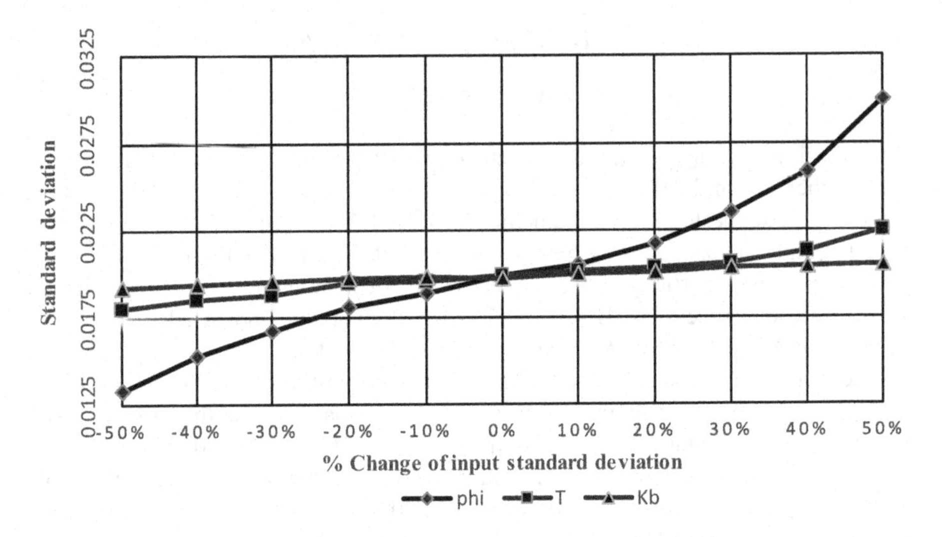

FIGURE 4.2 The result of sensitivity analysis for the proposed model by Aberoumand et al. [50].

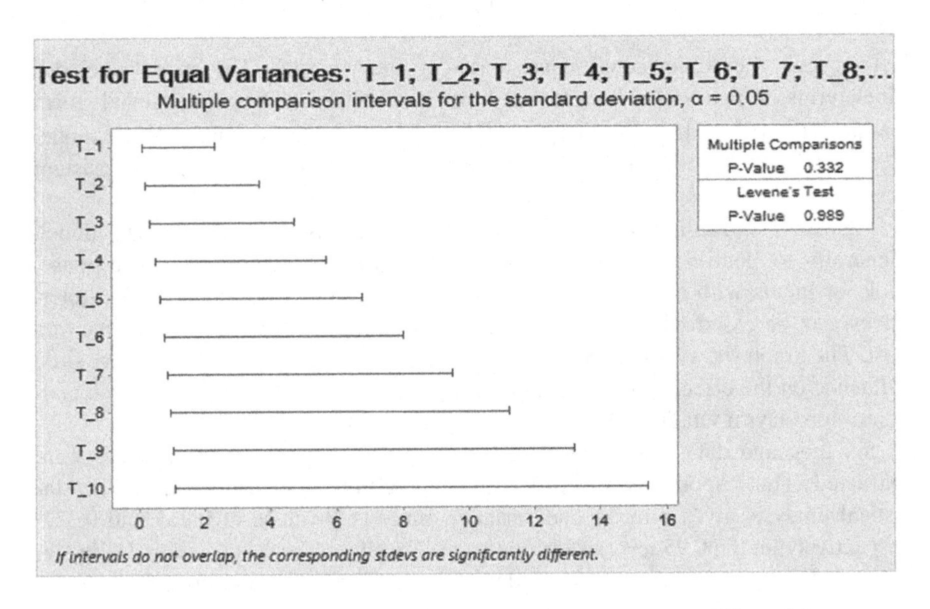

FIGURE 4.3 The result of variance analysis of the Aberoumand et al. [50] model.

this correlation is simple and accurate. It also gives effective thermal conductivity equal to the thermal conductivity of base fluid when the nanoparticle volume concentration is zero.

$$k_{\text{eff}} = k_{\text{b}}\left(1 + 1.25536\varphi^{0.29397}\right) \tag{4.7}$$

The average absolute error of this correlation from the experimental data is 3.62%. Therefore, this correlation has an appropriate goodness-of-fit and gives a reliable estimation of experimental data. Figures 4.4 and 4.5 show the acceptable accuracy of the correlation. The predicted value of the effective thermal conductivity of nanofluids is approximately equal to the thermal conductivity of the base fluid when the nanoparticle volume concentration is zero.

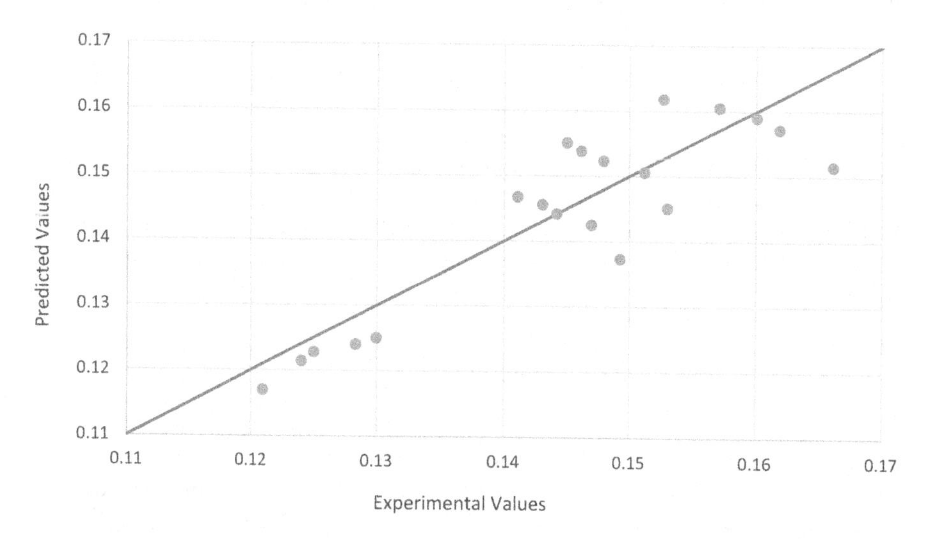

FIGURE 4.4 Agreement between experimental values and predicted values of effective thermal conductivity.

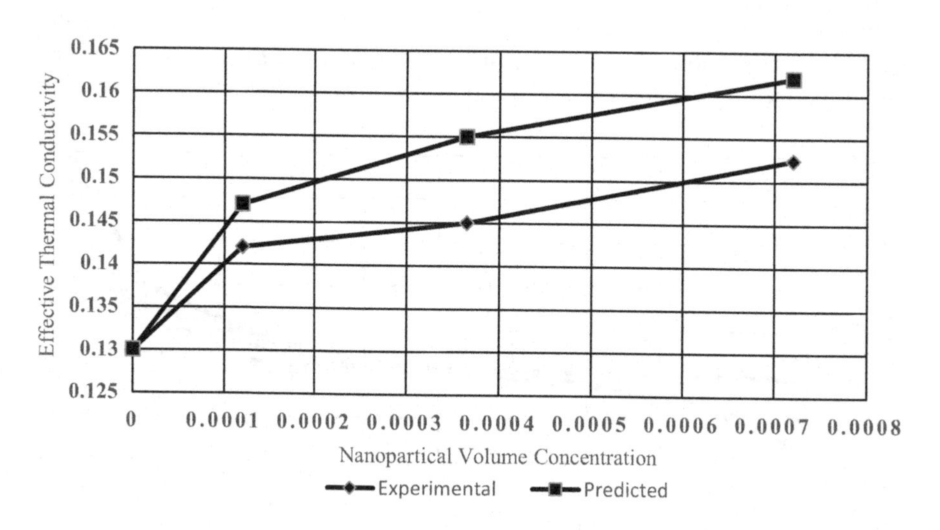

FIGURE 4.5 Comparison of experimental and predicted values of effective thermal conductivity.

4.5 MODEL BY FAKOOR PAKDAMAN ET AL.

An experimental investigation of the heat transfer performance of oil-based- nanofluids inside the vertical helically coiled tube was conducted by Fakoor Pakdaman et al. [51]. They dispersed MWCNT nanotubes 5–20 nm in diameter in heat transfer oil with a range of 0%–2% volume concentration. They introduced a correlation for the effective thermal conductivity of nanofluids (Eq. 4.8).

$$\frac{k_{\text{eff}}}{k_{\text{b}}} = 1 + 304.47(1+\varphi)^{136.35}\exp(-0.021T)\left(\frac{1}{d_{\text{p}}}\right)^{0.369}\left(\frac{T^{1.2321}}{10^{610.6287/(T-140)}}\right) \tag{4.8}$$

First of all, their model gives greater effective thermal conductivity than the base fluid thermal conductivity when the nanotube volume concentration is zero. When the nanotube volume concentration is assumed equal to zero (Eq. 4.9), the response will be 0.13530 W/m K. However, they measured the effective thermal conductivity of the base fluid at about 0.13250 W/m K at the temperature of 40°C.

$$\frac{k_{\text{eff}}}{k_{\text{b}}} = 1 + 304.47\exp(-0.021T)\left(\frac{1}{d_{\text{p}}}\right)^{0.369}\left(\frac{T^{1.2321}}{10^{610.6287/(T-140)}}\right) \tag{4.9}$$

Apart from these problems, the dimensional form shows that this model is dimensionally inconsistent, and the two sides of the equation do not give consistent dimensions. The left side is dimensionless, and the right side is dimensional. Figure 4.6 demonstrates the result of sensitivity analysis of their model. It can be concluded from the lines that the most critical variables in this model are nanoparticles' mean diameter and nanofluids temperature, respectively.

Eight thousand and thirty-six responses were extracted from their model in the valid ranges of variables. The responses using different temperatures in seven groups are categorized. Our statistical analysis provides P-values of 1.000 and 1.000 for

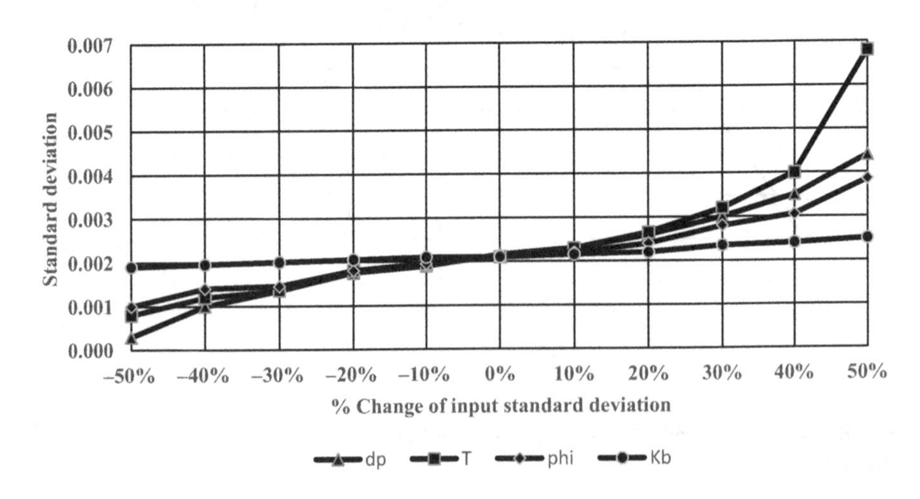

FIGURE 4.6 The result of sensitivity analysis for the proposed model by Fakoor Pakdaman et al. [51].

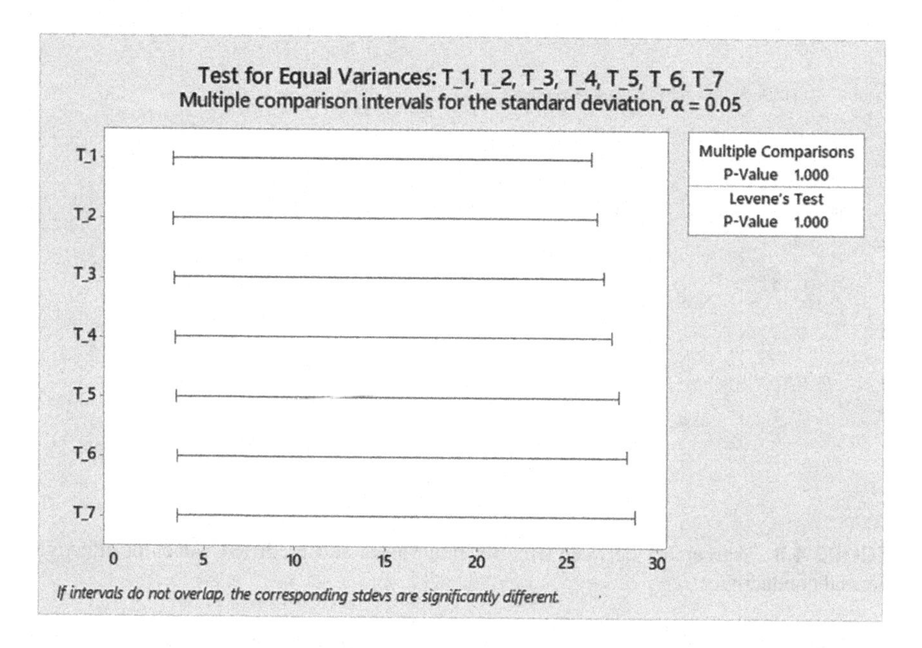

FIGURE 4.7 The result of variance analysis of the Fakoor Pakdaman et al. [51] model.

means and variances analysis, respectively (at the 0.95 level of significance). So, there is no statistically significant difference between different datasets at different temperatures. Statistically speaking, the responses are the same, despite the considerable temperature change. It can be concluded that the temperature does not play a vital role in the effective thermal conductivity prediction in their model with more than 95% probability, and so easily, the role of temperature can be neglected. Figure 4.7 shows the result of the variance analysis of the model of Fakoor Pakdaman et al. [51].

Then an effort is made to propose a correlation to predict the effective thermal conductivity of nanofluids, neglecting the role of temperature using the nonlinear regression method. Equation (4.10) is our proposed correlation with an average absolute error of 2.41%.

$$k_{\text{eff}} = k_{\text{b}} \left(1 + \left(\frac{2.734650\varphi}{d_{\text{p}}} \right)^{0.270822} \right) \tag{4.10}$$

Therefore, this correlation has excellent goodness-of-fit and gives a highly reliable estimation of response. Figures 4.8 and 4.9 show the acceptable accuracy of the correlation. Our proposed equation (Eq. 4.10) gives an effective thermal conductivity of 0.1235 W/m K when the nanoparticle volume concentration is zero. This is equal to the thermal conductivity of the base fluid. It means that when there is no nanoparticle in the liquid, we have the pure base fluid. However, the correlation proposed by Fakoor Pakdaman et al. [51] model (Eq. 4.8) gives 0.1358 W/m K for effective thermal conductivity in this situation. It means that although there is no nanoparticle in the liquid, the thermal conductivity value is 9.96% greater than the base fluid at the same conditions!

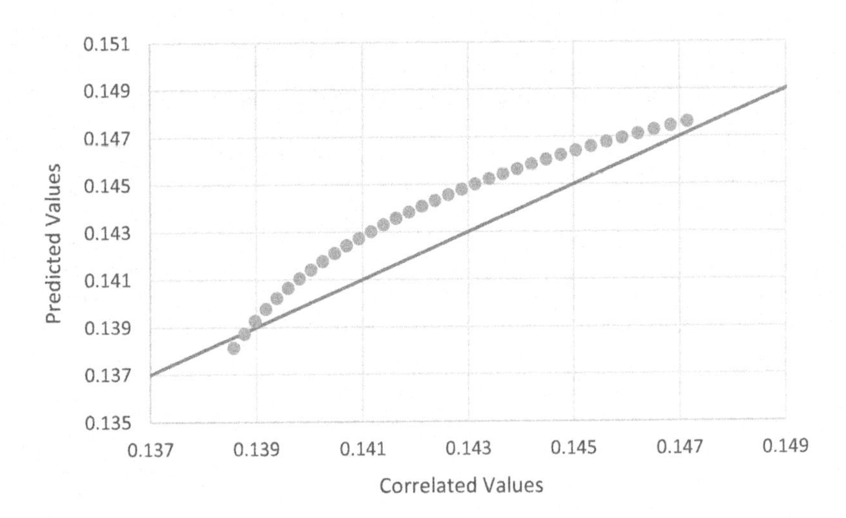

FIGURE 4.8 Agreement between experimental values and predicted values of effective thermal conductivity.

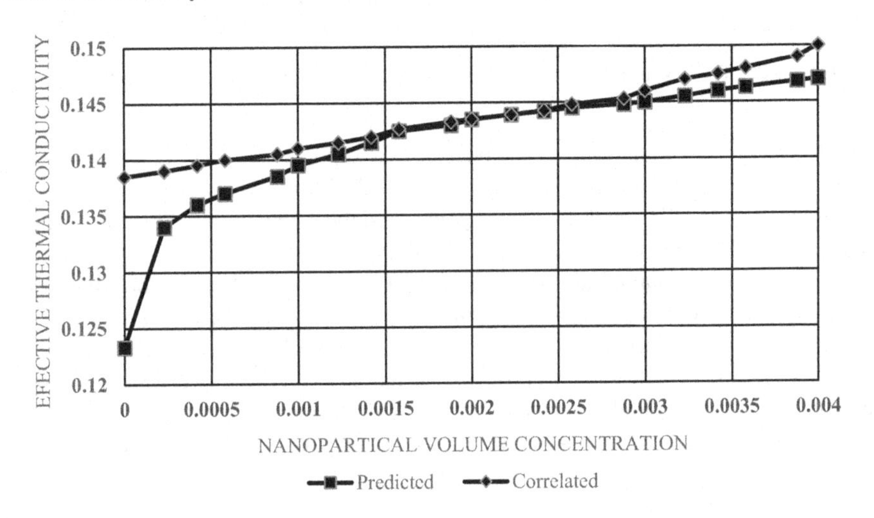

FIGURE 4.9 Comparison of experimental and predicted values of effective thermal conductivity.

4.6 MODEL BY AHAMMED ET AL.

Ahammed et al. [52] investigated heat transfer, and fluid flow of graphene–water nanofluids, experimentally implementing a transient hot-wire technique at temperatures below and above ambient conditions ranging from 10°C to 50°C. The author mentioned that the valid range of nanoparticle volume concentration is between 0.05% and 0.15%. An empirical correlation has been developed in the form of Eq. (4.11) (dimensionally consistent). Figure 4.10 shows the result of the sensitivity

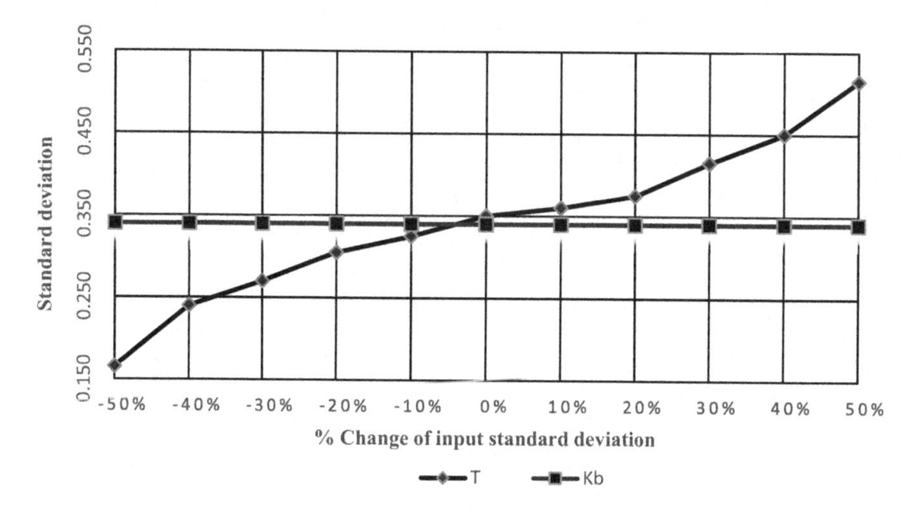

FIGURE 4.10 The result of sensitivity analysis of the proposed model by Ahammed et al. [52].

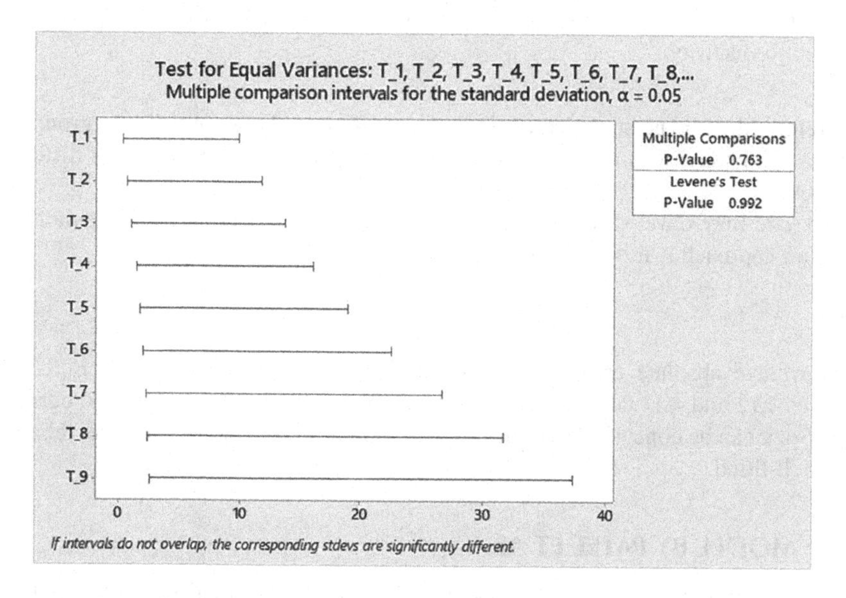

FIGURE 4.11 Test for equal variances of Ahammed et al. [52] model.

analysis of the model. It is shown that the temperature of the nanofluid plays a vital role in the prediction of the effective thermal conductivity of nanofluids.

$$\frac{k_{\text{eff}}}{k_{\text{b}}} = 2.5T_{\text{r}}^{1.032}\varphi^{0.108} \tag{4.11}$$

Nevertheless, a decision was made to investigate the case in detail, so 243 responses were extracted in different variables using their model and categorized into nine groups for different nanofluids temperature. The statistical analysis for mean and variance gives P-values of 0.772 and 0.762, respectively (Figure 4.11).

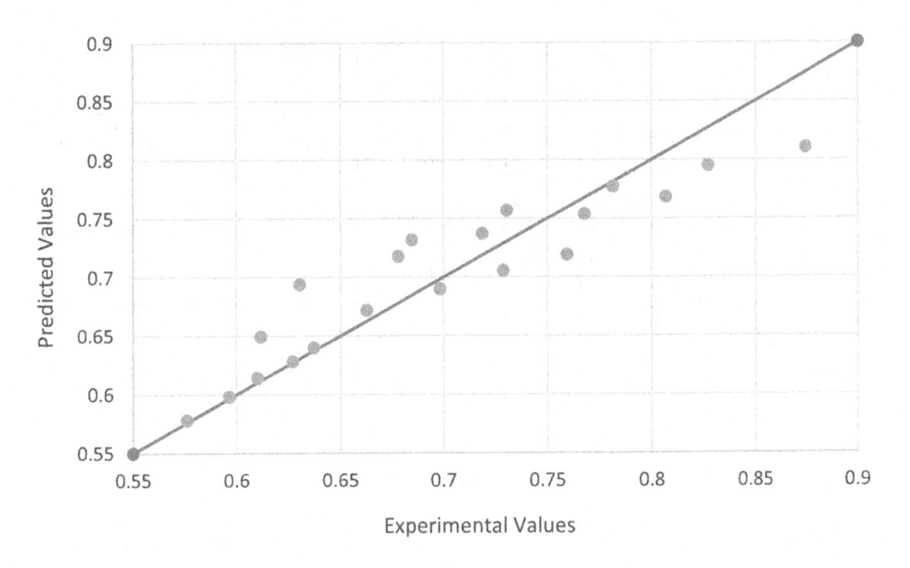

FIGURE 4.12 Agreement between experimental values and predicted values of effective thermal conductivity.

Therefore, there is no statistically significant difference between the nine groups. It means that changing temperature does not affect the mean and variances of different datasets.

So, one may develop a correlation neglecting the temperature of the nanofluid. Such an approach is followed, and Eq. (4.12) has resulted.

$$k_{\text{eff}} = k_{\text{b}}\left(1+\varphi^{0.698441}\right) \tag{4.12}$$

The average absolute error of this correlation from experimental data is 3.31%. Figures 4.12 and 4.13 demonstrate the acceptable accuracy of the proposed correlation. So, it can be concluded that the proposed simple correlation is accurate, reliable, and well-fitted.

4.7 MODEL BY PATEL ET AL.

A series of experiments have been conducted to measure the thermal conductivity of different oxide and metal-based nanofluids by Patel et al. [53]. They dispersed silver, alumina, copper, and aluminum nanoparticles in water, ethylene glycol, and transformer oil with volume concentrations in the range of 0.1%–3% using the sonication method. The mean diameter of nanoparticles was in the range of 10–50 nm, and the minimum and maximum nanofluids temperatures were 20°C and 50°C, respectively. They also proposed a correlation for the prediction of effective thermal conductivity of nanofluids (Eq. 4.13).

$$k_{\text{eff}} = k_{\text{b}}\left(1+0.135\times\left(\frac{k_{\text{p}}}{k_{\text{b}}}\right)^{0.273}\varphi^{0.467}\left(\frac{T}{20}\right)^{0.547}\left(\frac{100}{d_{\text{p}}}\right)^{0.234}\right) \tag{4.13}$$

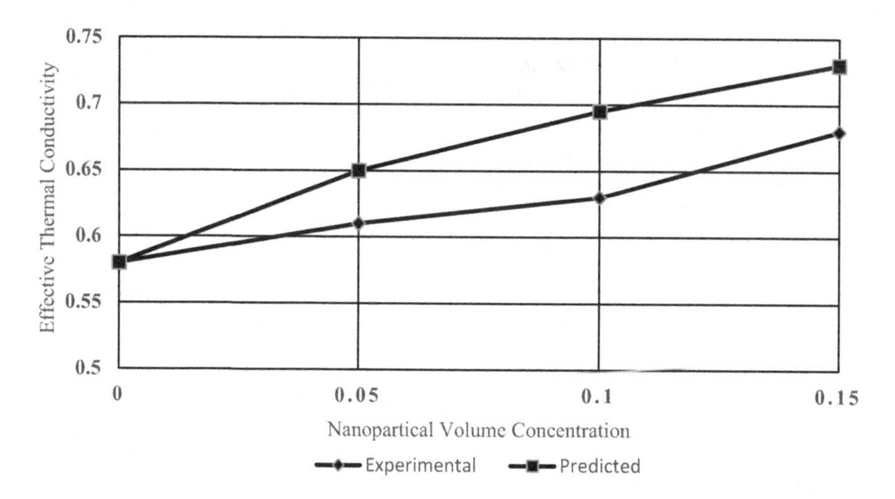

FIGURE 4.13 Comparison of experimental and predicted values of effective thermal conductivity.

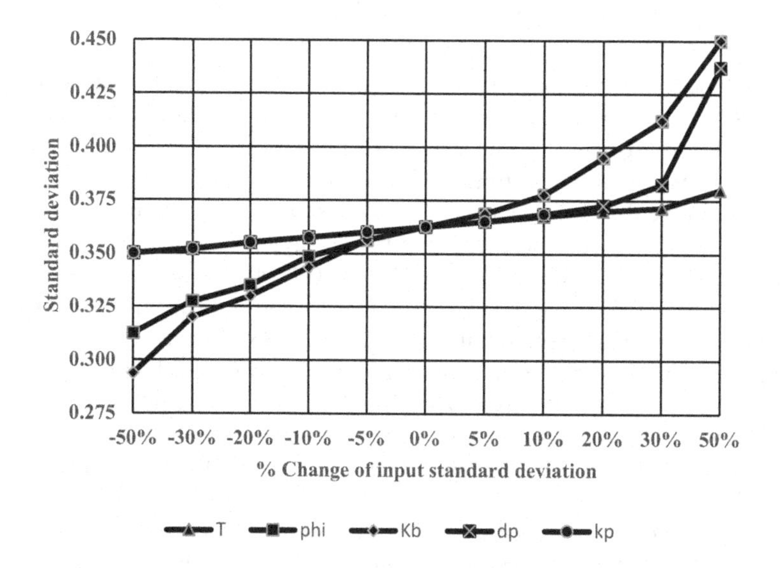

FIGURE 4.14 The result of sensitivity analysis of the proposed model by Patel et al. [53].

Let's check the dimensional consistency of their model. The dimensional form $\left([M][L][T]^{-3}[\theta]^{-1}=[M][L][T]^{-3}[\theta]^{-1}[\theta]^{0.547}[L]^{-0.234}\right)$ shows that this model is dimensionally inconsistent. Two sides of the equation do not give consistent dimensions. Figure 4.14 shows that the most and minor affective variables in the proposed model based on a sensitivity analysis are the thermal conductivity of the base fluid and nanofluids temperature, respectively.

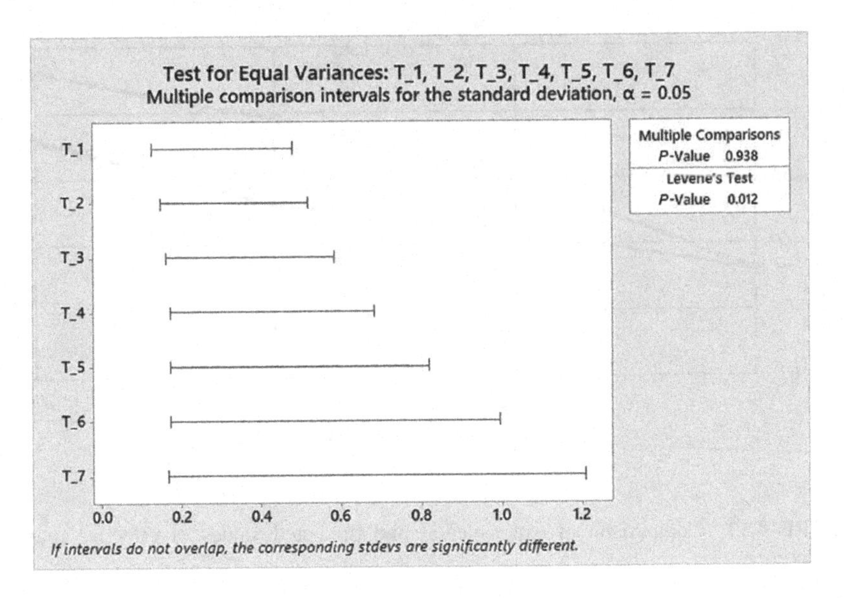

FIGURE 4.15 Test for equal variances of Patel et al. [53].

For detailed investigation, 105,847 responses from their model within the valid ranges of variables created seven unique datasets using different nanofluids temperatures. The statistical analysis shows that the *P*-values for the means and variances are 0 and 0.938, respectively. Although the *P*-value for means is zero, this test fails to prove the null hypothesis, the values of responses in different datasets were compared. It should be noted that the average response values for datasets with temperatures of 20°C, 25°C, 30°C, 35°C, 40°C, 45°C, and 50°C, are 0.49772, 0.50430, 0.51031, 0.51588, 0.52110, 0.52604, and 0.53072, respectively. This means that when the nanofluid temperature from 20°C to 25°C increases, only a 1.32% enhancement in effective thermal conductivity is experienced. Increasing the temperature from 45°C to 50°C gives only a 0.89% enhancement in effective thermal conductivity. Therefore, there is no significant difference between these seven groups of responses can be concluded. Figure 4.15 shows the result of the test for equal variances.

A simple correlation for predicting the effective thermal conductivity of nanofluids is developed, which neglects the role of nanofluids temperature, employing the nonlinear regression method. Equation (4.14) shows our proposed correlation.

$$k_{\text{eff}} = k_{\text{b}}\left(1 + \frac{0.232706\varphi^{0.402458} \times k_{\text{p}}}{d_{\text{p}}}\right) \tag{4.14}$$

The average absolute error from the experimental data is 3.44%. Figures 4.16 and 4.17 represent the accuracy and fitness of the proposed correlation. It can be concluded that Eq. (4.14) is accurate, reliable, and well-fitted.

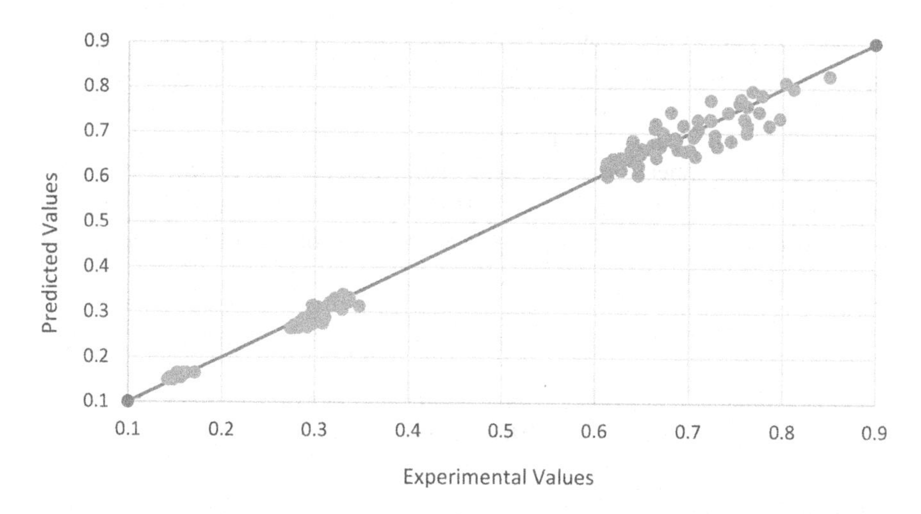

FIGURE 4.16 The fitness of predicted values by the proposed correlation on the experimental values.

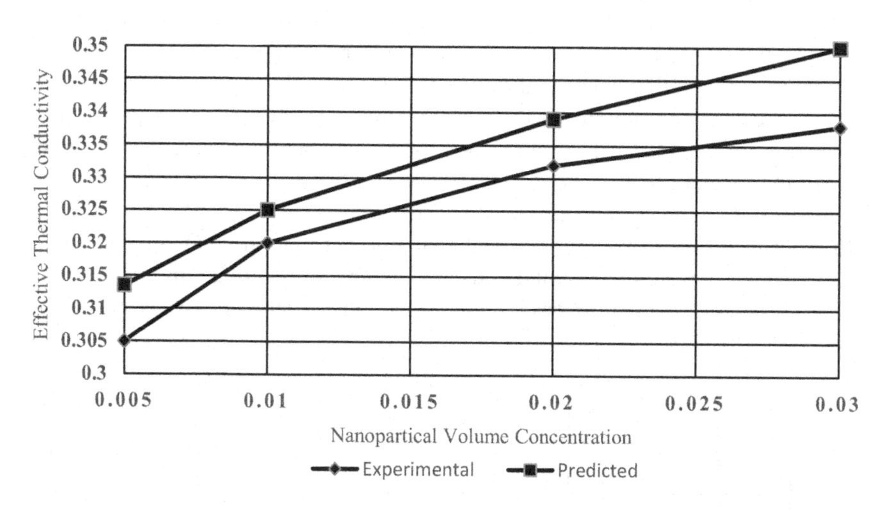

FIGURE 4.17 Comparison of experimental and predicted values for effective thermal conductivity.

4.8 MODELS BY LI AND PETERSON

Li and Peterson [54] introduced the first correlation for effective thermal conductivity of nanofluids, including the temperature of the nanofluid based on experimental data. They synthesized two categories of nanofluids in nanoparticle volume concentration in the range of 2%–10%: Al_2O_3–water and CuO–water. The mean diameters of alumina and copper oxide were around 36 and 20 nm, respectively. The experiments have been performed in nanofluids bulk temperatures of 27.5°C–34.7°C. Generally, they observed a significant enhancement in thermal conductivity up to 52% compared to the base fluid. Finally, they developed two correlations to predict effective thermal conductivity (Eqs. 4.15 and 4.16).

Let's check the dimensional consistency of their model. The dimensional form of two proposed correlations $\left(\left([M][L][T]^{-3}[\theta]^{-1}/[M][L][T]^{-3}[\theta]^{-1}\right)=[\theta]\right)$ shows that these models are dimensionally inconsistent. Two sides of the equations do not give consistent dimensions.

$$\frac{k_{\text{eff}}}{k_{\text{b}}}=0.7644815\varphi+0.018689T+0.537853, \quad \text{for } Al_2O_3 \tag{4.15}$$

$$\frac{k_{\text{eff}}}{k_{\text{b}}}=3.761088\varphi+0.017924T+0.69266, \quad \text{for CuO} \tag{4.16}$$

Figure 4.18 shows the result of the sensitivity analysis of their two proposed correlations. It has shown that nanofluids' temperature does not play a vital role in predicting the effective thermal conductivity of nanofluids.

For more investigation, 1862 responses from their correlations were extracted. They are categorized into seven unique datasets (for each correlation) with different temperatures. The statistical analysis gives P-Vales of 0.736 and 0.927 for means and variances, respectively for Eq. (4.15) and 0.755 and 0.925 for means and variances, respectively for Eq. (4.16). Therefore, there is no statistically meaningful difference between the means and variances of different datasets. Figure 4.19 shows the result of statistical analysis for variances of Eq. (4.15).

In order to develop two simple correlations for effective thermal conductivity, an effort was made, neglecting the nanofluids' bulk temperature. Equations (4.20 and 4.21) show the proposed correlation for alumina–water and copper oxide–water nanofluids.

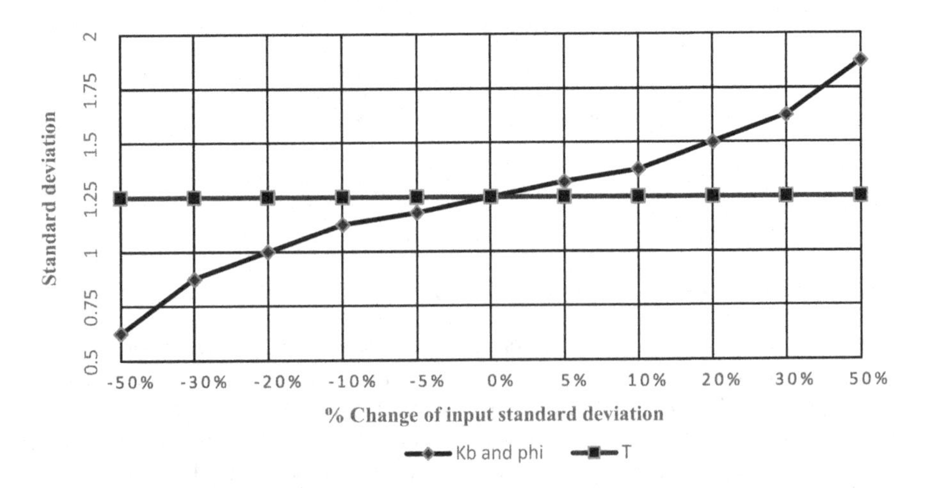

FIGURE 4.18 The result of the sensitivity analysis of two proposed correlations by Li and Peterson [54].

$$k_{\text{eff}} = k_{\text{b}}(1 + 0.81874\varphi^{0.56}) \quad \text{for } Al_2O_3 \tag{4.17}$$

$$k_{\text{eff}} = k_{\text{b}}(1 + \varphi^{0.255643}) \quad \text{for CuO} \tag{4.18}$$

Equations (4.17 and 4.18) give predicted values with an average absolute error of 4.88% and 2.79% from experimental data. Therefore, these two simple correlations are valid, accurate, and well-fitted. Figures 4.20–4.23 show the accuracy of proposed correlations for alumina–water and copper oxide–water nanofluids.

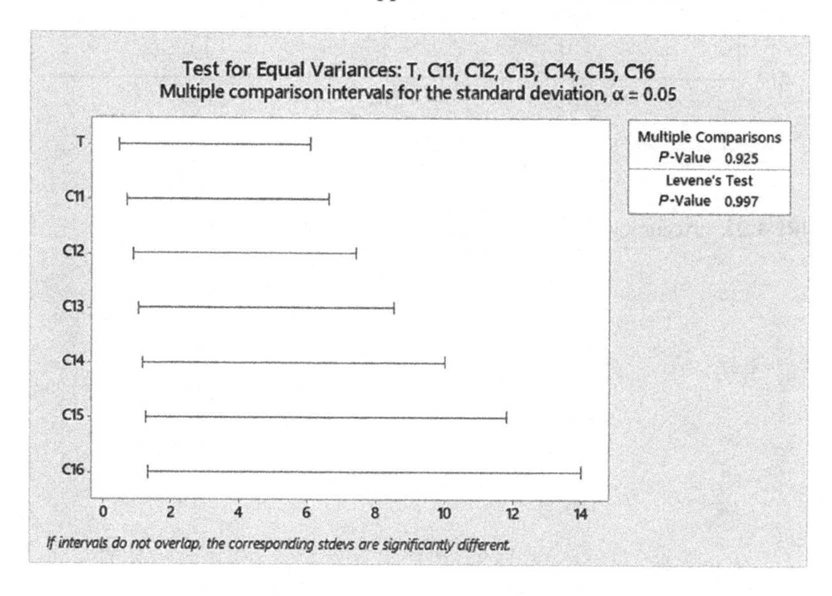

FIGURE 4.19 Test for equal variances of correlation for alumina (Eq. 4.15).

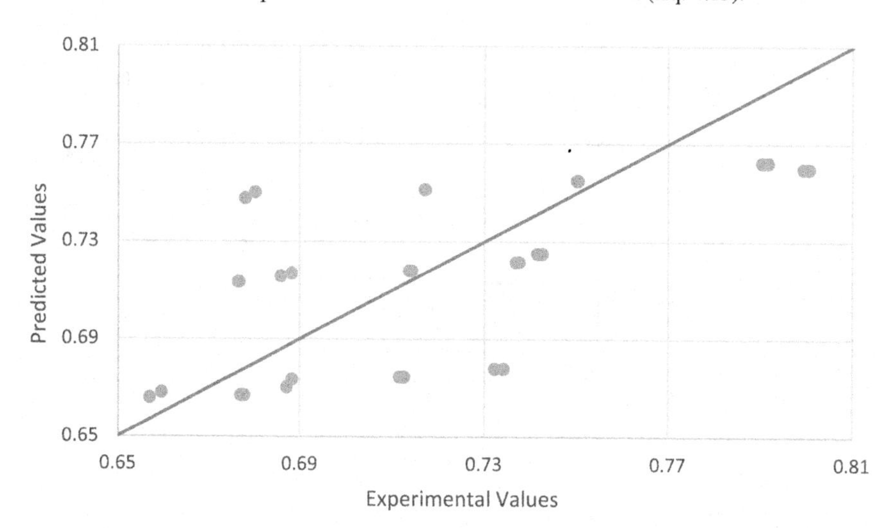

FIGURE 4.20 Fitness of predicted values by Eq. (4.17) versus experimental values.

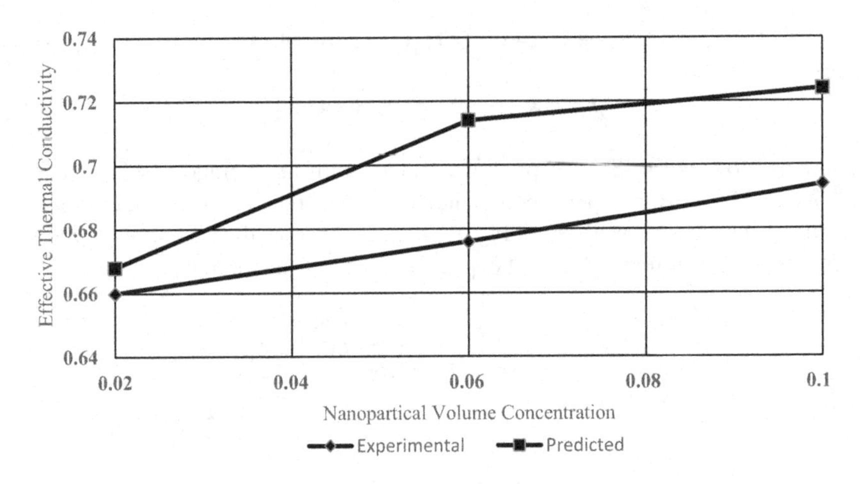

FIGURE 4.21 Accuracy of proposed correlation for alumina–water nanofluids.

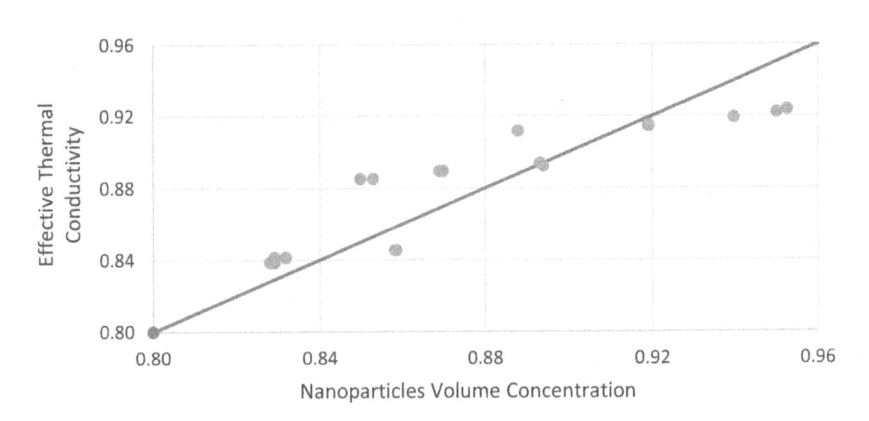

FIGURE 4.22 Fitness of predicted values by Eq. (4.18) versus experimental values.

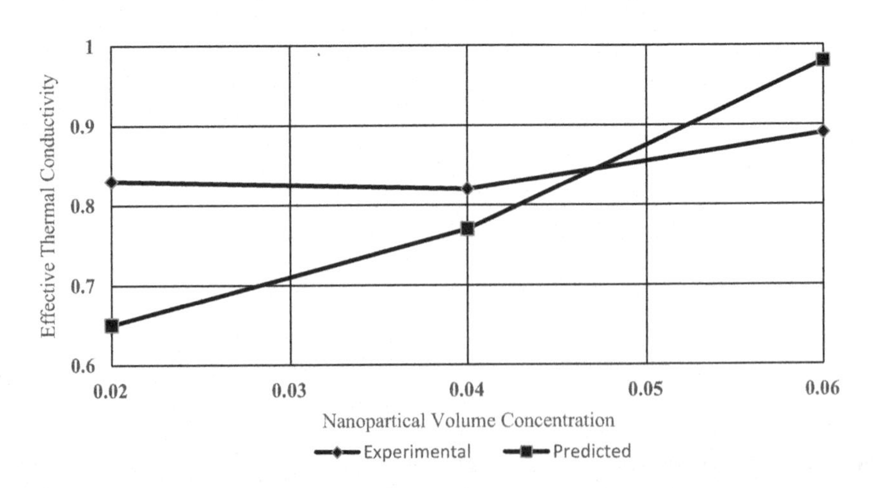

FIGURE 4.23 The accuracy of the proposed correlation for copper oxide–water nanofluids.

4.9 MODEL BY HEMMAT ESFE ET AL.

An experimental study has been performed on the thermal conductivity of MgO-based nanofluids by Hemmat Esfe et al. [55]. They dispersed 40 nm MgO nanoparticles in a binary mixture of water and ethylene glycol (60:40) with a volume concentration in the range of 0.1%–3%, using an ultrasonic homogenizer. The minimum and maximum nanofluids' bulk temperatures were 20°C and 50°C, respectively. They proposed a correlation for effective thermal conductivity (Eq. 4.19) using experimental data.

$$k_{\text{eff}} = 0.4 + 0.0332\varphi + 0.00101T + 0.000619\varphi T + 0.0687\varphi^3 + 0.0148\varphi^5 - 0.00218\varphi^6 - 0.0419\varphi^4 - 0.0604\varphi^2 \quad (4.19)$$

The proposed correlation has ten terms, including only three-dimensional terms. The dimensional form of the proposed correlation $\left([M][L][T]^{-3}[\theta]^{-1} = [\theta] + [\theta]\right)$ shows that this model is dimensionally inconsistent. Two sides of the equation do not give consistent dimensions. Figure 4.24 represents the result of the sensitivity analysis of their proposed model, indicating that the temperature of the nanofluid does not considerably affect the effective thermal conductivity of nanofluids.

For detailed investigation, 784 responses were extracted from their model in the validation ranges of variables and were categorized into seven unique datasets with different temperatures. The statistical analysis gives P-values of 0.995 and 0.927 for means and variances, respectively. Therefore, the test fails to reject the null hypothesis. There is no statistically significant difference between these seven groups, despite considerably changing the nanofluids' bulk temperature. Figure 4.25 shows the result of the test for equal variances.

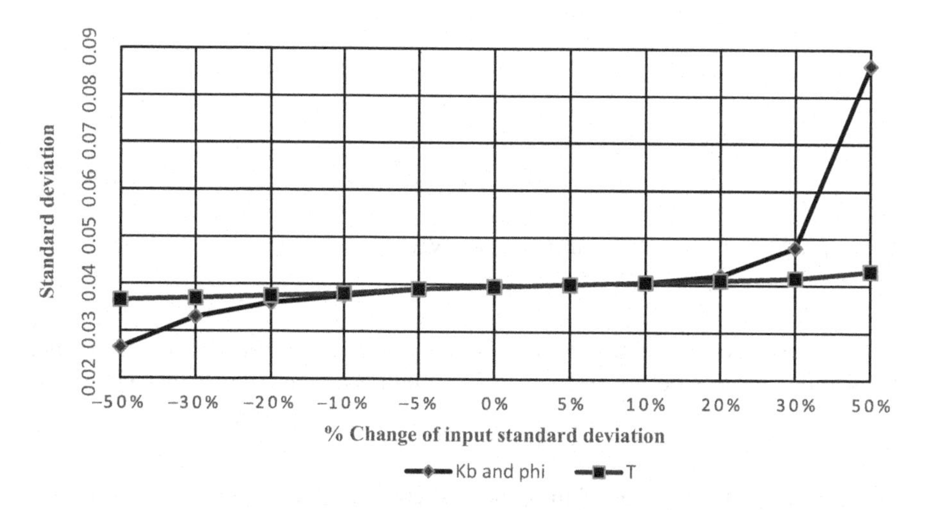

FIGURE 4.24 The result of the sensitivity analysis of the proposed model by Hemmat Esfe et al. [55].

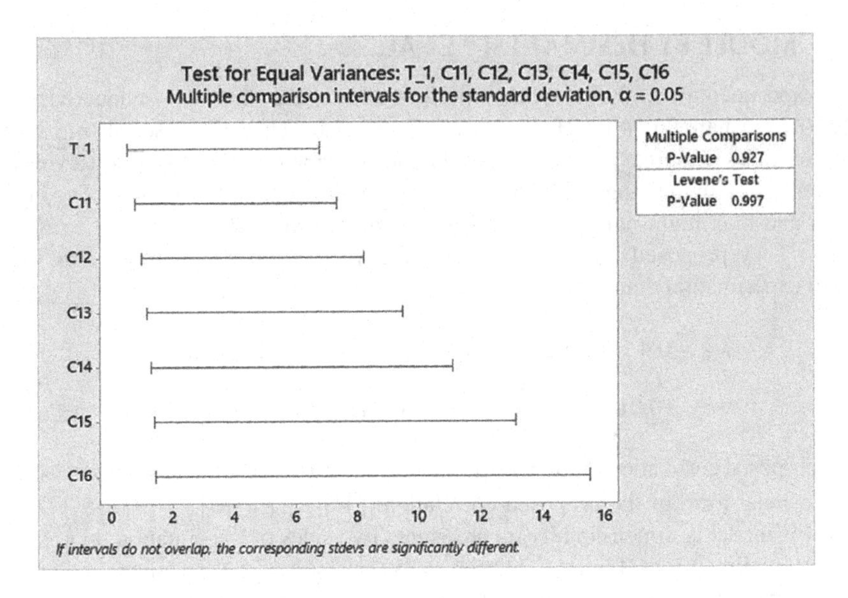

FIGURE 4.25 The result of the test for equal variances.

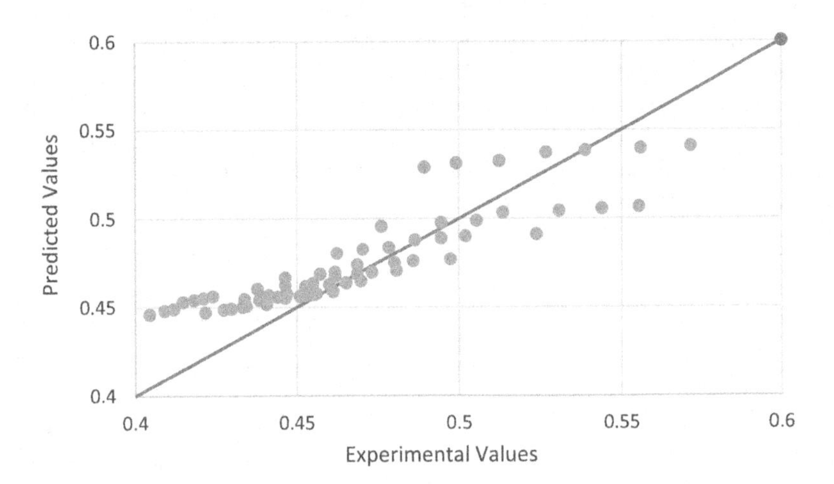

FIGURE 4.26 The fitness of predicted values by the proposed correlation versus the experimental values.

An effort was made to develop a simple correlation neglecting the bulk nanofluids temperature using experimental data and the nonlinear regression method (Eq. 4.20).

$$k_{\mathrm{eff}} = k_{\mathrm{b}}(1 + 15.9106\varphi^{1.26896}) \tag{4.20}$$

The average absolute error of the proposed correlation from experimental data is 3.5%. Figures 4.26 and 4.27 show the accuracy of the proposed correlation. Therefore, the proposed simple correlation is accurate, reliable, and well-fitted.

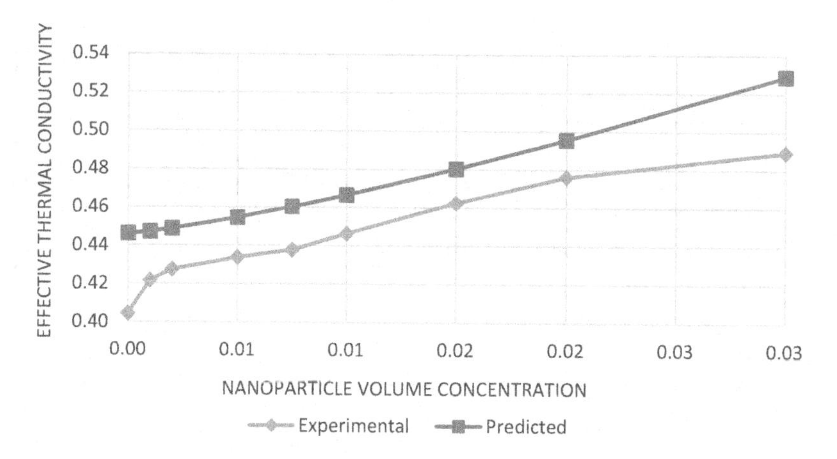

FIGURE 4.27 The comparison of experimental and predicted values of effective thermal conductivity.

4.10 MODELS BY HEMMAT ESFE ET AL.

Hemmat Esfe et al. [56] conducted an experimental study on the thermal conductivity of alumina-ethylene glycol nanofluids. The experiments were performed at a temperature ranging from 24°C to 50°C with volume concentration up to 5%. They proposed two correlations for predicting the effective thermal conductivity of nanofluids (Eqs. 4.21 and 4.22).

The dimensional form of their proposed correlations $\left(\left([M][L][T]^{-3}[\theta]^{-1}/[M][L][T]^{-3}[\theta]^{-1}\right)=[\theta]+[\theta]+[\theta]\right)$ represents that these models are dimensionally inconsistent. Two sides of the equations do not give consistent dimensions.

$$\begin{aligned}\frac{k_{\text{eff}}}{k_{\text{b}}} &= 1.04+5.91\times10^{-5}T+0.00154T\varphi+0.0195\varphi^2 \\ &-0.014\varphi-0.00253\varphi^3-0.000104T\varphi^2-0.0357 \\ &\times\sin(1.72+0.407\varphi^2-1.67\varphi)\end{aligned} \tag{4.21}$$

$$\begin{aligned}\frac{k_{\text{eff}}}{k_{\text{b}}} &= 0.999+9.581\times10^{-5}T+0.00142T\varphi+0.0519\varphi^2 \\ &+0.00208\varphi^2+0.00208\varphi^4-0.00719\varphi-0.0193\varphi^3 \\ &-8.21\times10^{-5}T\varphi^2\end{aligned} \tag{4.22}$$

As shown in Figure 4.28, the sensitivity analysis results show that the role of changing temperature on the effective thermal conductivity is negligible.

Two thousand five hundred and forty-eight responses were extracted from these two correlations in their valid performance range and categorized into seven unique

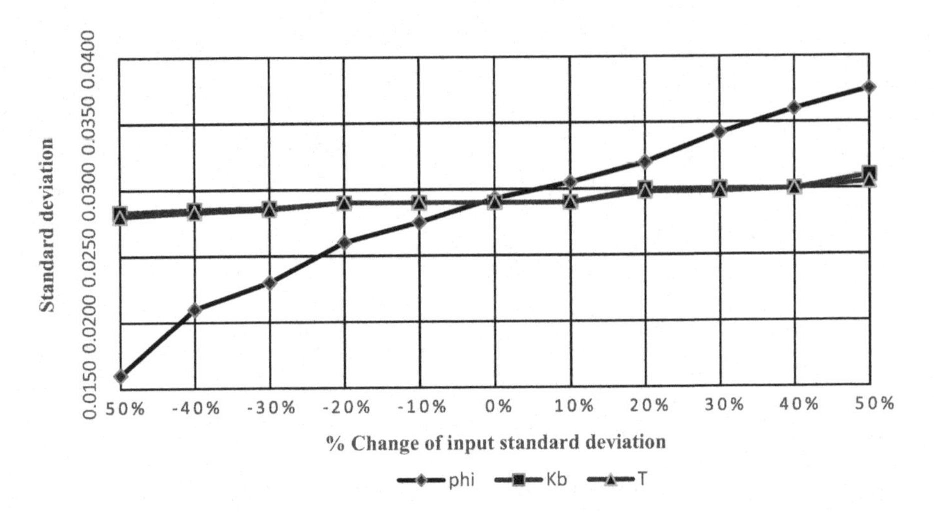

FIGURE 4.28 The result of the sensitivity analysis of two proposed models by Hemmat Esfe et al. [56].

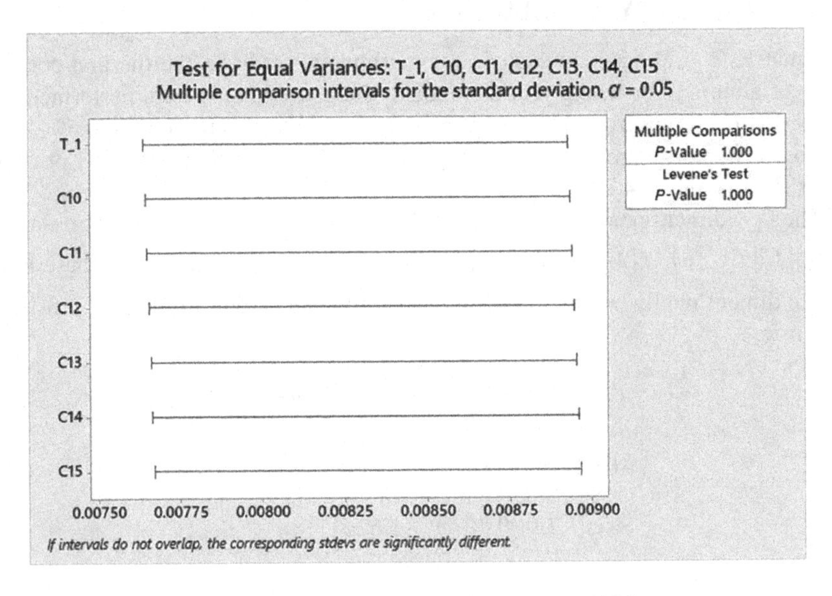

FIGURE 4.29 The result of the test for equal variances of Eq. (4.21).

datasets (seven datasets for each correlation). Statistical analysis for means and variances gives *P*-values of 0.971 and 1.000, respectively, for Eq. (4.21) and 0.880 and 1.000, respectively for Eq. (4.22). Therefore, it can be concluded that there is no statistically significant difference between datasets. Figures 4.29 and 4.30 show the results of tests for equal variances of Eqs. (4.21 and 4.22).

For more investigation, an effort was made to develop a more precise correlation neglecting the nanofluids' bulk temperature using experimental data and a nonlinear

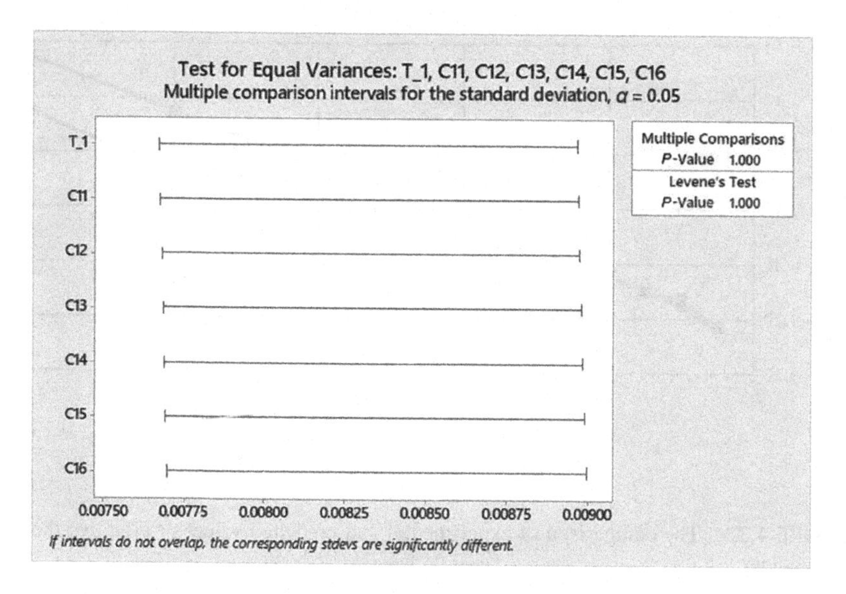

FIGURE 4.30 The result of the test for equal variances of Eq. (4.22).

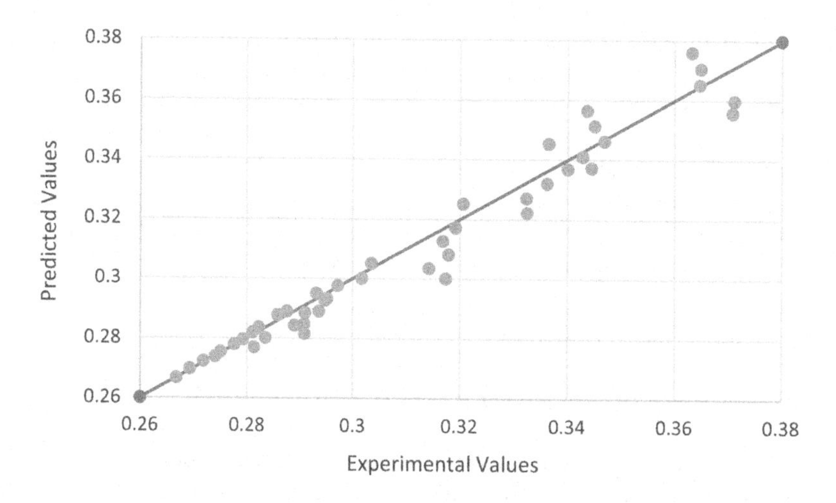

FIGURE 4.31 Fitness of predicted and experimental values.

regression model (Eq. 4.23). This correlation predicts the effective thermal conductivity of nanofluids (covering all 14 datasets) with an average absolute error of 1.69%. Figures 4.31 and 4.32 show the accuracy of the proposed correlation for the prediction of effective thermal conductivity.

This simple correlation is accurate, reliable, and well-fitted.

$$k_{\text{eff}} = k_{\text{b}}(1 + 7.05149\varphi) \tag{4.23}$$

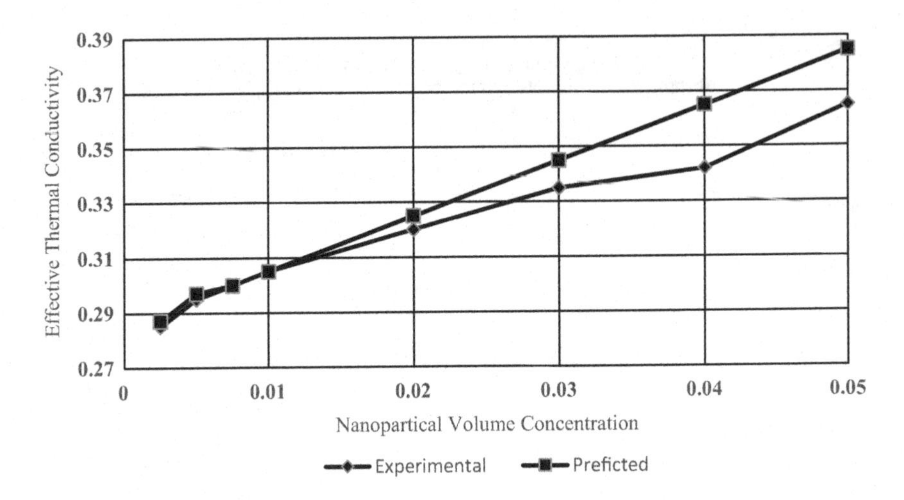

FIGURE 4.32 The comparison of experimental and predicted values of effective thermal conductivity.

4.11 MODEL BY HEMMAT ESFE ET AL.

An interesting study on the thermal conductivity of hybrid nanofluids has been performed by Hemmat Esfe et al. [57]. They used experimental data about the dispersion of Cu and TiO_2 nanoparticles with mean diameters of 70 and 40 nm in a binary water–ethylene glycol (60:40) base fluid with a volume concentration of 0.1%–2%. They implemented an artificial neural network (ANN) to correlate the result in a mathematical model. They considered temperature and volume concentration as input layers and relative thermal conductivity (the proportion of effective thermal conductivity of nanofluids to the thermal conductivity of base fluid) as the output layer. Finally, they developed a trigonometric correlation:

$$\frac{k_{\text{eff}}}{k_{\text{b}}} = 1.04 + 0.000589T + \frac{-0.000184}{T\varphi} \times \cos(6.11 + 0.00673T + 4.41T\varphi - 0.0414\sin(T)) - 32.5\varphi \quad (4.24)$$

Apart from this ambiguity, which is why a trigonometric correlation is used, how can temperature (a dimensional variable) and volume concentration as sine and cosine arguments be used? The sine of an angle is defined as the ratio of the lengths of two respective sides of a triangle. Thus, the dimensions of the sine are [L]/[L], or 1. Nevertheless, the result of sensitivity analysis shows that the most and most minor affective variables in predicting the effective thermal conductivity of nanofluids based on their model are nanoparticle volume concentration and nanofluids temperature, respectively (Figure 4.33). Using trigonometric functions leads to having a periodical behavior of volume concentration, shown in Figure 4.33.

Nine hundred and eighty values were extracted from their model in the mentioned variables and then categorized into seven unique datasets with different temperatures for detailed investigation. The statistical analysis gives P-values of 0.999 and

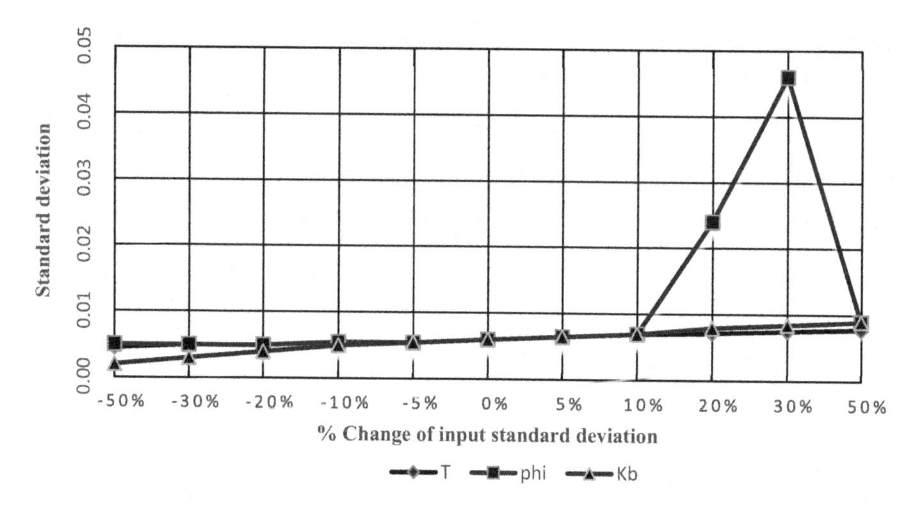

FIGURE 4.33 The result of the sensitivity analysis of the proposed model by Hemmat Esfe et al. [57].

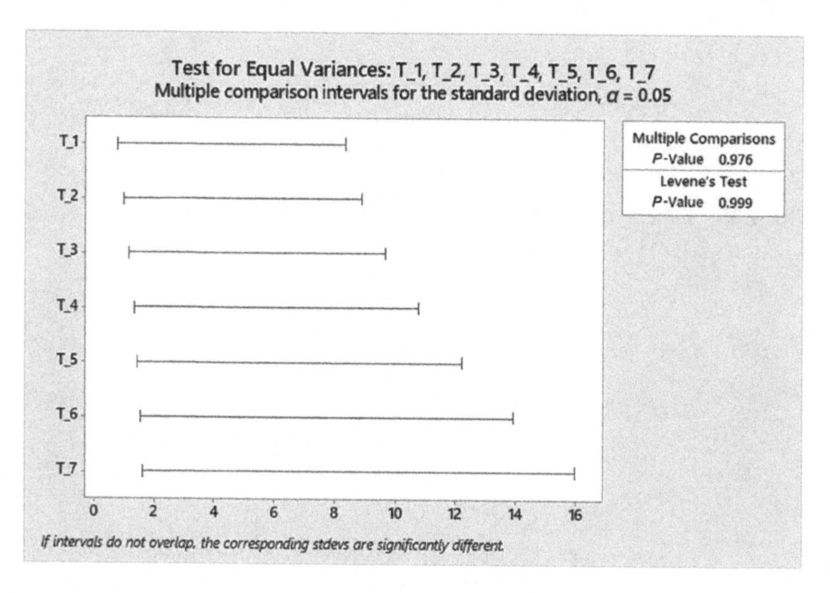

FIGURE 4.34 The result of the test for equal variances for seven unique datasets.

0.976 for means and variances of the model, respectively. This test fails to reject the null hypothesis, and therefore, there is no statistically significant difference between these seven datasets. This means that temperature does not affect the responses considerably, despite changing temperature considerably. Figure 4.34 shows the result of the test for equal variances of seven datasets.

Then, an effort was made to develop a simple correlation to predict effective thermal conductivity based on experimental data neglecting the role of temperature.

The nonlinear regression method to propose Eq. (4.25) was implemented. The proposed correlation gives an average absolute error of 3.86%.

$$k_{\text{eff}} = k_{\text{b}}(1+\varphi^{0.30409}) \tag{4.25}$$

So, the proposed correlation is accurate, reliable, and well-fitted. Figures 4.35 and 4.36 show the fitness and accuracy of the proposed correlation.

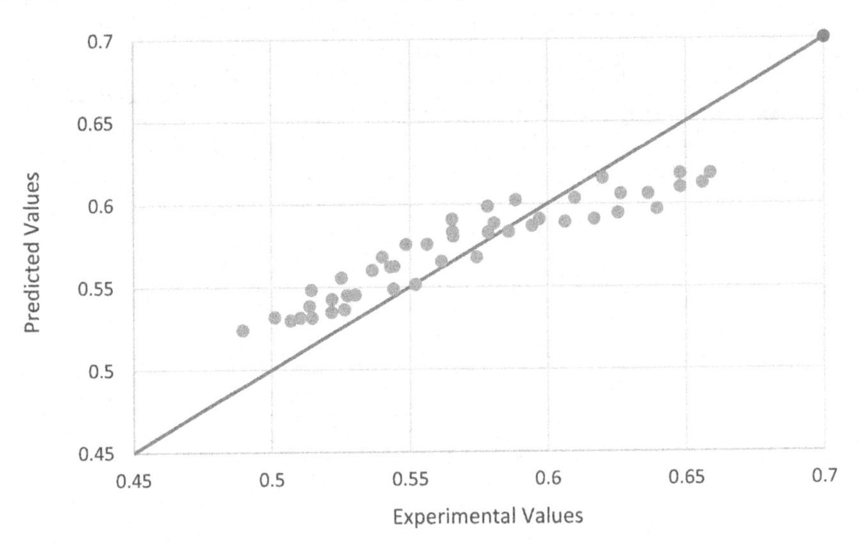

FIGURE 4.35 Fitness of predicted values on the experimental values.

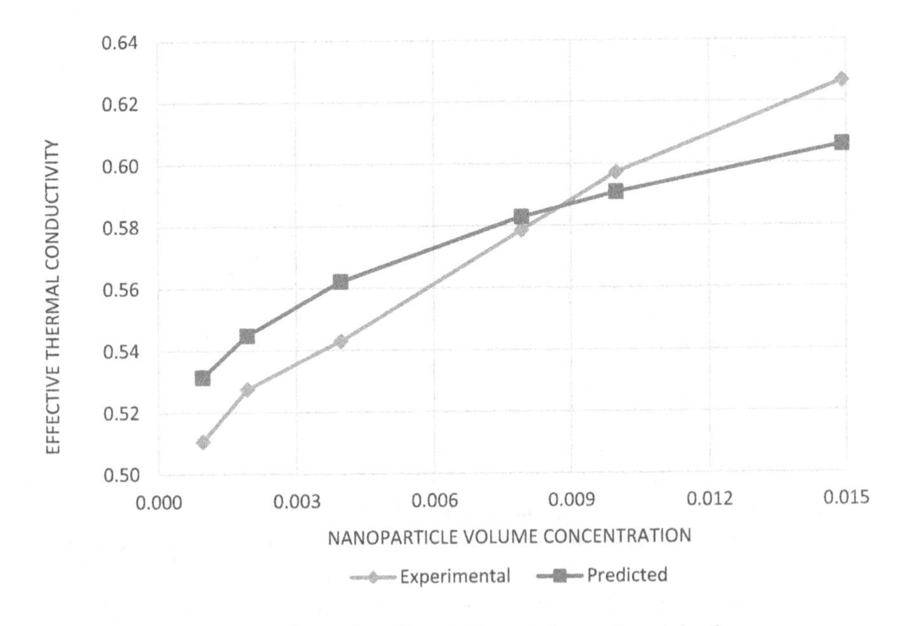

FIGURE 4.36 The comparison of predicted values and experimental values.

4.12 MODEL BY HEMMAT ESFE ET AL.

Hemmat Esfe et al. [58] dispersed carbon nanotubes with a mean average of 5–15 nm and alumina nanoparticles with a mean diameter of 20 nm into the water using an ultrasonic vibrator and a magnetic mixer to measure the thermal conductivity of nanofluids experimentally. Experiments were conducted with various solid volume concentrations of 0.02%, 0.04%, 0.1%, 0.2%, 0.4%, 0.8%, and 1.0% and various fluid temperatures of 303, 314, 323, and 332 K. Measured data reveal that the thermal conductivity of nanofluids highly depends on the solid volume concentration. They finally proposed Eq. (4.26) to predict the effective thermal conductivity of nanofluids. It should be noted that their proposed correlation is dimensionally consistent. Figure 4.37 shows the result of sensitivity analysis indicating that the nanofluid temperature is the least influential variable on the prediction of effective thermal conductivity.

$$\frac{k_{\text{eff}}}{k_{\text{b}}} = \frac{-214.83 + T}{346.58 - 106.98\varphi} + \frac{227.69}{T} \tag{4.26}$$

Thousand twenty-nine different responses were extracted from their model within their valid ranges of variables and then categorized into seven unique datasets. The statistical analysis gives *P*-values of 1.000 and 1.000 for mean and variances, respectively. This means that there is no statistically significant difference between the seven datasets. So, changing the considerable temperature does not affect responses considerably. Figure 4.38 shows the results of statistical analysis of variances of different datasets.

By using the nonlinear regression method, a simple correlation is proposed to predict effective thermal conductivity based on experimental data. Eq. (4.27) gives

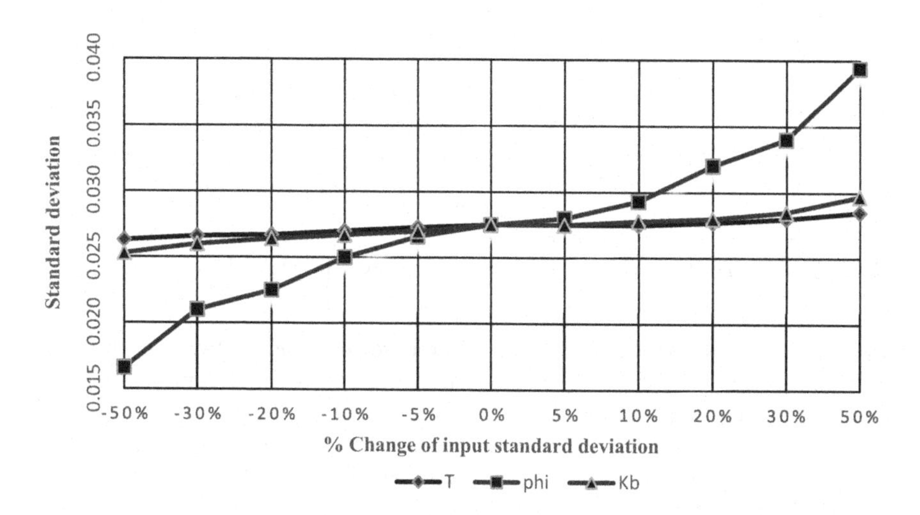

FIGURE 4.37 The result of sensitivity analysis on the proposed model by Hemmat Esfe et al. [58].

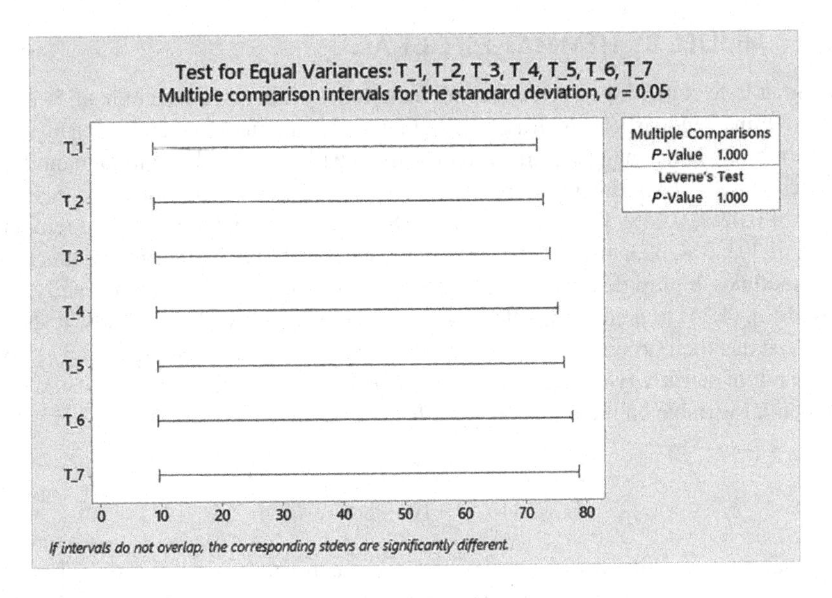

FIGURE 4.38 The result of the test for equal variances of seven datasets.

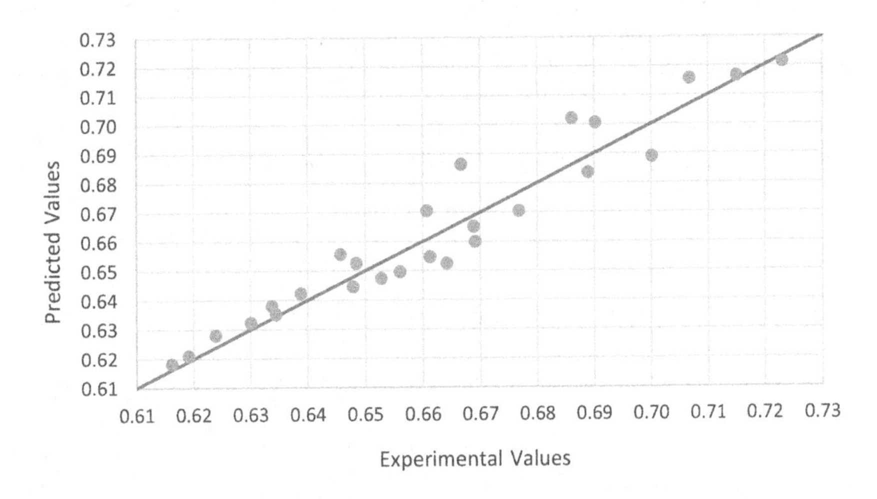

FIGURE 4.39 The fitness of predicted values on the experimental values.

an average absolute error of 1.11%. So, the proposed correlation is very accurate and well-fitted. Figures 4.39 and 4.40 demonstrate the fitness and accuracy of the proposed correlation (Eq. 4.27).

$$k_{eff} = k_b(1 + 5.72977\,\varphi^{0.806027}) \tag{4.27}$$

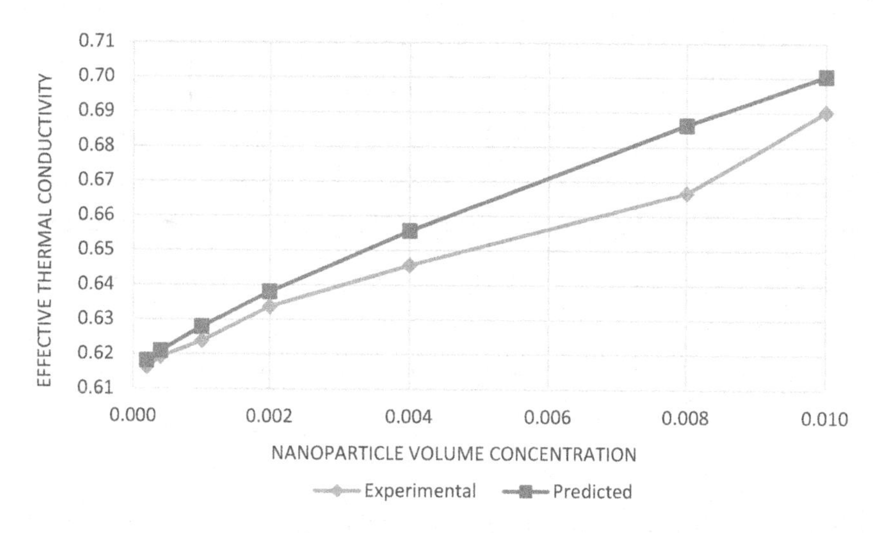

FIGURE 4.40 The comparison of predicted values and experimental values.

4.13 MODEL BY HARANDI ET AL.

Harandi et al. [59] conducted an experimental study on the thermal conductivity of hybrid nanofluids. They considered the effect of nanofluids' bulk temperature and volume concentration. They dispersed F-MWCNT-Fe_3O_4 nanoparticles in ethylene glycol using an ultrasonic processor. The experiments were carried out for a solid volume concentration range of 0%–2.3% in temperatures ranging from 25°C to 50°C. They finally developed a correlation (Eq. 4.28) using their experimental data to predict effective thermal conductivity.

$$\frac{k_{\text{eff}}}{k_{\text{b}}} = 1 + 0.0162\,\varphi^{0.7038} T^{0.6009} \tag{4.28}$$

The dimensional form of their proposed correlation $\left([M][L][T]^{-3}[\theta]^{-1}\right) = [\theta]^{0.6009}$ represents that this model is dimensionally inconsistent. Two sides of the equations do not give consistent dimensions. A sensitivity analysis of their proposed model is conducted, and it is concluded that the role of nanofluids temperature on the response can be neglected (Figure 4.41).

Eight hundred and twenty-eight responses were extracted from their model in the valid range of variables and then categorized into six unique datasets. The statistical analysis gives P-values of 0.997 and 1.000 for the mean and variances of seven unique datasets. So, the temperature does not play a vital role in predicting effective thermal conductivity in their model. Figure 4.42 shows the result of the test for equal variances of six unique datasets.

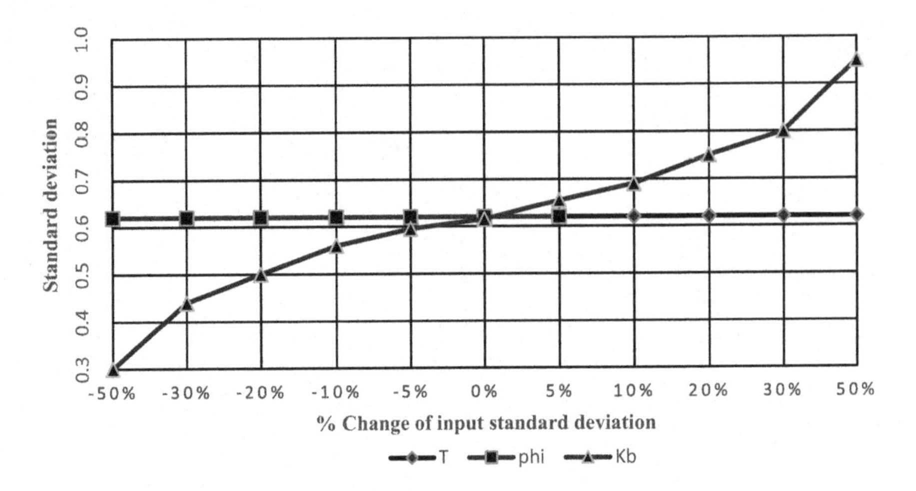

FIGURE 4.41 The result of sensitivity analysis on the proposed model by Harandi et al. [59].

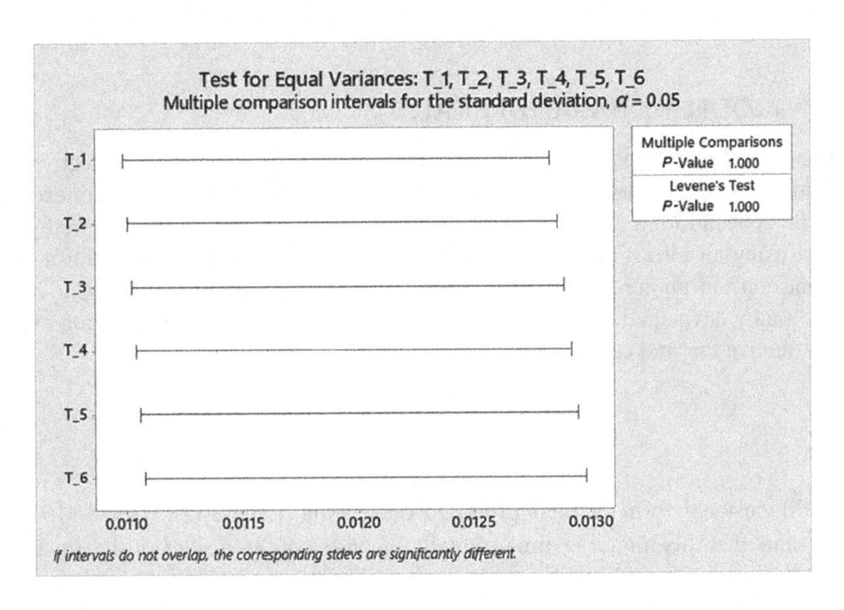

FIGURE 4.42 The result of the test for equal variances.

A simple correlation is proposed based on experimental data using the nonlinear regression method (Eq. 4.29).

$$k_{\text{eff}} = k_{\text{b}}(1 - 0.526495\varphi^{-0.0539237}) \tag{4.29}$$

The average absolute error of the proposed correlation is 2.7%. Figures 4.43 and 4.44 show the fitness and accuracy of the proposed correlation. It can be concluded that the proposed correlation is accurate and well-fitted.

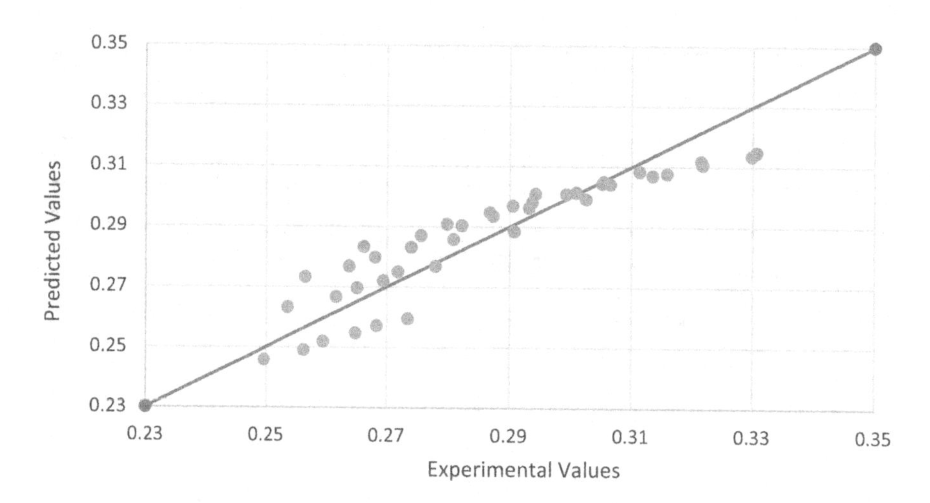

FIGURE 4.43 The fitness of predicted values on the experimental values.

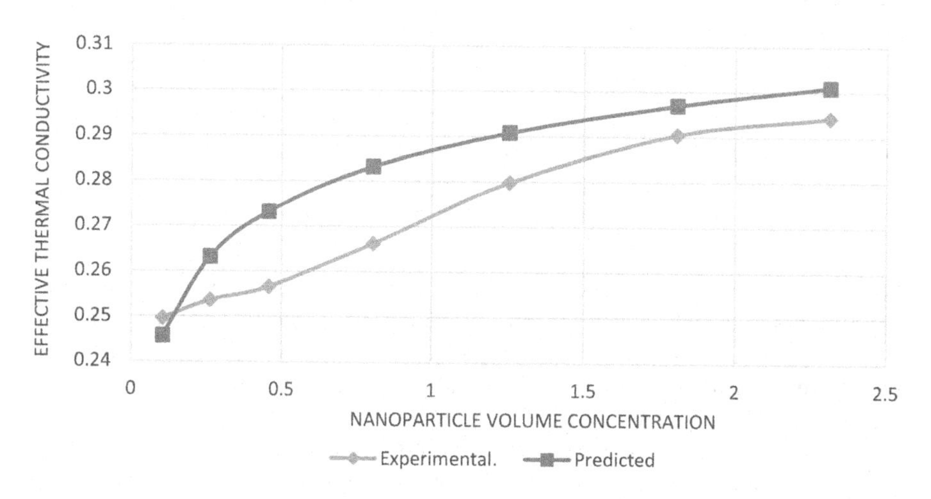

FIGURE 4.44 Comparison of the predicted values and experimental values.

4.14 MODEL BY ZADKHAST ET AL.

Another experimental study on the thermal conductivity of hybrid nanofluids has been performed by Zadkhast et al. [60]. They used an ultrasonic processor to disperse MWCNTs and CuO nanoparticles into the water with a volume concentration of 0.05%–0.6%. All thermal conductivity measurements are repeated three times in the range of 25°C–50°C. During the measurements, a hot water bath is used to stabilize the temperature at 25°C, 30°C, 35°C, 40°C, 45°C, and 50°C. They developed a mathematical model (Eq. 4.30) to predict the effective thermal conductivity of hybrid nanofluids using experimental data. Figure 4.45 represents the result of sensitivity analysis on their model. It can be concluded that the most and most minor affective

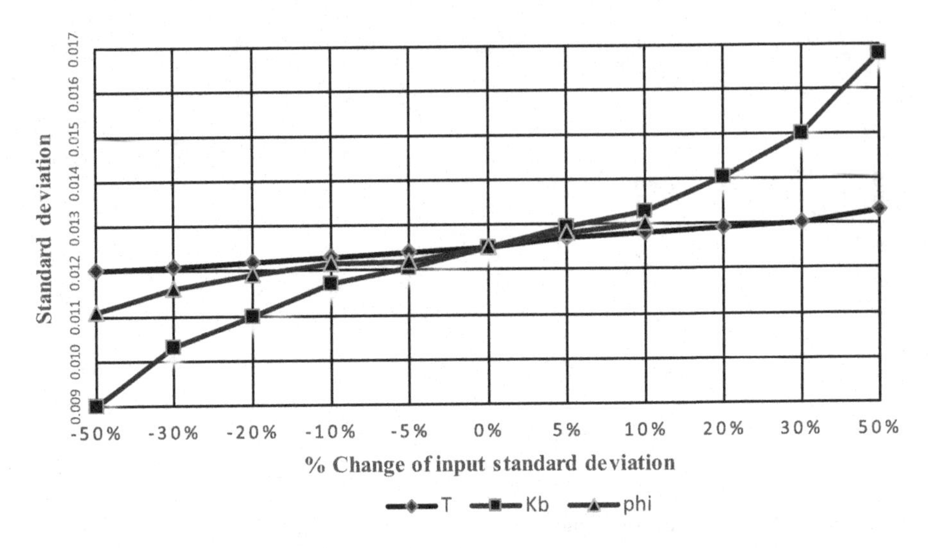

FIGURE 4.45 The result of the sensitivity analysis of the proposed model by Zadkhast et al. [60].

variables on the prediction of the effective thermal conductivity are the thermal conductivity of the base fluid and the temperature of the nanofluid.

$$\frac{k_{\mathrm{eff}}}{k_{\mathrm{b}}} = 0.907\exp(0.36\varphi^{0.3111} + 0.000956T) \tag{4.30}$$

Four hundred thirty-two responses were extracted from their model within their valid ranges and categorized into six unique datasets with different temperatures. The statistical analysis gives P-values of 0.065 and 1.000 for means and variances, respectively. Therefore, there is no statistically significant difference between datasets. Figure 4.46 shows the result of the test for equal variances.

The effort is made to propose a simple correlation using the nonlinear regression model based on experimental data. Equation (4.31) gives an average absolute error of 4.71%. Figures 4.47 and 4.48 also show the fitness and accuracy of the proposed correlation. So, the proposed correlation is accurate and well-fitted.

$$k_{\mathrm{eff}} = k_{\mathrm{b}}(1 + 0.574519\varphi) \tag{4.31}$$

4.15 MODEL BY KAKAVANDI AND AKBARI

Kakavandi and Akbari [61] measured the thermal conductivity of SiC-MWCNT/water–EG hybrid nanofluids experimentally. The volume concentration was in the range of 0%–0.75%. They finally introduced a correlation to predict the effective thermal conductivity of hybrid nanofluids (Eq. 4.32).

$$\frac{k_{\mathrm{eff}}}{k_{\mathrm{b}}} = 0.0017 \times \varphi^{0.698} \times T^{1.386} + 0.981 \tag{4.32}$$

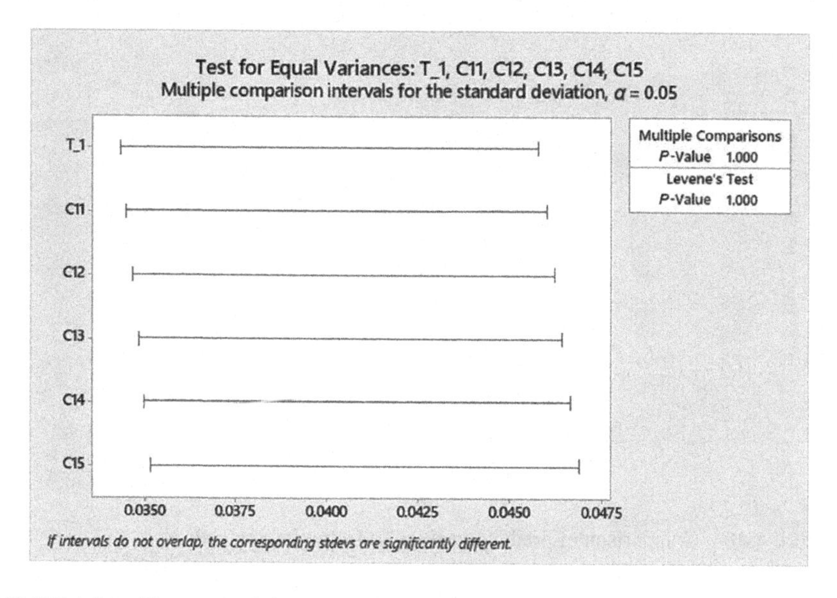

FIGURE 4.46 The result of the test for equal variances.

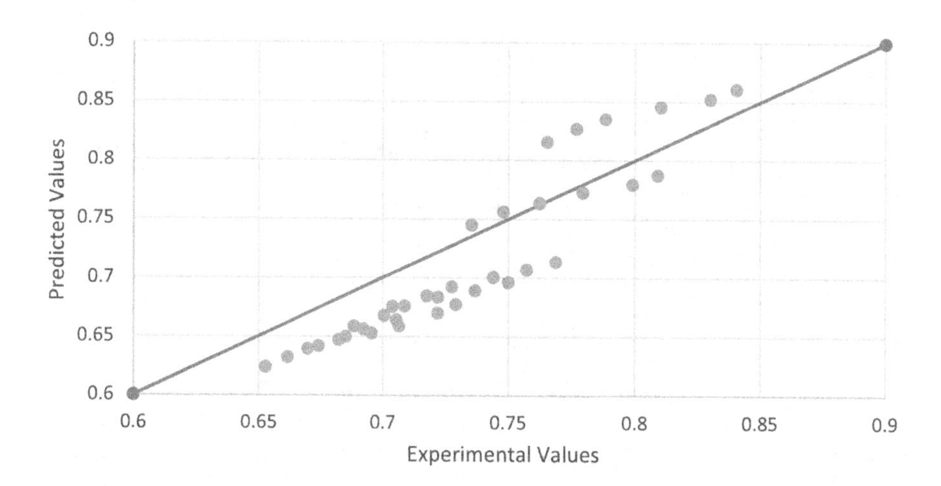

FIGURE 4.47 The fitness of predicted values versus experimental data.

Let's check the dimensional consistency of the model. The dimensional form of their proposed correlation $\left(\left([M][L][T]^{-3}[\theta]^{-1}/[M][L][T]^{-3}[\theta]^{-1}\right)=[\theta]^{1.386}\right)$ represents that this model is dimensionally inconsistent. Two sides of the equations do not give consistent dimensions. A sensitivity analysis of their proposed model is performed, indicating that the temperature of the nanofluid does not play a vital role in the prediction of the effective thermal conductivity (Figure 4.49).

The effort was made to extract as many as possible responses from their model. It categorized 576 responses into six unique datasets. The statistical analysis gives

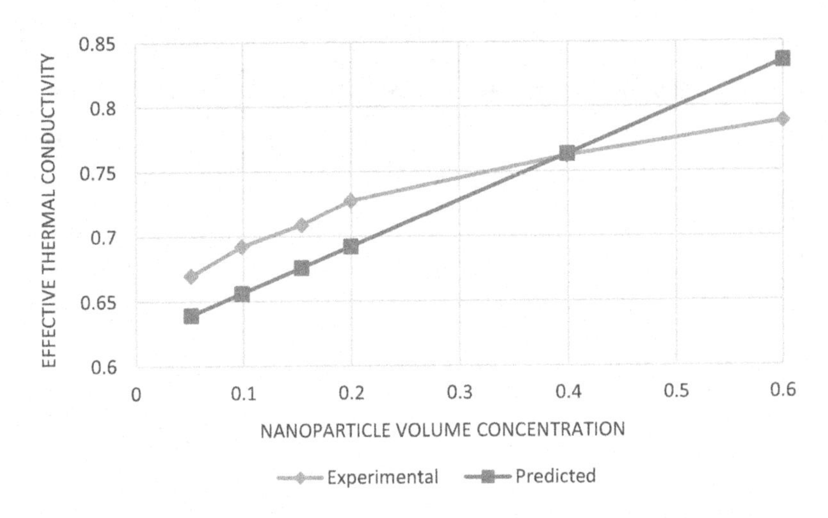

FIGURE 4.48 Comparison of predicted values and experimental values.

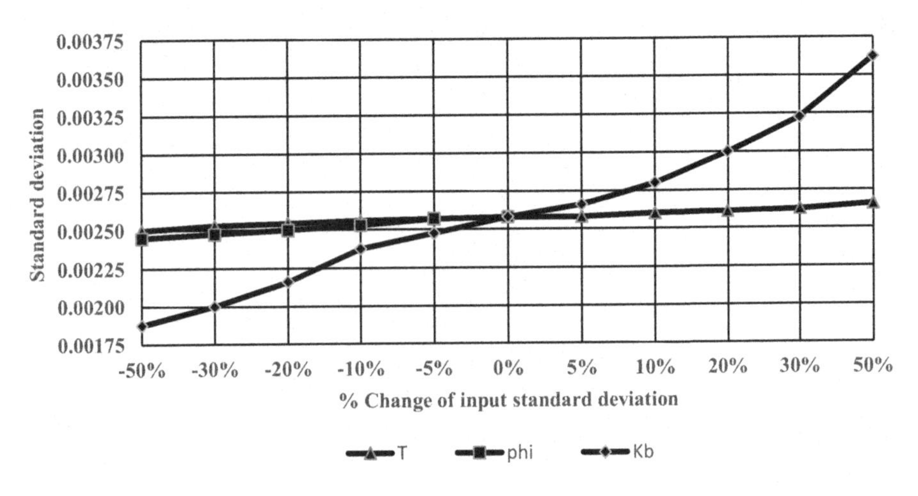

FIGURE 4.49 The result of the sensitivity analysis of the proposed model by Kakavandi and Akbari [61].

P-values of 0.998 and 0.632 for the means and variances of these datasets, respectively. Therefore, the test failed to reject the null hypothesis, so the temperature does not affect the response considerably. Figure 4.50 shows the result of the test for equal variances.

A simple correlation is proposed, which neglects the role of temperature using the nonlinear regression model based on experimental data with an average absolute error of 6.27% from experimental data since the proposed correlation (Eq. 4.33) is accurate and well-fitted enough. Figures 4.51 and 4.52 demonstrate the fitness and accuracy of the proposed correlation.

$$k_{\text{eff}} = k_{\text{b}}(1 + 39842\varphi^{2.61172}) \tag{4.33}$$

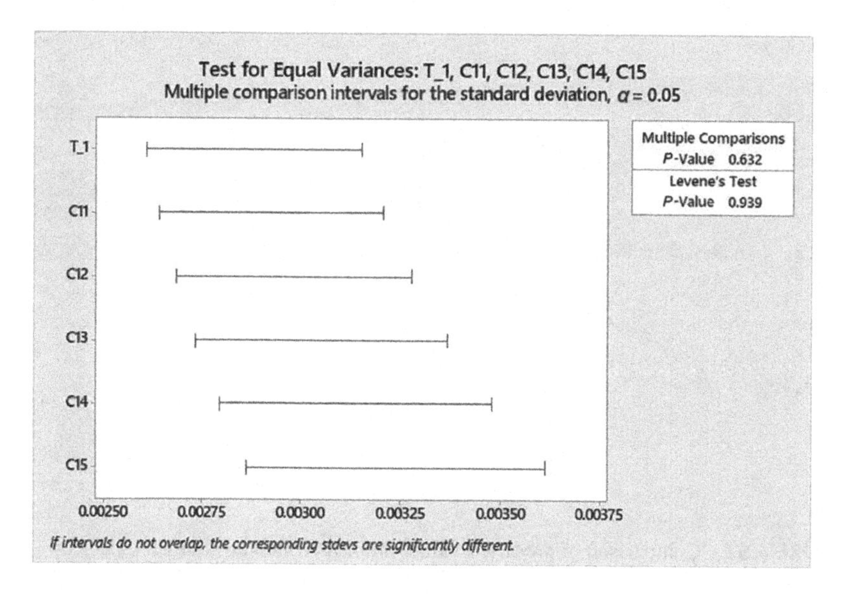

FIGURE 4.50 The result of the test for equal variances.

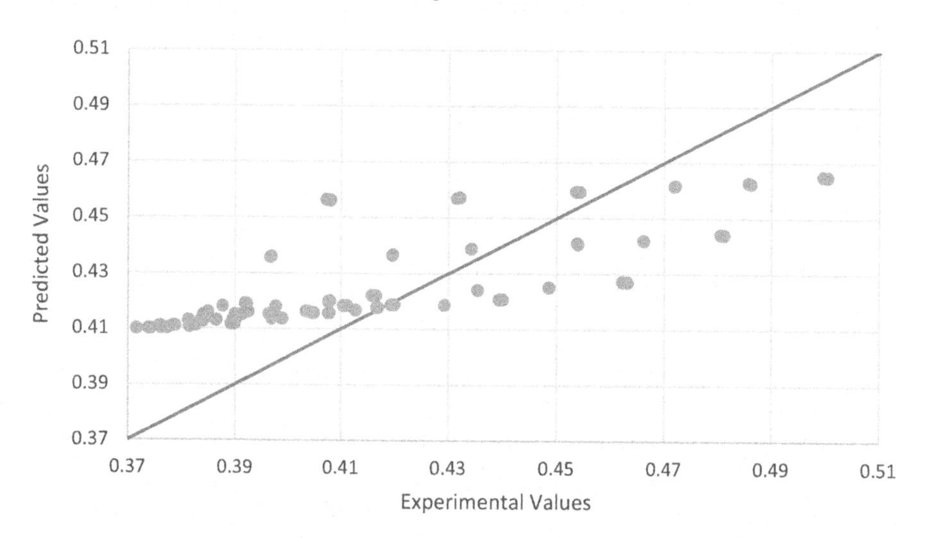

FIGURE 4.51 The fitness of predicted values versus experimental data.

4.16 MODEL BY KARIMI ET AL.

An experimental investigation of the thermal conductivity of Fe_3O_4 and $CoFe_2O_4$ – magnetic water nanofluids has been performed by Karimi et al. [62]. They synthesized nanoparticles using the co-precipitation method. The X-ray diffraction, electronic transmission microscopy, and vibration sample magnetometer are implemented to characterize nanoparticles' structure, size, and magnetic properties. They measured the thermal conductivity of nanofluids with volume concentrations in the

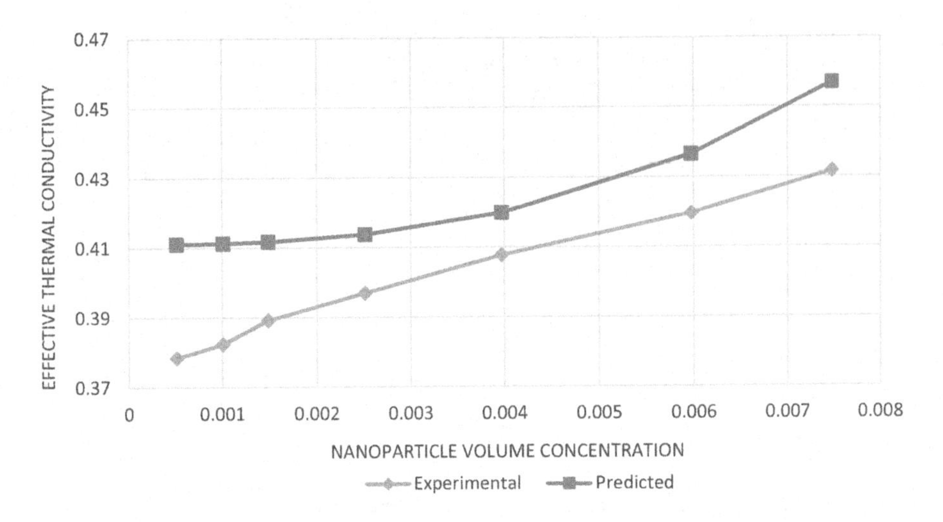

FIGURE 4.52 Comparison of predicted values and experimental values.

range of 0%–4.8%. Finally, they developed two correlations to predict the effective thermal conductivity of different nanofluids (Eqs. 4.34 and 4.35).

$$k_{eff} = \left(1 + 359.28\varphi\left(\frac{T}{T_{max}}\right)\right)^{0.06418} \times (-6.58\times10^{-6}T^2 + 0.0018T + 0.5694) \text{ for } Fe_3O_4 \tag{4.34}$$

$$k_{eff} = \left(1 + 418.81\varphi\left(\frac{T}{T_{max}}\right)\right)^{0.06748} \times (-6.58\times10^{-6}T^2 + 0.0018T + 0.5694) \text{ for } CoFe_2O_4 \tag{4.35}$$

Let's check the dimensional consistency of the models. The dimensional form of their proposed correlations $\left([M][L][T]^{-3}[\theta]^{-1} = [\theta]^2 + [\theta]\right)$ represents that these models are dimensionally inconsistent. Two sides of the equations do not give consistent dimensions. Figure 4.53 also demonstrates the result of the sensitivity analysis of their proposed correlations. As shown, the most and most minor affective variables on the response are the thermal conductivity of the base fluid and nanofluids temperature, respectively.

Four thousand eight hundred and fifty-one responses were extracted from these two models and categorized into nine unique datasets. The statistical analysis of these responses gives P-values of 0.982 and 0.930 for means and variances. It means that changing temperature does not affect responses considerably. Figure 4.54 presents the result of the test for equal variances.

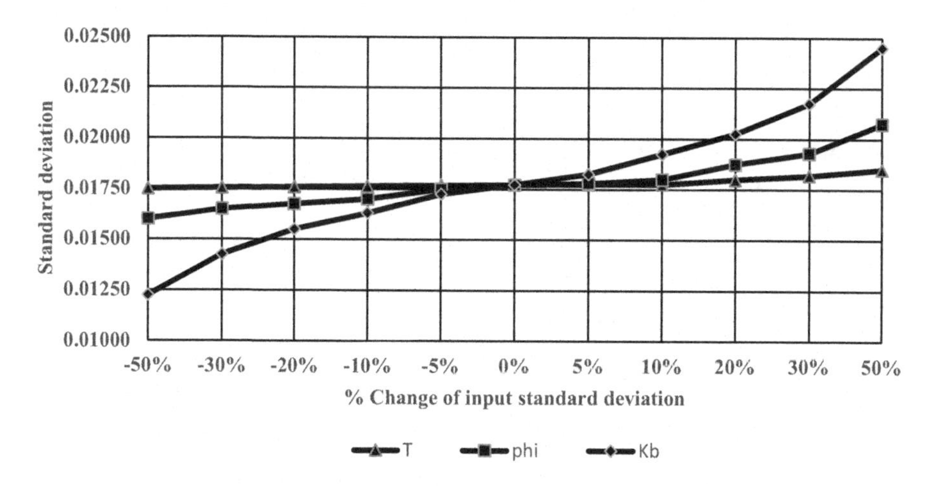

FIGURE 4.53 The result of sensitivity analysis of the proposed model by Karimi et al. [62].

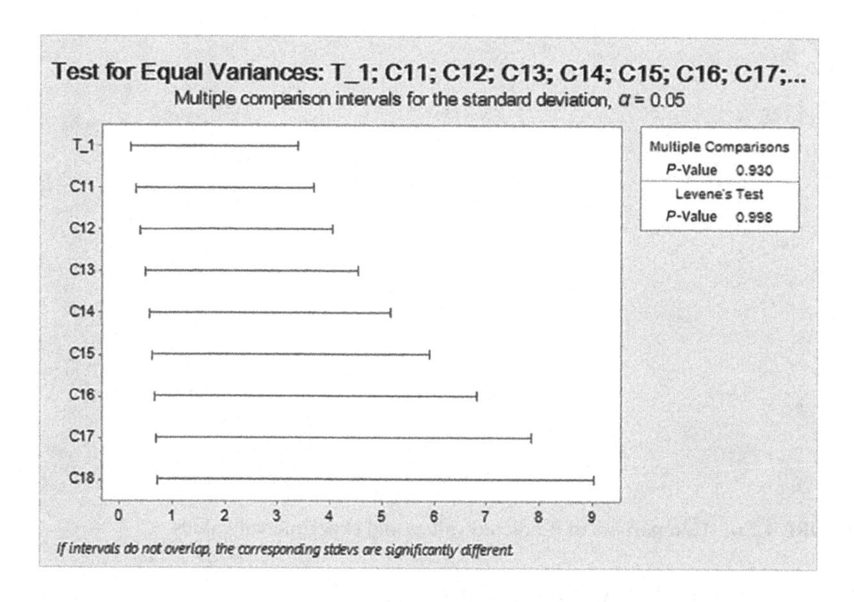

FIGURE 4.54 The result of the test for equal variances.

A simple correlation is proposed for the prediction of the effective thermal conductivity of nanofluids using the nonlinear regression method based on experimental data (Eq. 4.36).

$$k_{\text{eff}} = k_{\text{b}}(1 + 0.148349\varphi^{0.0614858}) \tag{4.36}$$

This correlation gives an average absolute error of 3.06% from experimental data. Figures 4.55 and 4.56 show the fitness and accuracy of the proposed correlation. It can be concluded that this correlation is accurate and well-fitted.

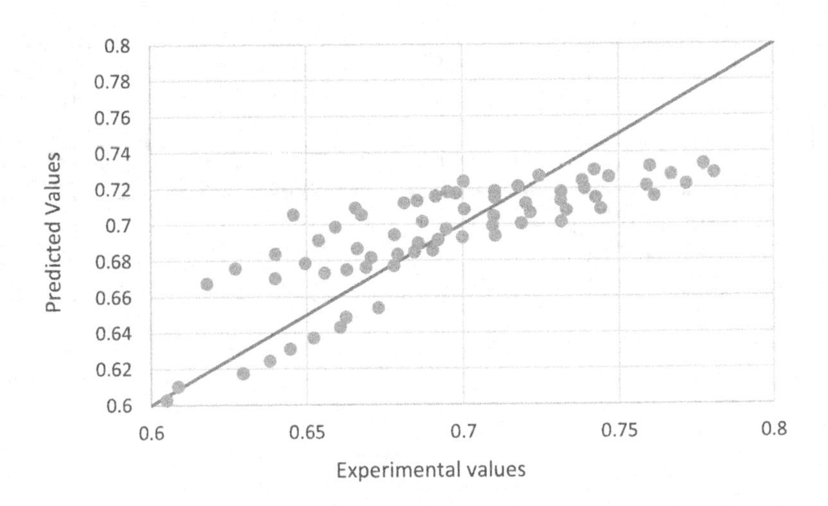

FIGURE 4.55 The fitness of predicted values versus experimental data.

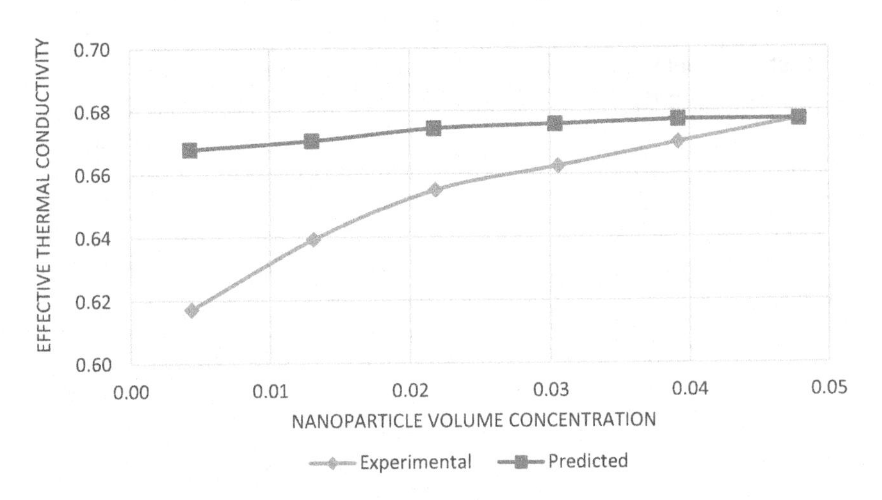

FIGURE 4.56 Comparison of predicted values and experimental values.

4.17 MODEL BY KARIMIPOUR ET AL.

Karimipour et al. [63] measured the thermal conductivity of CuO/liquid paraffin nanofluids in a different range of volume concentration (0.25 to 6%wt.) and bulk temperature (25°C–100°C), experimentally. They introduced a mathematical model to predict the effective thermal conductivity of nanofluids (Eq. 4.37). Figure 4.57 shows that the proposed model has a medium sensitivity to the temperature of the nanofluid. In comparison, it showed the highest sensitivity on the thermal conductivity of the base fluid.

$$\frac{k_{\text{eff}}}{k_{\text{b}}} = 0.792194 + 0.0547913\varphi + 0.00998805T + 0.000730423\varphi T - 0.00421237\varphi^2 - 0.0000643292T^2 \tag{4.37}$$

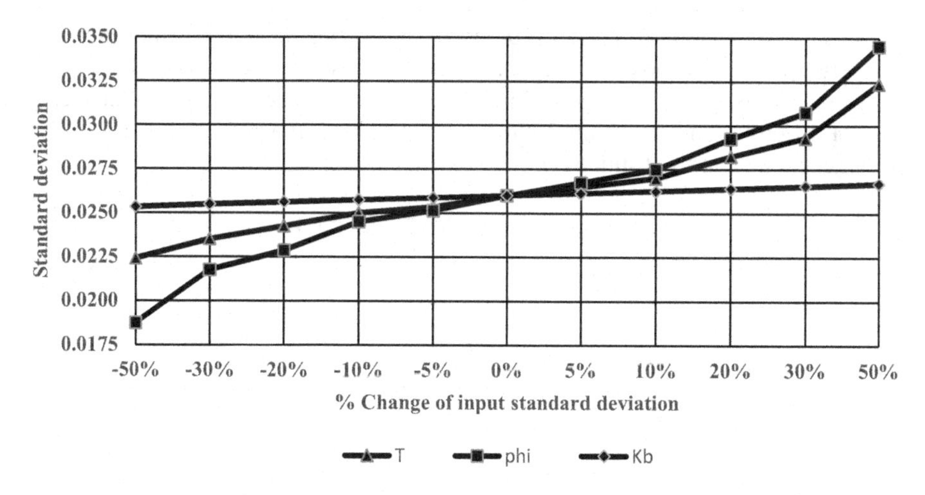

FIGURE 4.57 The result of the sensitivity analysis of the proposed model by Karimipour et al. [63].

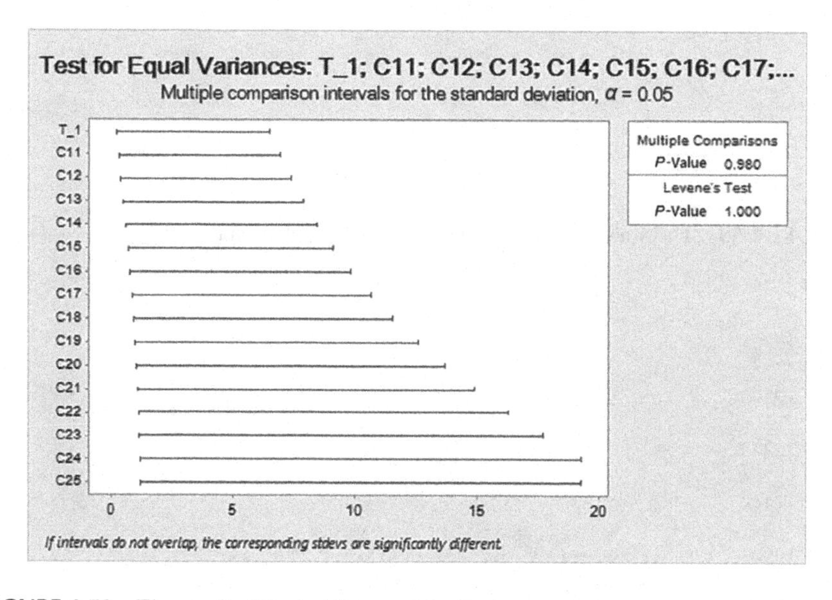

FIGURE 4.58 The result of the test for equal variances.

This case needs investigation, and therefore, 6144 responses were extracted from their model in the valid ranges of variables and categorized into 16 unique datasets. Our statistical analysis gives P-values of 1.000 and 0.980 for the means and variances of these datasets. Therefore, nanofluids' temperature does not play a vital role in the estimation of effective thermal conductivity. Figure 4.58 demonstrates the result of the test for equal variances.

A simple correlation (Eq. 4.38) is proposed, neglecting the role of bulk nanofluids temperature using the nonlinear regression method based on experimental data with an average absolute error of 7.34%. Figures 4.59 and 4.60 show the fitness and accuracy of the proposed correlation. It can be concluded that the proposed correlation is accurate enough and well-fitted.

$$k_{\text{eff}} = k_b(1+2.13017\varphi^{0.500216}) \tag{4.38}$$

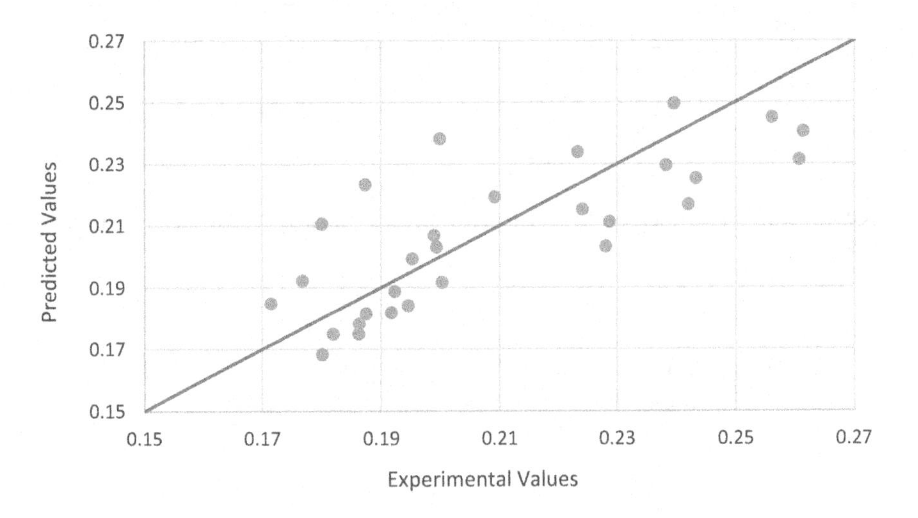

FIGURE 4.59 The fitness of predicted values versus experimental values.

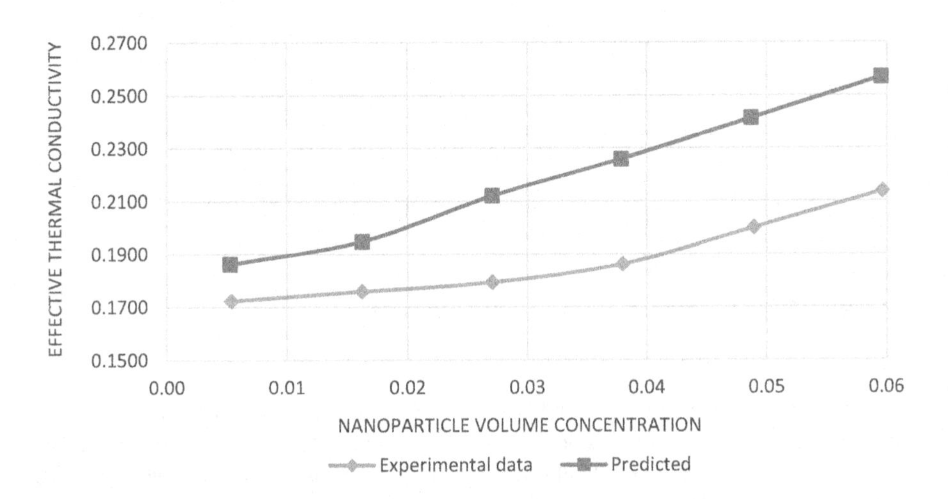

FIGURE 4.60 Comparison of predicted values and experimental values.

4.18 MODEL BY RANJBARZADEH ET AL.

Ranjbarzadeh et al. [64] conducted an experimental study on the water-silica nanofluids with an eco-friendly method of nanoparticles production recently. Their temperature and nanoparticle volume concentration ranges are 20°C–55°C and 0.1%–3%, respectively. They proposed a correlation using Levenberg–Marquardt algorithm, based on experimental data (Eq. 4.39):

$$\frac{k_{\text{eff}}}{k_{\text{b}}} = 1 + 0.4281\left(\frac{T}{100}\right)^{1.707} \varphi^{0.8449} \tag{4.39}$$

Figure 4.61 shows the result of sensitivity analysis for their correlation. It can be concluded that the thermal conductivity of the base fluid is the most influential variable in effective thermal conductivity prediction, and the response is not much sensitive to the temperature of the nanofluid. For further investigation, 441 different responses were extracted from their correlation in the range mentioned above of variables. They categorized them into seven unique datasets with different temperatures. *P*-values of 0.998 and 0.960 are computed for means and variances, respectively, using statistical analysis. It means that when considering the temperature of the nanofluid is changed, the responses do not change significantly. Figure 4.62 represents the result of the test for equal variances. So, a simple correlation is proposed, using the nonlinear regression method, using experimental data (Eq. 4.40). This correlation predicts the effective thermal conductivity of nanofluids with an average absolute error of 3.44%. Figures 4.63 and 4.64 also demonstrate the fitness and accuracy of the proposed correlation (Eq. 4.40). It can be concluded that the proposed correlation is simple, accurate, and well-fitted.

$$k_{\text{eff}} = k_{\text{b}}(1 + 5.02743\varphi^{0.867218}) \tag{4.40}$$

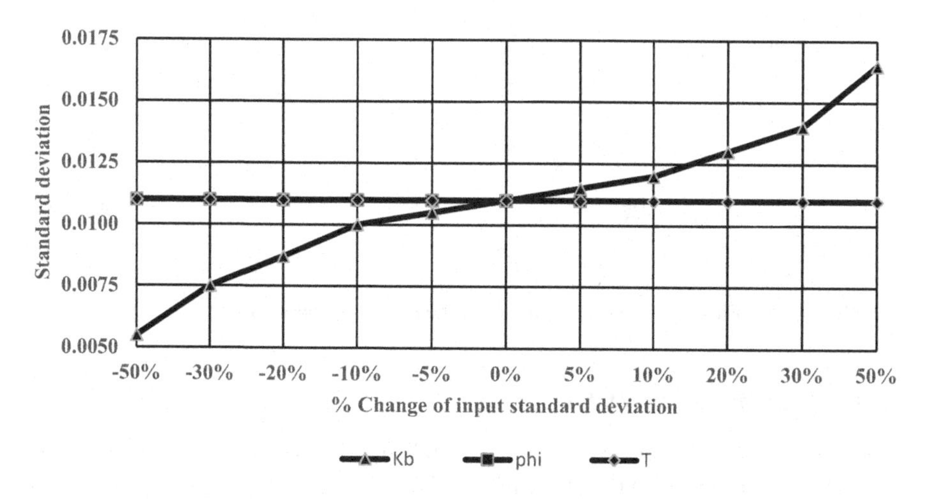

FIGURE 4.61 The result of the sensitivity analysis of the proposed correlation by Ranjbarzadeh et al. [64].

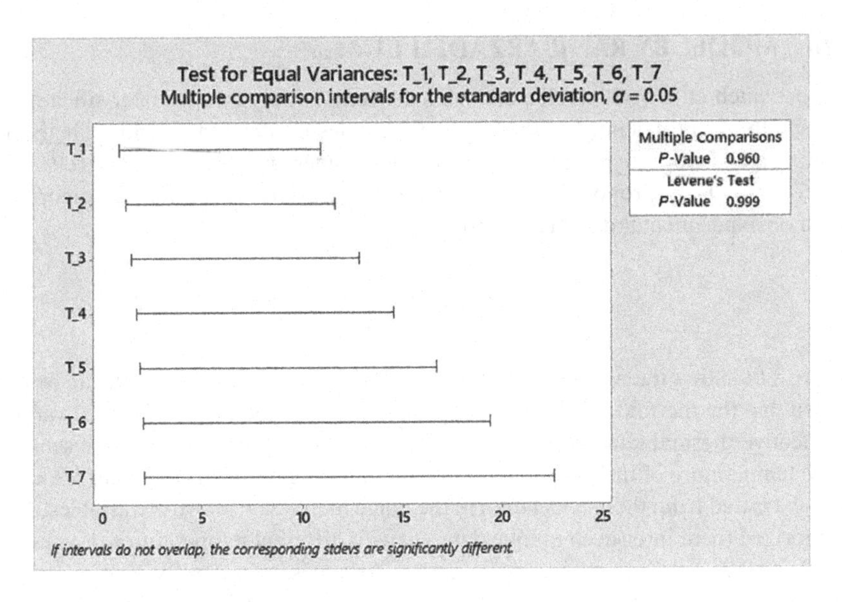

FIGURE 4.62 The result of the test for equal variances.

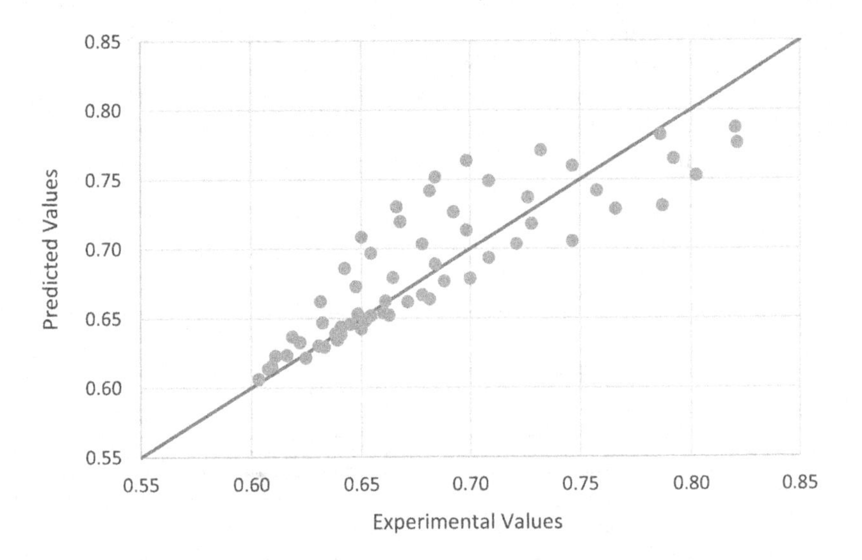

FIGURE 4.63 Fitness of predicted values on the experimental values.

4.19 MODEL BY KEYVANI ET AL.

Keyvani et al. [65] conducted an experimental study on the measurement of cerium (CeO_2)-ethylene glycol nanofluids, probably for the first time. The samples were made in the volume concentration range of 0.25%–2.5% using a two-step method. Measurements were done for all samples at temperatures ranging from 25°C to 50°C.

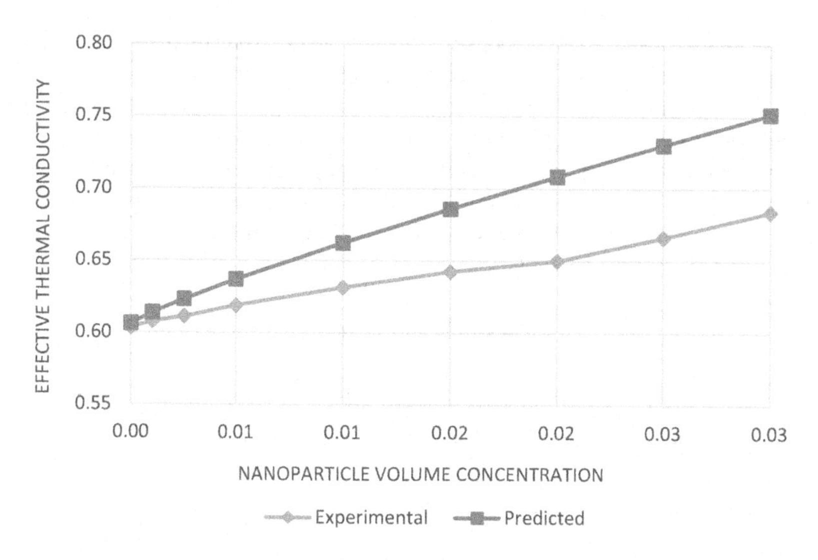

FIGURE 4.64 Comparison of predicted values and experimental values.

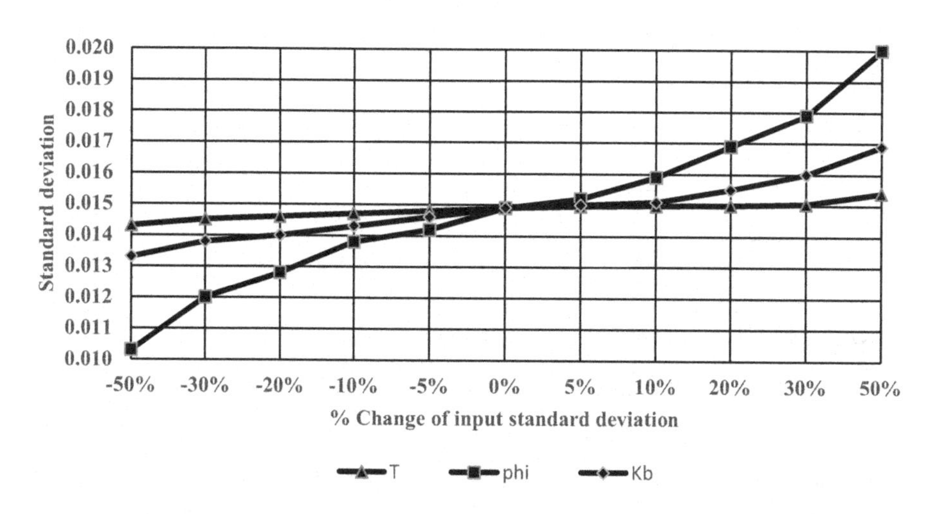

FIGURE 4.65 The result of the sensitivity analysis of the model of Keyvani et al. [65].

They eventually proposed a new correlation to predict the effective thermal conductivity of nanofluids (Eq. 4.41):

$$\frac{k_{\text{eff}}}{k_{\text{b}}} = 0.9320 + 0.0673\varphi + 0.0021T \tag{4.41}$$

Figure 4.65 shows the result of sensitivity analysis of their correlation. As shown, the response is a susceptible function of the nanoparticle volume concentration and relatively insensitive to the temperature of the nanofluid.

Three hundred and sixty responses were extracted from their model within the valid range of variables and categorized into six unique datasets for further investigation. The statistical analysis of the means and variances gives 0.997 and 0.984, respectively (Figure 4.66). This shows that there is no significant difference between datasets. A more straightforward correlation is proposed, neglecting the temperature of the nanofluid using experimental data (Eq. 4.42) with an average absolute error of 5.66% from experimental values. Figures 4.67 and 4.68 represent the fitness and accuracy of the proposed correlation.

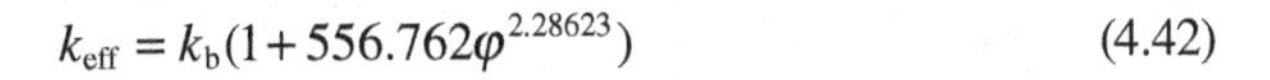

$$k_{eff} = k_b(1 + 556.762\varphi^{2.28623}) \tag{4.42}$$

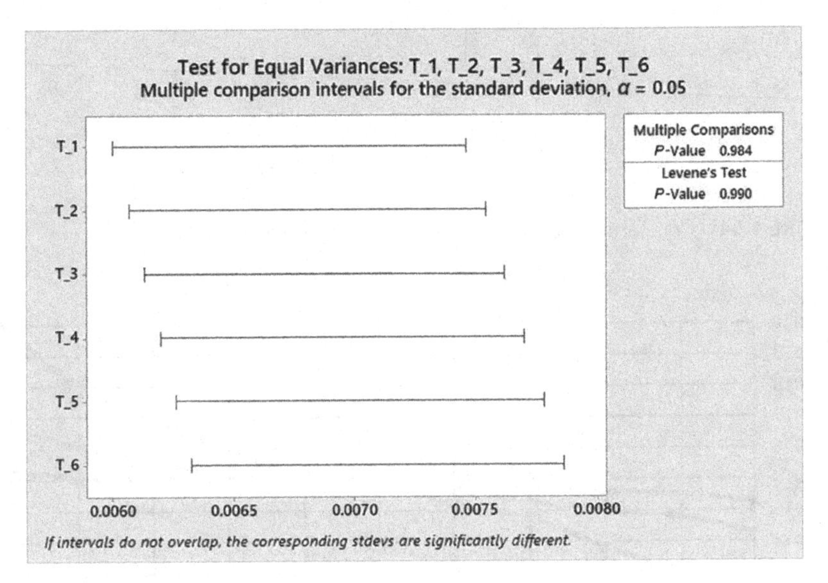

FIGURE 4.66 The result of the test for equal variances.

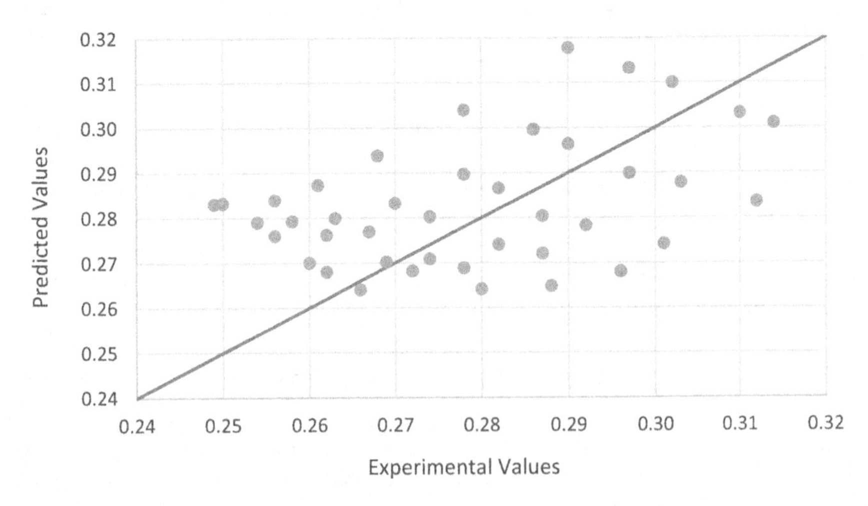

FIGURE 4.67 The fitness of predicted values on the experimental values.

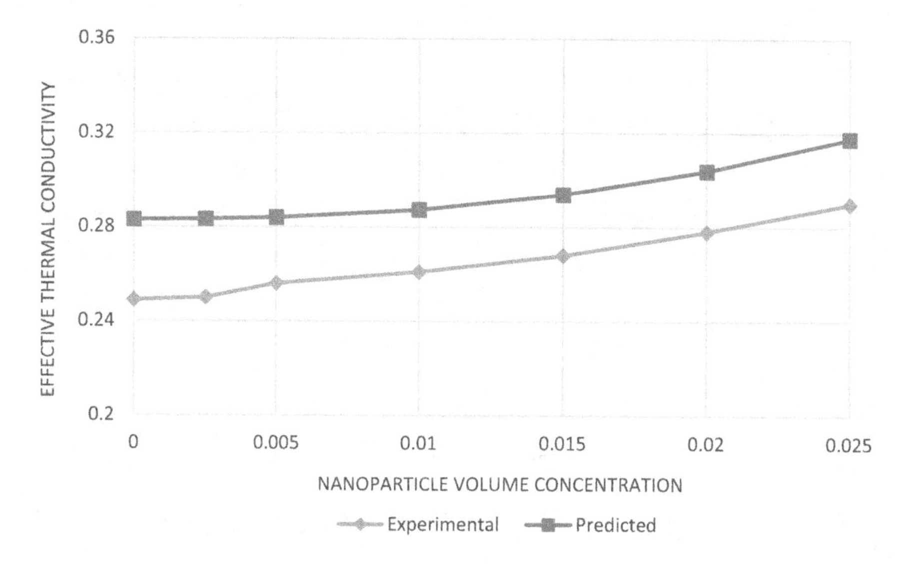

FIGURE 4.68 Comparison of experimental data and predicted values.

4.20 MODEL BY AFRAND ET AL.

An experimental study on the thermal conductivity of Fe_3O_4–water nanofluids has been performed by Afrand et al. [66] with the nanoparticle volume concentrations of 0.1%, 0.2%, 0.4%, 1%, 2%, and 3% under temperatures ranging from 20°C to 55°C. They introduced a new correlation for the prediction of effective thermal conductivity of nanofluids (Eq. 4.43).

$$\frac{k_{\text{eff}}}{k_{\text{b}}} = 0.9320 + 0.0673\varphi^{0.323}T^{0.245} \tag{4.43}$$

Figure 4.69 shows the result of the sensitivity analysis of their proposed correlation. The most and most minor affective variables on effective thermal conductivity are nanoparticle volume concentration and nanofluids temperature.

1920 responses were extracted from their model and categorized into eight unique datasets. Our statistical analysis gives *P*-values of 0.937 and 0.926 for means and variances of datasets, respectively (Figure 4.70). An effort was made to propose replacing more direct correlation ignoring the role of nanofluids temperature. The proposed correlation gives an average absolute error of 5.98%. Figures 4.71 and 4.72 show the fitness and accuracy of the proposed correlation.

$$k_{\text{eff}} = k_{\text{b}}(1 + 2.42534\varphi^{0.241448}) \tag{4.44}$$

Table 4.2 lists the published experimental model for the effective thermal conductivity of nanofluids considering the nanofluids' bulk temperature. *P*-values for means and variances are listed based on statistical analysis. The proposed replacing

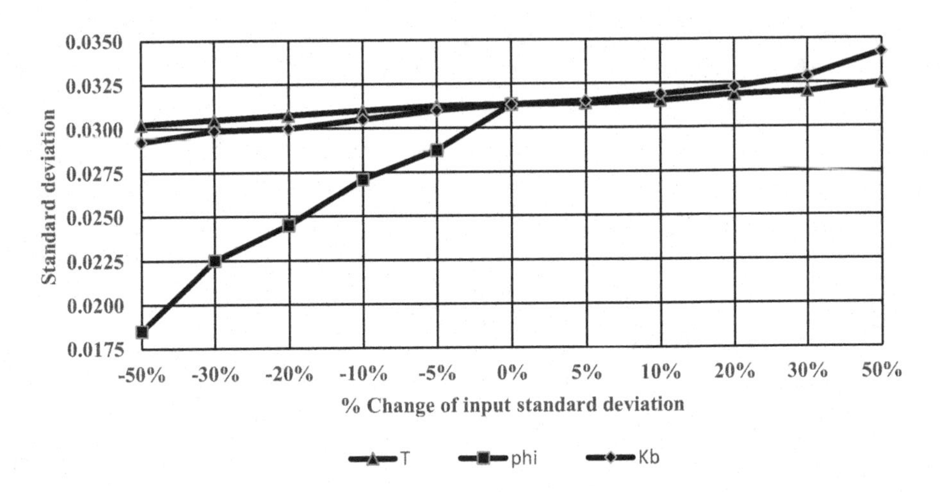

FIGURE 4.69 The result of the sensitivity analysis of the model introduced by Afrand et al. [66].

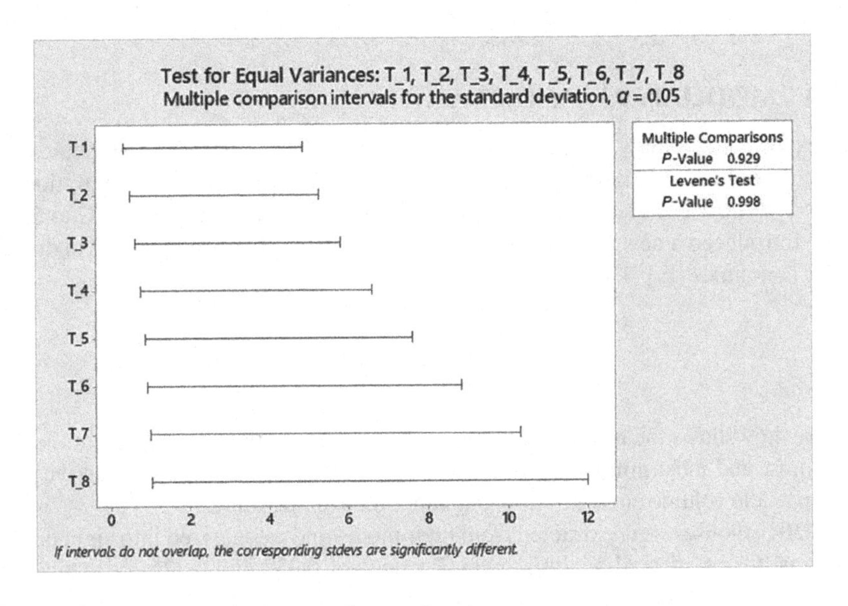

FIGURE 4.70 The result of the test for equal variances.

correlation (neglecting the role of temperature) for each model is shown in a separate column. The average absolute error and the maximum deviation for each correlation also are mentioned. All proposed correlations have been proven simple, accurate, reliable, and well-fitted.

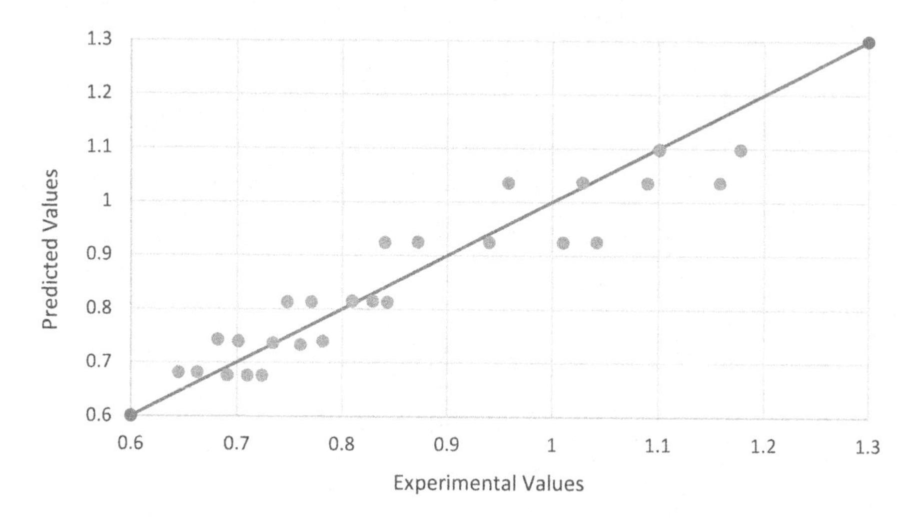

FIGURE 4.71 The fitness of predicted values on the experimental values.

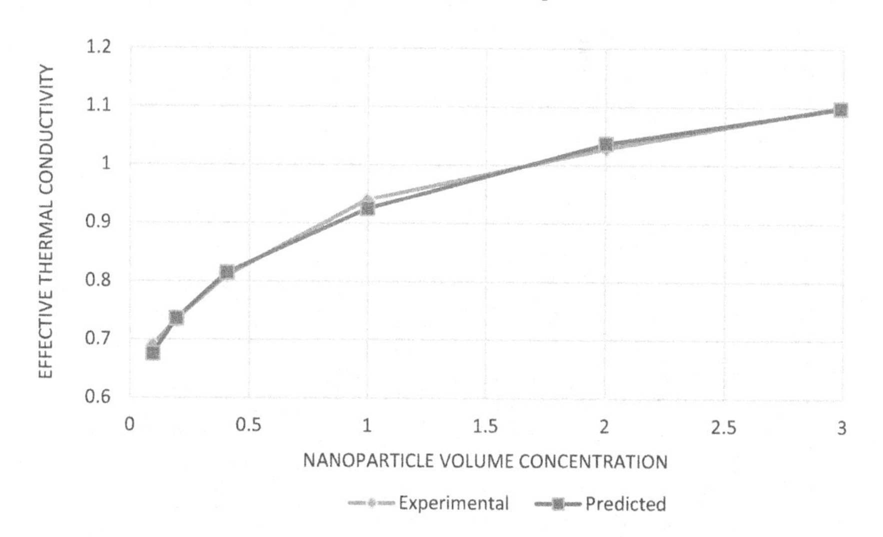

FIGURE 4.72 Comparison of predicted values and experimental values.

TABLE 4.2
The Summary of Statistical Analysis on the Published Models and Their Proposed Replacing Correlations

Ref.		Model	*P*-Value for Means	*P*-Value for Variances	Proposed Correlation	Average Error (%)	Maximum Error (%)
1	Aberoumand et al. [50]	$k_{eff} = (3.9\times10^{-5}T - 0.0305)\varphi^2 + (0.086 - 1.6\times10^{-4}T)\varphi + 3.1\times10^{-4}T + 0.129 - 5.77\times10^{-6}k_p - 40\times10^{-4}$	0.963	0.332	$k_{eff} = k_b(1 + 1.25536\varphi^{0.29397})$	3.62	8.82
2	Fakoor Pakdaman et al. [51]	$\frac{k_{eff}}{k_b} = 1 + 304.47(1+\varphi)^{136.35}\exp(-0.021T)\left(\frac{1}{d_p}\right)^{0.369}\left(\frac{T^{1.2321}}{10^{610.6287/(T-140)}}\right)$	1.000	1.000	$k_{eff} = k_b\left(1 + \left(\frac{2.73465\varphi}{d_p}\right)^{0.270822}\right)$	2.41	14.04
3	Ahammed et al. [52]	$\frac{k_{eff}}{k_b} = 2.5T_r^{1.032}\varphi^{0.108}$	0.772	0.762	$\frac{k_{eff}}{k_b} = 2.5T_r^{1.032}\varphi^{0.108}$	3.31	7.37
4	Patel et al. [53]	$k_{eff} = k_b\left(1 + 0.135\times\left(\frac{k_p}{k_b}\right)^{0.273}\varphi^{0.467}\left(\frac{T}{20}\right)^{0.547}\left(\frac{100}{d_p}\right)^{0.234}\right)$	0.000	0.938	$k_{eff} = k_b\left(1 + \frac{0.232706\varphi^{0.402458}\times k_p}{d_p}\right)$	3.44	10.68
5	Li and Peterson [54]	$\frac{k_{eff}}{k_b} = 0.7644815\varphi + 0.018689T + 0.537853$, for Al_2O_3	0.736	0.927	$k_{eff} = k_b(1 + 0.81874\varphi^{0.56})$ for Al_2O_3	4.88	11.41
		$\frac{k_{eff}}{k_b} = 3.761088\varphi + 0.017924T + 0.69266$, for CuO	0.755	0.925	$k_{eff} = k_b(1 + \varphi^{0.255643})$ for CuO	2.79	7.74
6	Hemmat Esfe et al. [55]	$k_{eff} = 0.4 + 0.0332\varphi + 0.00101T + 0.000619\varphi T + 0.0687\varphi^3 + 0.0148\varphi^5 - 0.00218\varphi^6 - 0.0419\varphi^4 - 0.0604\varphi^2$	0.995	0.927	$k_{eff} = k_b(1 + 15.9106\varphi^{1.26896})$	3.5	10.26

(*Continued*)

TABLE 4.2 (Continued)
The Summary of Statistical Analysis on the Published Models and Their Proposed Replacing Correlations

Ref.	Model	*P*-Value for Means	*P*-Value for Variances	Proposed Correlation	Average Error (%)	Maximum Error (%)
7 Hemmat Esfe et al. [56]	$\frac{k_{eff}}{k_b}=1.04+5.91\times10^{-5}T+0.00154T\varphi+0.0195\varphi^2 -0.014\varphi-0.00253\varphi^3-0.000104T\varphi^2-0.0357 \times\sin(1.72+0.407\varphi^2-1.67\varphi)$	0.971	1.000	$k_{eff}=k_b(1+7.05149\varphi)$	1.69	6.84
	$\frac{k_{eff}}{k_b}=0.999+9.581\times10^{-5}T+0.00142T\varphi+0.0519\varphi^2 +0.00208\varphi^2+0.00208\varphi^4-0.00719\varphi-0.0193\varphi^3 -8.21\times10^{-5}T\varphi^2$	0.880	1.000			
8 Hemmat Esfe et al. [57]	$\frac{k_{eff}}{k_b}=1.04+0.000589T+\frac{-0.000184}{T\varphi} \times\cos(6.11+0.00673T+4.41T\varphi-0.0414\sin(T))-32.5\varphi$	0.999	0.976	$k_{eff}=k_b(1+\varphi^{0.30409})$	3.86	8.53
9 Hemmat Esfe et al. [58]	$\frac{k_{eff}}{k_b}=\frac{-214.83+T}{346.58-106.98\varphi}+\frac{227.69}{T}$	1.000	1.000	$k_{eff}=k_b(1+5.72977\varphi^{0.806027})$	1.11	3.43
10 Harandi et al. [59]	$\frac{k_{eff}}{k_b}=1+0.0162\varphi^{0.7038}T^{0.6009}$	0.997	1.000	$k_{eff}=k_b(1-0.526495\varphi^{-0.0539237})$	2.70	7.59
11 Zadkhast et al. [60]	$\frac{k_{eff}}{k_b}=0.907\exp(0.36\varphi^{0.3111}+0.000956T)$	0.065	1.000	$k_{eff}=k_b(1+0.574519\varphi)$	4.71	7.20

(Continued)

TABLE 4.2 (*Continued*)
The Summary of Statistical Analysis on the Published Models and Their Proposed Replacing Correlations

Ref.	Model	*P*-Value for Means	*P*-Value for Variances	Proposed Correlation	Average Error (%)	Maximum Error (%)
12 Kakavandi and Akbari [61]	$\frac{k_{eff}}{k_b} = 0.0017 \times \varphi^{0.698} \times T^{1.386} + 0.981$	0.998	0.632	$k_{eff} = k_b(1 + 39842\varphi^{2.61172})$	6.27	32.51
13 Karimi et al. [62]	$k_{eff} = \left(1 + 359.28\varphi\left(\frac{T}{T_{max}}\right)\right)^{0.06418} \times(-6.58 \times 10^{-6}T^2 + 0.0018T + 0.5694)$ for Fe_3O_4 $k_{eff} = \left(1 + 418.81\varphi\left(\frac{T}{T_{max}}\right)\right)^{0.06748} \times(-6.58 \times 10^{-6}T^2 + 0.0018T + 0.5694)$ for $CoFe_2O_4$	0.982	0.930	$k_{eff} = k_b(1 + 0.148349\varphi^{0.0614858})$	3.06	9.21
14 Karimipour et al. [63]	$\frac{k_{eff}}{k_b} = 0.792194 + 0.0547913\varphi + 0.00998805T + 0.000730423\varphi T - 0.00421237\varphi^2 - 0.0000643292T^2$	1.000	0.980	$k_{eff} = k_b(1 + 2.13017\varphi^{0.500216})$	7.34	19.13
15 Ranjbarzadeh et al. [64]	$\frac{k_{eff}}{k_b} = 1 + 0.4281\left(\frac{T}{100}\right)^{1.707}\varphi^{0.8449}$	0.998	0.960	$k_{eff} = k_b(1 + 5.02743\varphi^{0.867218})$	3.44	10.24
16 Keyvani et al. [65]	$\frac{k_{eff}}{k_b} = 0.9320 + 0.0673\varphi + 0.0021T$	0.997	0.984	$k_{eff} = k_b(1 + 556.762\varphi^{2.28623})$	5.66	13.65
17 Afrand et al. [66]	$\frac{k_{eff}}{k_b} = 0.9320 + 0.0673\varphi^{0.323}T^{0.245}$	0.937	0.926	$k_{eff} = k_b(1 + 2.42534\varphi^{0.241448})$	5.98	14.16

4.21 A COMPREHENSIVE CORRELATION

In order to propose a comprehensive correlation, a statistical look at the sensitivity analysis result of published models is needed. As shown in Table 4.3, the nanofluid temperature is the most frequent variable with the title "Least Affective Variable." Since the role of nanofluids temperature is ignored, 603 experimental data are used, extracted from ten published experimental papers (all mentioned papers except those related to hybrid nanofluids) to correlate a comprehensive mathematical model, named MAG (Molana-Armaghani-Ghasemiasl) correlation. Equation (4.39) has been derived employing the nonlinear regression method to predict the effective thermal conductivity of nanofluids:

$$k_{\mathrm{eff}} = k_{\mathrm{b}}(1+1.9647\varphi^{0.410499}\, k_{\mathrm{Pr}}{}^{0.169384}\, d_{\mathrm{Pr}}{}^{-0.244673})$$

$$k_{\mathrm{Pr}} = \frac{k_{\mathrm{p}}}{k_{\mathrm{Pref}}}, \quad d_{\mathrm{Pr}} = \frac{d_{\mathrm{p}}}{d_{\mathrm{Pref}}} \tag{4.39}$$

TABLE 4.3
The Summary of Sensitivity Analysis of the Published Models

Ref.	Most Affective Variable	Least Affective Variable
Aberoumand et al. [50]	Volume concentration	Thermal conductivity of nanoparticles
Fakoor Pakdaman et al. [51]	Mean diameter of nanoparticles	Thermal conductivity of the base fluid
Ahammed et al. [52]	Nanofluids temperature	Thermal conductivity of the base fluid, volume concentration
Patel et al. [53]	Thermal conductivity of the base fluid	Nanofluids temperature, thermal conductivity of nanoparticles
Li and Peterson [54]	Volume concentration	Nanofluids temperature, thermal conductivity of the base fluid
Hemmat Esfe et al. [55]	Volume concentration	Nanofluids temperature
Hemmat Esfe et al. [56]	Volume concentration	Nanofluids temperature
Hemmat Esfe et al. [57]	Volume concentration	Nanofluids temperature
Hemmat Esfe et al. [58]	Volume concentration	Nanofluids temperature
Harandi et al. [59]	Thermal conductivity of the base fluid	Nanofluids temperature, volume concentration
Zadkhast et al. [60]	Thermal conductivity of the base fluid	Nanofluids temperature
Kakavandi and Akbari [61]	Thermal conductivity of the base fluid	Nanofluids temperature
Karimi et al. [62]	Thermal conductivity of the base fluid	Nanofluids temperature
Karimipour et al. [63]	Volume concentration	Thermal conductivity of the base fluid
Ranjbarzadeh et al. [64]	Thermal conductivity of the base fluid	Nanofluids temperature, volume concentration
Keyvani et al. [65]	Volume concentration	Nanofluids temperature
Afrand et al. [66]	Volume concentration	Nanofluids temperature

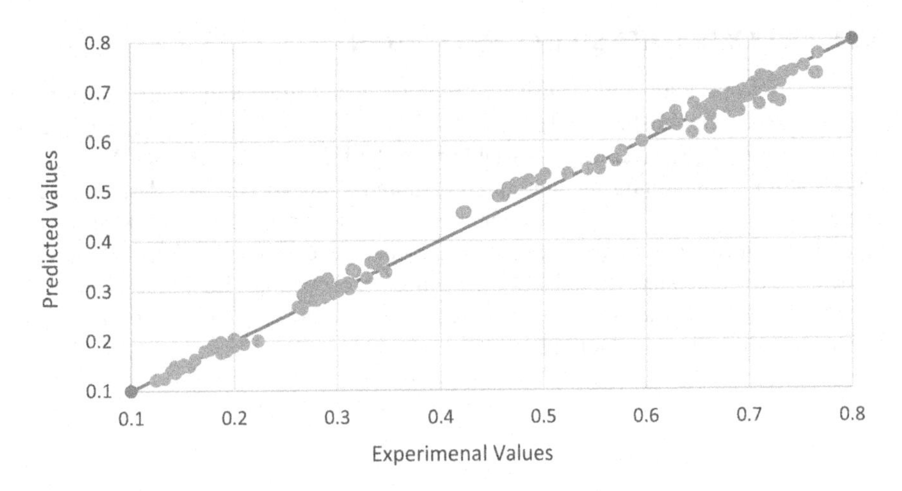

FIGURE 4.73 The fitness of responses predicted by the proposed correlation on the experimental values.

where k_b, k_{pr}, φ, and d_{pr} are thermal conductivity of the base fluid (W/m K), thermal conductivity ratio of nanoparticles (W/m K), nanoparticle volume concentration, and mean diameter ratio of nanoparticles (nm). The thermal conductivity ratio of nanoparticles is assumed thermal conductivity (k_{Pref}), which is 1500 (W/m K) for MWCNTs. The mean diameter ratio of nanoparticles also is the proportion of the mean diameter of used nanoparticles to the reference value of mean diameter ($d_{Pref} = 10$). Equation (4.39) is dimensionless, and its average absolute error of proposed compared to the experimental data is 5.39%. This correlation is more straightforward than the other correlations involving trigonometric (sinus and cosine), exponential, and logarithmic.

Furthermore, it covers all measured experimental data in volume concentrations ranging from 0% to 4% and the mean diameter of nanoparticles ranging from 10 to 150 nm. Figure 4.73 demonstrates the fitness of predicted values versus experimental data. Figure 4.74 also shows the accuracy of the proposed correlation. Table 4.4 shows the predicted values of effective thermal conductivity versus the experimental data and their deviations. The two last columns list the correlated values by the models provided by the other researchers and their deviations. As shown, our proposed correlation gives more accurate values. It finally can be concluded that the proposed correlation is comprehensive, reliable in the mentioned range of variables, more straightforward than the other correlations, accurate enough, and well-fitted on the experimental values. Therefore, it is recommended to use this correlation to predict nanofluids' thermal conductivity in studies employing the single-phase approach.

4.22 CONCLUSION

This section takes a skeptical look at the published mathematical model of the effective thermal conductivity of nanofluids. There are a couple of correlations considering different variables with different roles. Nevertheless, there is no global agreement on the role of different variables on the nanofluids' thermal conductivity. It is started

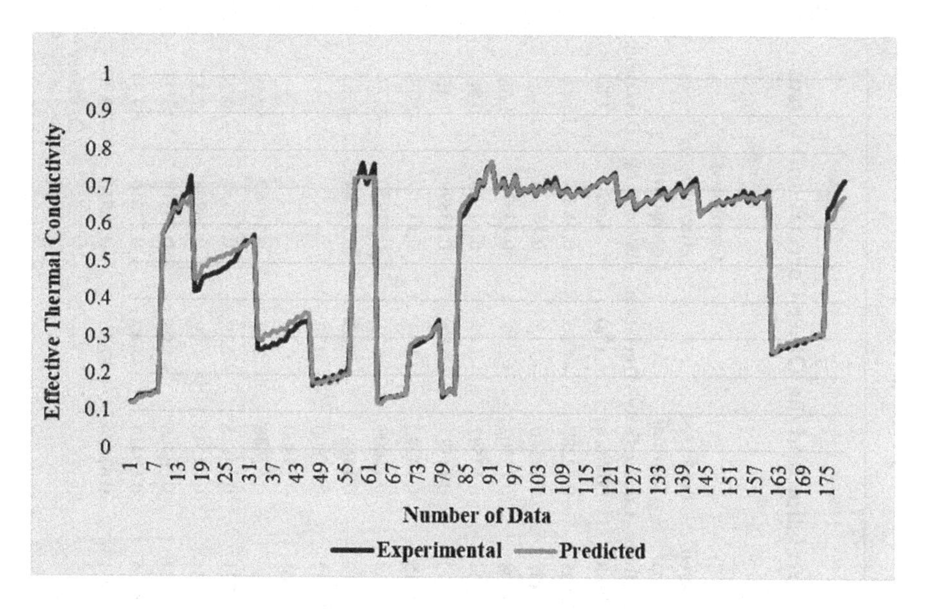

FIGURE 4.74 The accuracy of the proposed correlation.

with the models, which driven by an experimental study. Our sensitivity analysis demonstrated that in most published models, the role of nanofluids' bulk temperature is negligible. A big data bank is extracted from the models in the valid ranges of variables for detailed investigation. The next step was performing statistical analysis of the variances and means of different datasets (data populations). The results showed that there is no significant statistical difference between different datasets. It means that changing the temperature of the nanofluid does not considerably affect the nanofluid's thermal conductivity.

This is not claimed that nanofluids' bulk temperature does not affect the effective thermal conductivity. Nonetheless, the proposed mathematical models for effective thermal conductivity of nanofluids did not devote a considerable role to temperature. Since it is believed that the processes of developing a correlation should be more careful, the correlation must satisfy the physical aspects of the phenomena. Moreover, it was shown that most of the proposed correlations for effective thermal conductivity are dimensionally inconsistent.

A more straightforward correlation (neglecting the nanofluid temperature) is introduced for each published model with a low average absolute error from experimental data. In the end, a comprehensive correlation (MAG correlation) is proposed to estimate all available experimental data. In this correlation, the role of nanofluids' bulk temperature has been ignored. However, the predicted values of the effective thermal conductivity are in good agreement with experimental results.

It is recommended that editors and reviewers scrutinize any correlation to predict thermophysical properties. For instance, it is assumed that any predictive correlation for nanofluids should give the effective thermal conductivity equal to the thermal conductivity of the base fluid when the nanoparticle volume concentration is zero.

TABLE 4.4
The Predicted Values of Effective Thermal Conductivity (by Our Proposed Correlation and the Other Correlations) versus the Experimental Data and Their Deviations

Ref.	Temperature (°C)	Nanoparticle Volume Concentration	Nanoparticle Thermal Conductivity (W/m K)	Base Fluid Thermal Conductivity (W/m K)	Nanoparticle Mean Diameter (nm)	Measured Thermal Conductivity (W/m K)	Predicted Thermal Conductivity (W/m K)	Deviation (%)	Correlated Thermal Conductivity (W/m K) by Ref.	Deviation (%)
Aberoumand	60	0	403	0.1228	20	0.12505	0.12280	−1.80	0.14483	15.82
et al. [50]	70	0	403	0.1214	20	0.12407	0.12140	−2.15	0.14793	19.23
	40	0.0012	403	0.125	20	0.14108	0.13702	−2.87	0.13871	−1.68
	50	0.0012	403	0.124	20	0.14304	0.13592	−4.97	0.14178	−0.88
	40	0.0036	403	0.125	20	0.14500	0.14387	−0.78	0.13899	−4.15
	50	0.0036	403	0.124	20	0.14615	0.14272	−2.34	0.14209	−2.78
	40	0.0072	403	0.125	20	0.15269	0.15008	−1.71	0.13930	−8.77
	50	0.0072	403	0.124	20	0.15710	0.14888	−5.23	0.14234	−9.40
Ahammed	10	0	1500	0.578	20	0.57627	0.57800	0.30	-	-
et al. [52]	20	0	1500	0.598	20	0.59661	0.59800	0.23	-	-
	10	0.0005	1500	0.578	20	0.61186	0.62648	2.39	0.24967	−59.19
	20	0.0005	1500	0.598	20	0.66271	0.64815	−2.20	0.52537	−20.72
	10	0.001	1500	0.578	20	0.63051	0.64243	1.89	0.26908	−57.32
	20	0.001	1500	0.598	20	0.67797	0.66466	−1.96	0.56315	−16.93
	10	0.0015	1500	0.578	20	0.68475	0.65410	−4.47	0.27810	−59.39
	20	0.0015	1500	0.598	20	0.73051	0.67674	−7.36	0.58516	−19.90

(Continued)

TABLE 4.4 (*Continued*)
The Predicted Values of Effective Thermal Conductivity (by Our Proposed Correlation and the Other Correlations) versus the Experimental Data and Their Deviations

Ref.	Temperature (°C)	Nanoparticle Volume Concentration	Nanoparticle Thermal Conductivity (W/m K)	Base Fluid Thermal Conductivity (W/m K)	Nanoparticle Mean Diameter (nm)	Measured Thermal Conductivity (W/m K)	Predicted Thermal Conductivity (W/m K)	Deviation (%)	Correlated Thermal Conductivity (W/m K) by Ref.	Deviation (%)
Hemmat Esfe et al. [55]	45	0	69.036	0.455	40	0.42113	0.45500	8.04	0.44563	5.82
	50	0	69.036	0.456	40	0.42400	0.45600	7.55	0.45057	6.27
	45	0.002	69.036	0.455	40	0.45612	0.48878	7.16	0.44568	−2.29
	50	0.002	69.036	0.456	40	0.46128	0.48985	6.19	0.45071	−2.29
	45	0.005	69.036	0.455	40	0.46530	0.50420	8.36	0.44586	−4.18
	50	0.005	69.036	0.456	40	0.46989	0.50531	7.54	0.45099	−4.02
	45	0.0075	69.036	0.455	40	0.47333	0.51311	8.41	0.44602	−5.77
	50	0.0075	69.036	0.456	40	0.48078	0.51424	6.96	0.45106	−6.18
	45	0.01	69.036	0.455	40	0.48595	0.52040	7.09	0.44617	−8.18
	50	0.01	69.036	0.456	40	0.49742	0.52154	4.85	0.45123	−9.29
	45	0.015	69.036	0.455	40	0.50201	0.53224	6.02	0.44647	−11.06
	50	0.015	69.036	0.456	40	0.52381	0.53341	1.83	0.45146	−13.81
	45	0.02	69.036	0.455	40	0.54388	0.54192	−0.36	0.44678	−17.85
	50	0.02	69.036	0.456	40	0.55535	0.54312	−2.20	0.45187	−18.63
	45	0.03	69.036	0.455	40	0.55593	0.55767	0.31	0.44737	−19.53
	50	0.03	69.036	0.456	40	0.57142	0.55889	−2.19	0.45240	−20.83
	45	0.002	38	0.266	5	0.26932	0.29568	9.79	0.26799	−0.49

(*Continued*)

TABLE 4.4 (*Continued*)
The Predicted Values of Effective Thermal Conductivity (by Our Proposed Correlation and the Other Correlations) versus the Experimental Data and Their Deviations

Ref.	Temperature (°C)	Nanoparticle Volume Concentration	Nanoparticle Thermal Conductivity (W/m K)	Base Fluid Thermal Conductivity (W/m K)	Nanoparticle Mean Diameter (nm)	Measured Thermal Conductivity (W/m K)	Predicted Thermal Conductivity (W/m K)	Deviation (%)	Correlated Thermal Conductivity (W/m K) by Ref.	Deviation (%)
	50	0.002	38	0.263	5	0.26676	0.29235	9.59	0.26505	−0.64
	45	0.005	38	0.266	5	0.27501	0.30924	12.45	0.26803	−2.54
	50	0.005	38	0.263	5	0.27182	0.30575	12.48	0.26509	−2.48
	45	0.0075	38	0.266	5	0.28355	0.31707	11.82	0.26806	−5.46
	50	0.0075	38	0.263	5	0.28139	0.31350	11.41	0.26513	−5.78
	45	0.01	38	0.266	5	0.29095	0.32347	11.18	0.26809	−7.86
	50	0.01	38	0.263	5	0.29096	0.31983	9.92	0.26517	−8.87
	45	0.02	38	0.266	5	0.31429	0.34239	8.94	0.26822	−14.66
	50	0.02	38	0.263	5	0.31742	0.33853	6.65	0.26532	−16.41
	45	0.03	38	0.266	5	0.33250	0.35623	7.14	0.26836	−19.29
	50	0.03	38	0.263	5	0.33824	0.35221	4.13	0.26547	−21.51
	45	0.04	38	0.266	5	0.34275	0.36754	7.23	0.26850	−21.66
	50	0.04	38	0.263	5	0.34443	0.36339	5.51	0.26563	−22.88
Karimipour et al. [63]	25	0.0025	18	0.167	30	0.17148	0.17861	4.16	0.16760	−2.26
	40	0.0025	18	0.164	30	0.18746	0.17540	−6.44	0.17866	−4.69
	25	0.005	18	0.167	30	0.17678	0.18243	3.20	0.16732	−5.35
	40	0.005	18	0.164	30	0.19228	0.17915	−6.83	0.17858	−7.13
	25	0.015	18	0.167	30	0.18006	0.19122	6.20	0.16766	−6.89

(*Continued*)

TABLE 4.4 (*Continued*)
The Predicted Values of Effective Thermal Conductivity (by Our Proposed Correlation and the Other Correlations) versus the Experimental Data and Their Deviations

Ref.	Temperature (°C)	Nanoparticle Volume Concentration	Nanoparticle Thermal Conductivity (W/m K)	Base Fluid Thermal Conductivity (W/m K)	Nanoparticle Mean Diameter (nm)	Measured Thermal Conductivity (W/m K)	Predicted Thermal Conductivity (W/m K)	Deviation (%)	Correlated Thermal Conductivity (W/m K) by Ref.	Deviation (%)
	40	0.015	18	0.164	30	0.19893	0.18779	−5.60	0.17886	−10.09
	25	0.025	18	0.167	30	0.18734	0.19687	5.09	0.16769	−10.49
	40	0.025	18	0.164	30	0.20918	0.19334	−7.57	0.17897	−14.44
	25	0.04	18	0.167	30	0.19988	0.20323	1.68	0.16789	−16.00
	40	0.04	18	0.164	30	0.22329	0.19958	−10.62	0.17910	−19.77
Li and Peterson [54]	33	0.02	38	0.618	36	0.71265	0.72750	2.08	0.72097	1.17
	35	0.02	38	0.621	36	0.73419	0.73103	−0.43	0.75043	2.21
	35	0.02	38	0.622	36	0.76644	0.73221	−4.47	0.76006	−0.83
	33	0.02	38	0.618	36	0.71166	0.72750	2.23	0.72083	1.29
	35	0.02	38	0.621	36	0.73227	0.73103	−0.17	0.75000	2.42
	35	0.02	38	0.622	36	0.76418	0.73221	−4.18	0.75996	−0.55
	29	0.04	18	0.613	29	0.82073	0.74711	−8.97	0.83482	1.72
	29	0.04	18	0.613	29	0.81876	0.74711	−8.75	0.83460	1.93
	31	0.04	18	0.615	29	0.84978	0.74955	−11.80	0.86321	1.58

(*Continued*)

TABLE 4.4 (*Continued*)
The Predicted Values of Effective Thermal Conductivity (by Our Proposed Correlation and the Other Correlations) versus the Experimental Data and Their Deviations

Ref.	Temperature (°C)	Nanoparticle Volume Concentration	Nanoparticle Thermal Conductivity (W/m K)	Base Fluid Thermal Conductivity (W/m K)	Nanoparticle Mean Diameter (nm)	Measured Thermal Conductivity (W/m K)	Predicted Thermal Conductivity (W/m K)	Deviation (%)	Correlated Thermal Conductivity (W/m K) by Ref.	Deviation (%)
Fakoor	40	0	1500	0.125	20	0.13250	0.12500	−5.67	0.13117	−1.00
Pakdaman	50	0	1500	0.124	20	0.13208	0.12400	−6.12	0.13205	−0.03
et al. [51]	40	0.001	1500	0.125	20	0.13890	0.13893	0.03	0.13208	−4.91
	50	0.001	1500	0.124	20	0.14143	0.13782	−2.55	0.13322	−5.80
	40	0.002	1500	0.125	20	0.14185	0.14352	1.18	0.13311	−6.16
	50	0.002	1500	0.124	20	0.14487	0.14237	−1.72	0.13457	−7.11
	50	0.004	1500	0.124	20	0.14733	0.14842	0.74	0.13787	−6.42
	60	0.004	1500	0.1228	20	0.14936	0.14698	−1.60	0.14002	−6.26
Patel et al.	50	0.005	38	0.263	150	0.27422	0.28160	2.69	0.28045	2.27
[53]	50	0.01	38	0.263	150	0.27743	0.28772	3.71	0.28717	3.51
	50	0.02	38	0.263	150	0.28368	0.29586	4.29	0.29637	4.47
	50	0.03	38	0.263	150	0.29186	0.30181	3.41	0.30329	3.91
	50	0.005	38	0.263	11	0.29960	0.29825	−0.45	0.29515	−1.48
	50	0.01	38	0.263	11	0.30478	0.30986	1.67	0.30750	0.89
	50	0.02	38	0.263	11	0.32875	0.32528	−1.05	0.32452	−1.29
	50	0.03	38	0.263	11	0.34721	0.33656	−3.07	0.33727	−2.86
	20	0.00495	18	0.136	31	0.14257	0.14841	4.10	0.14368	0.78
	30	0.010057	18	0.136	31	0.15138	0.15260	0.81	0.14935	−1.34

(*Continued*)

TABLE 4.4 (*Continued*)
The Predicted Values of Effective Thermal Conductivity (by Our Proposed Correlation and the Other Correlations) versus the Experimental Data and Their Deviations

Ref.	Temperature (°C)	Nanoparticle Volume Concentration	Nanoparticle Thermal Conductivity (W/m K)	Base Fluid Thermal Conductivity (W/m K)	Nanoparticle Mean Diameter (nm)	Measured Thermal Conductivity (W/m K)	Predicted Thermal Conductivity (W/m K)	Deviation (%)	Correlated Thermal Conductivity (W/m K) by Ref.	Deviation (%)
	40	0.030122	18	0.136	31	0.16200	0.16205	0.03	0.16208	0.05
	50	0.005021	18	0.136	31	0.14902	0.14848	−0.36	0.14876	−0.18
	20	0.005032	38	0.598	150	0.62118	0.64041	3.09	0.61726	−0.63
	30	0.004792	38	0.614	150	0.62903	0.65668	4.40	0.63796	1.42
	40	0.005032	38	0.628	150	0.64670	0.67253	4.00	0.65716	1.62
	50	0.005032	38	0.64	150	0.66769	0.68539	2.65	0.67341	0.86
	50	0.001	413	0.64	80	0.67529	0.68085	0.82	0.67489	−0.06
	50	0.005	413	0.64	80	0.72263	0.71909	−0.49	0.71399	−1.20
	30	0.01	413	0.614	80	0.70921	0.71485	0.80	0.68898	−2.85
	30	0.02	413	0.614	80	0.75343	0.74804	−0.71	0.71782	−4.73
	30	0.03	413	0.614	80	0.76717	0.77232	0.67	0.73969	−3.58
	20	0.009993	237	0.598	80	0.70054	0.68738	−1.88	0.64867	−7.40
	30	0.005028	237	0.64	80	0.70476	0.71215	1.05	0.68822	−2.35
	40	0.010064	237	0.628	80	0.72752	0.72213	−0.74	0.70498	−3.10
	50	0.005024	237	0.614	80	0.68785	0.68320	−0.68	0.67585	−1.74

(*Continued*)

TABLE 4.4 (*Continued*)
The Predicted Values of Effective Thermal Conductivity (by Our Proposed Correlation and the Other Correlations) versus the Experimental Data and Their Deviations

Ref.	Temperature (°C)	Nanoparticle Volume Concentration	Nanoparticle Thermal Conductivity (W/m K)	Base Fluid Thermal Conductivity (W/m K)	Nanoparticle Mean Diameter (nm)	Measured Thermal Conductivity (W/m K)	Predicted Thermal Conductivity (W/m K)	Deviation (%)	Correlated Thermal Conductivity (W/m K) by Ref.	Deviation (%)
Zadkhast et al. [60]	25	0.002	1500	0.606	15	0.70389	0.70225	−0.23	0.59300	−15.75
	25	0.00399	1500	0.606	15	0.73538	0.73394	−0.20	0.60048	−18.34
	30	0.00101	1500	0.614	15	0.68513	0.68776	0.38	0.59775	−12.75
	30	0.00152	1500	0.614	15	0.70055	0.70127	0.10	0.60118	−14.18
	35	0.00099	1500	0.621	15	0.69226	0.69482	0.37	0.60726	−12.28
	35	0.00154	1500	0.621	15	0.70879	0.70957	0.11	0.61102	−13.79
	40	0.00052	1500	0.628	15	0.68234	0.68563	0.48	0.61247	−10.24
	40	0.001	1500	0.628	15	0.70545	0.70305	−0.34	0.61715	−12.52
	45	0.00051	1500	0.634	15	0.69552	0.69160	−0.56	0.62112	−10.70
	45	0.00099	1500	0.634	15	0.72192	0.70936	−1.74	0.62592	−13.30
	50	0.00051	1500	0.64	15	0.70650	0.69814	−1.18	0.63000	−10.83
	50	0.00101	1500	0.64	15	0.72906	0.71688	−1.67	0.63509	−12.89
Karimi et al. [62]	50	0.005	0.58	0.64295	10	0.68133	0.68639	0.74	0.68185	0.08
	55	0.005	0.58	0.64849	10	0.69525	0.69231	−0.42	0.69033	−0.71
	60	0.005	0.58	0.65371	10	0.70032	0.69788	−0.35	0.69832	−0.28
	35	0.01	0.58	0.62434	10	0.67822	0.68040	0.32	0.67134	−1.01
	40	0.01	0.58	0.63087	10	0.68714	0.68752	0.06	0.68237	−0.69
	45	0.01	0.58	0.63707	10	0.70046	0.69428	−0.88	0.69286	−1.09

(*Continued*)

TABLE 4.4 (*Continued*)
The Predicted Values of Effective Thermal Conductivity (by Our Proposed Correlation and the Other Correlations) versus the Experimental Data and Their Deviations

Ref.	Temperature (°C)	Nanoparticle Volume Concentration	Nanoparticle Thermal Conductivity (W/m K)	Base Fluid Thermal Conductivity (W/m K)	Nanoparticle Mean Diameter (nm)	Measured Thermal Conductivity (W/m K)	Predicted Thermal Conductivity (W/m K)	Deviation (%)	Correlated Thermal Conductivity (W/m K) by Ref.	Deviation (%)
	30	0.02	0.58	0.61747	10	0.68583	0.69118	0.78	0.68105	−0.70
	35	0.02	0.58	0.62434	10	0.69476	0.69886	0.59	0.69400	−0.11
	40	0.02	0.58	0.63087	10	0.70992	0.70617	−0.53	0.70615	−0.53
	35	0.03	0.58	0.62434	10	0.70949	0.71236	0.40	0.70928	−0.03
	40	0.03	0.58	0.63087	10	0.72152	0.71981	−0.24	0.72202	0.07
	45	0.03	0.58	0.63707	10	0.73136	0.72689	−0.61	0.73389	0.35
	35	0.04	0.58	0.62434	10	0.71867	0.72339	0.66	0.72098	0.32
	40	0.04	0.58	0.63087	10	0.73313	0.73096	−0.30	0.73419	0.14
	45	0.04	0.58	0.63707	10	0.74263	0.73815	−0.60	0.74639	0.51
	35	0.005	0.635	0.62434	15	0.65833	0.66313	0.73	0.65369	−0.70
	40	0.005	0.635	0.63087	15	0.67335	0.67006	−0.49	0.66357	−1.45
	45	0.005	0.635	0.63707	15	0.68088	0.67665	−0.62	0.67287	−1.18
	55	0.005	0.635	0.64849	15	0.69448	0.68878	−0.82	0.69021	−0.62
	20	0.01	0.635	0.60276	15	0.64767	0.65254	0.75	0.63404	−2.10
	25	0.01	0.635	0.61028	15	0.65881	0.66068	0.28	0.64737	−1.74
	30	0.01	0.635	0.61747	15	0.66701	0.66847	0.22	0.65950	−1.13
	35	0.01	0.635	0.62434	15	0.67934	0.67589	−0.51	0.67125	−1.19
	20	0.02	0.635	0.60276	15	0.66234	0.66893	0.99	0.65212	−1.54

(*Continued*)

TABLE 4.4 (*Continued*)
The Predicted Values of Effective Thermal Conductivity (by Our Proposed Correlation and the Other Correlations) versus the Experimental Data and Their Deviations

Ref.	Temperature (°C)	Nanoparticle Volume Concentration	Nanoparticle Thermal Conductivity (W/m K)	Base Fluid Thermal Conductivity (W/m K)	Nanoparticle Mean Diameter (nm)	Measured Thermal Conductivity (W/m K)	Predicted Thermal Conductivity (W/m K)	Deviation (%)	Correlated Thermal Conductivity (W/m K) by Ref.	Deviation (%)
	25	0.02	0.635	0.61028	15	0.67759	0.67727	−0.05	0.66701	−1.56
	30	0.02	0.635	0.61747	15	0.69486	0.68525	−1.38	0.68091	−2.01
	35	0.02	0.635	0.62434	15	0.70348	0.69287	−1.51	0.69406	−1.34
	20	0.03	0.635	0.60276	15	0.67485	0.68091	0.90	0.66477	−1.49
	25	0.03	0.635	0.61028	15	0.68895	0.68940	0.07	0.68077	−1.19
	30	0.03	0.635	0.61747	15	0.70149	0.69752	−0.56	0.69559	−0.84
	35	0.03	0.635	0.62434	15	0.71599	0.70528	−1.50	0.70915	−0.95
	20	0.04	0.635	0.60276	15	0.68176	0.69070	1.31	0.67488	−1.01
	25	0.04	0.635	0.61028	15	0.70075	0.69932	−0.20	0.69152	−1.32
	30	0.04	0.635	0.61747	15	0.71563	0.70756	−1.13	0.70675	−1.24
	35	0.04	0.635	0.62434	15	0.73074	0.71542	−2.10	0.72085	−1.35
Ranjbarzadeh et al. [64]	40	0	1.4	0.63040	40.77	0.63049	0.63040	−0.01	0.63040	0.0149
	55	0	1.4	0.64570	40.77	0.64489	0.64570	0.12	0.64570	−0.1248
	45	0.001	1.4	0.63450	40.77	0.64998	0.65273	0.42	0.63470	2.4063
	55	0.001	1.4	0.64570	40.77	0.66014	0.66425	0.62	0.64599	2.1905
	45	0.0025	1.4	0.63450	40.77	0.66268	0.66105	−0.25	0.63494	4.3692
	50	0.0025	1.4	0.64380	40.77	0.67115	0.67074	−0.06	0.64433	4.1622

(*Continued*)

TABLE 4.4 (*Continued*)
The Predicted Values of Effective Thermal Conductivity (by Our Proposed Correlation and the Other Correlations) versus the Experimental Data and Their Deviations

Ref.	Temperature (°C)	Nanoparticle Volume Concentration	Nanoparticle Thermal Conductivity (W/m K)	Base Fluid Thermal Conductivity (W/m K)	Nanoparticle Mean Diameter (nm)	Measured Thermal Conductivity (W/m K)	Predicted Thermal Conductivity (W/m K)	Deviation (%)	Correlated Thermal Conductivity (W/m K) by Ref.	Deviation (%)
	40	0.005	1.4	0.63040	40.77	0.66099	0.66546	0.68	0.63104	4.7454
	45	0.005	1.4	0.63450	40.77	0.67793	0.66979	−1.20	0.63529	6.7117
	35	0.01	1.4	0.62150	40.77	0.66438	0.66745	0.46	0.62241	6.7433
	40	0.01	1.4	0.63040	40.77	0.68386	0.67700	−1.00	0.63155	8.282
	35	0.015	1.4	0.62150	40.77	0.67793	0.67577	−0.32	0.62278	8.8561
	40	0.015	1.4	0.63040	40.77	0.69826	0.68544	−1.84	0.63203	10.48
	30	0.02	1.4	0.61560	40.77	0.66777	0.67609	1.25	0.61684	8.2561
	35	0.02	1.4	0.62150	40.77	0.69233	0.68257	−1.41	0.62313	11.106
	25	0.025	1.4	0.60600	40.77	0.66607	0.67126	0.78	0.60708	9.7175
	30	0.025	1.4	0.61560	40.77	0.68132	0.68189	0.08	0.61710	10.407
	25	0.03	1.4	0.60600	40.77	0.68386	0.67633	−1.10	0.60726	12.614
	30	0.03	1.4	0.61560	40.77	0.69826	0.68704	−1.61	0.61734	13.107

(*Continued*)

TABLE 4.4 (*Continued*)
The Predicted Values of Effective Thermal Conductivity (by Our Proposed Correlation and the Other Correlations) versus the Experimental Data and Their Deviations

Ref.	Temperature (°C)	Nanoparticle Volume Concentration	Nanoparticle Thermal Conductivity (W/m K)	Base Fluid Thermal Conductivity (W/m K)	Nanoparticle Mean Diameter (nm)	Measured Thermal Conductivity (W/m K)	Predicted Thermal Conductivity (W/m K)	Deviation (%)	Correlated Thermal Conductivity (W/m K) by Ref.	Deviation (%)
Keyvani et al. [65]	45	0	11.71	0.268	30	0.262	0.26800	2.29	0.27510	5.00
	50	0	11.71	0.264	30	0.266	0.26400	−0.75	0.27377	2.92
	45	0.0025	11.71	0.268	30	0.272	0.28532	4.90	0.27961	2.80
	50	0.0025	11.71	0.264	30	0.28	0.28106	0.38	0.27821	−0.64
	45	0.005	11.71	0.268	30	0.278	0.29103	4.69	0.28412	2.20
	50	0.005	11.71	0.264	30	0.288	0.28668	−0.46	0.28265	−1.86
	45	0.01	11.71	0.268	30	0.287	0.29860	4.04	0.29314	2.14
	50	0.01	11.71	0.264	30	0.296	0.29415	−0.63	0.29154	−1.51
	45	0.015	11.71	0.268	30	0.292	0.30415	4.16	0.30216	3.48
	50	0.015	11.71	0.264	30	0.301	0.29961	−0.46	0.30042	−0.19
	45	0.02	11.71	0.268	30	0.303	0.30868	1.87	0.31117	2.70
	50	0.02	11.71	0.264	30	0.312	0.30407	−2.54	0.30930	−0.86
	40	0.025	11.71	0.27	30	0.31	0.31491	1.58	0.31975	3.14
	45	0.025	11.71	0.268	30	0.314	0.31258	−0.45	0.32019	1.97
Afrand et al. [66]	20	0.001044	0.58	0.59803	30	0.64539	0.614269	−4.82237	0.6141	−4.8491
	25	0.001044	0.58	0.606	30	0.6629	0.622455	−6.1009	0.62392	−5.8806
	35	0.000986	0.58	0.621	30	0.69091	0.637471	−7.73491	0.63611	−7.9322
	45	0.000986	0.58	0.634	30	0.71017	0.650816	−8.35754	0.6486	−8.669
	55	0.000986	0.58	0.645	30	0.72417	0.662108	−8.57058	0.65915	−8.9792

In this chapter, all the tables and some of the Figures 4.1–4.3, 4.6, 4.7, 4.73, and 4.74 are presented from the article "A different look at the effect of temperature on the nanofluids thermal conductivity: focus on the experimental-based models," in *Journal of Thermal Analysis and Calorimetry* by the authors Maysam Molana, Ramin Ghasemiasl, and Taher Armaghani [67], published in "Springer Nature" publications with the volume and page number 147 and 4553–4577, respectively.

REFERENCES

1. A. Moradikazerouni, et al., Investigation of a computer CPU heat sink under laminar forced convection using a structural stability method, *Int. J. Heat Mass Transfer* 134 (2019) 1218–1226.
2. T. Armaghani, et al., Forced convection heat transfer of nanofluids in a channel filled with porous media under local thermal non-equilibrium condition with three new models for absorbed heat flux, *J. Nanofluids* 6.2 (2017) 362–367.
3. A. Moradikazerouni, et al., Comparison of the effect of five different entrance channel shapes of a micro-channel heat sink in forced convection with application to cooling a supercomputer circuit board, *Appl. Therm. Eng.* 150 (2019) 1078–1089.
4. S. Banooni, H. Zarea, M. Molana, Thermodynamic and economic optimization of plate fin heat exchangers using the bees algorithm, *Heat Transfer-Asian Res.* 43.5 (2014) 427–446.
5. A. Qashqaei, R. Ghasemi Asl, Numerical modeling and simulation of copper oxide nanofluids used in compact heat exchangers, *Int. J. Mech. Eng.* 4.2 (2015) 1–8.
6. A. Yari, et al., Numerical simulation for thermal design of a gas water heater with turbulent combined convection, Proceedings of the ASME/JSME/KSME 2015 Joint Fluids Engineering Conference. Volume 1: Symposia. Seoul, South Korea. July 26–31, 2015. V001T03A007. ASME. https://doi.org/10.1115/AJKFluids2015-3305.
7. H. Esmaeili, et al. Turbulent combined forced and natural convection of nanofluid in a 3D rectangular channel using two-phase model approach, *J. Therm. Anal. Calorim.* 135.6 (2019) 3247–3257.
8. A.M. Rashad, et al., MHD mixed convection and entropy generation of nanofluid in a lid-driven U-shaped cavity with internal heat and partial slip, *Phy. Fluids* 31.4 042006 (2019). https://doi.org/10.1063/1.5079789.
9. T. Armaghani, et al., MHD mixed convection flow and heat transfer in an open C-shaped enclosure using water-copper oxide nanofluid, *Heat Mass Transfer* 54 (2018) 1791–1801.
10. T. Armaghani, et al., Mixed convection and entropy generation of an Ag-water nanofluid in an inclined L-shaped channel, *Energies* 12.6 (2019) 1150.
11. R. Ghasemiasl, S. Hoseinzadeh, M.A. Javadi, Numerical analysis of energy storage systems using two phase-change materials with nanoparticles, *J. Thermophys. Heat Transfer* 32.2 (2018) 440–448.
12. S. Hoseinzadeh, et al., Numerical investigation of rectangular thermal energy storage units with multiple phase change materials, *J. Mol. Liq.* 271 (2018) 655–660.
13. J.C. Maxwell, A Treatise on Electricity and Magnetism. Vol. 1. Clarendon Press, Oxford, 1873.
14. S. Choi, J.A. Eastman, Enhancing thermal conductivity of fluids with nanoparticles. No. ANL/MSD/CP-84938; CONF-951135-29. Argonne National Lab.(ANL), Argonne, IL (United States), 1995.

15. J.A. Eastman, et al., Anomalously increased effective thermal conductivities of ethylene glycol-based nanofluids containing copper nanoparticles, *Appl. Phys. Lett.* 78.6 (2001) 718–720.
16. M. Molana, A comprehensive review on the nanofluids application in the tubular heat exchangers, *Am. J. Heat Mass Transf* 3.5 (2016) 352–381.
17. M.H. Esfe, et al., Multi-objective optimization of nanofluid flow in double tube heat exchangers for applications in energy systems, *Energy* 137 (2017) 160–171.
18. J. Alsarraf, et al., Hydrothermal analysis of turbulent boehmite alumina nanofluid flow with different nanoparticle shapes in a minichannel heat exchanger using two-phase mixture model, *Physica A* 520 (2019) 275–288.
19. M. Molana, S. Banooni, Investigation of heat transfer processes involved liquid impingement jets: A review, *Braz. J. Chem. Eng.* 30 (2013) 413–435.
20. K. Wongcharee, V. Chuwattanakul, S. Eiamsa-Ard, Influence of CuO/water nanofluid concentration and swirling flow on jet impingement cooling, *Int. Commun. Heat Mass Transfer* 88 (2017) 277–283.
21. E. Khodabandeh, et al., Application of nanofluid to improve the thermal performance of horizontal spiral coil utilized in solar ponds: geometric study, *Renewable Energy* 122 (2018) 1–16.
22. O. Mahian, et al., A review of the applications of nanofluids in solar energy, *Int. J. Heat Mass Transfer* 57.2 (2013) 582–594.
23. A. Asadi, et al., Heat transfer efficiency of Al_2O_3-MWCNT/thermal oil hybrid nanofluid as a cooling fluid in thermal and energy management applications: An experimental and theoretical investigation, *Int. J. Heat Mass Transfer* 117 (2018) 474–486.
24. E. Hemmat, Mohammad, et al., Thermal conductivity and viscosity of Mg (OH) 2-ethylene glycol nanofluids: Finding a critical temperature, *J. Therm. Anal. Calorim.* 120 (2015) 1145–1149.
25. A. Moradikazerouni, et al., Assessment of thermal conductivity enhancement of nano-antifreeze containing single-walled carbon nanotubes: Optimal artificial neural network and curve-fitting, *Physica A* 521 (2019) 138–145.
26. A. Asadi, et al., Recent advances in preparation methods and thermophysical properties of oil-based nanofluids: A state-of-the-art review, *Powder Technol.* 352 (2019) 209–226.
27. M. Molana, On the nanofluids application in the automotive radiator to reach the enhanced thermal performance: A review, *Am. J. Heat Mass Transfer* 4.4 (2017) 168–187.
28. H. Kakavand, M. Molana, A numerical study of heat transfer charcteristics of a car radiator involved nanofluids, *Heat Transfer-Asian Res.* 47.1 (2018) 88–102.
29. A.J. Chamkha, et al., On the nanofluids applications in microchannels: A comprehensive review, *Powder Technol.* 332 (2018) 287–322.
30. S. Izadi, et al., A comprehensive review on mixed convection of nanofluids in various shapes of enclosures, *Powder Technol.* 343 (2019) 880–907.
31. P. Nitiapiruk, et al., Performance characteristics of a microchannel heat sink using TiO_2/water nanofluid and different thermophysical models, *Int. Commun. Heat Mass Transfer* 47 (2013) 98–104.
32. D.D. Vo, et al., Numerical investigation of γ-AlOOH nano-fluid convection performance in a wavy channel considering various shapes of nanoadditives, *Powder Technol* 345 (2019) 649–657.
33. S. Zeinali Heris, et al., Rheological behavior of zinc-oxide nanolubricants, *J. Dispersion Sci. Technol.* 36.8 (2015) 1073–1079.
34. M.H. Esfe, et al., Effects of temperature and concentration on rheological behavior of MWCNTs/SiO_2 (20-80)-SAE40 hybrid nano-lubricant, *Int. Commun. Heat Mass Transfer* 76 (2016) 133–138.

35. M.H. Esfe, et al., Examination of rheological behavior of MWCNTs/ZnO-SAE40 hybrid nano-lubricants under various temperatures and solid volume fractions, *Exp. Therm. Fluid Sci.* 80 (2017) 384–390.
36. D. Tripathi, O. Anwar Bég, A study on peristaltic flow of nanofluids: Application in drug delivery systems, *Int. J. Heat Mass Transfer* 70 (2014) 61–70.
37. M.A. Abbas, et al., Application of drug delivery in magnetohydrodynamics peristaltic blood flow of nanofluid in a non-uniform channel, *J, Mech. Med. Biol.* 16.04 (2016) 1650052.
38. S.E. Ghasemi, et al., Analytical and numerical investigation of nanoparticle effect on peristaltic fluid flow in drug delivery systems, *J. Mol. Liq.* 215 (2016) 88–97.
39. M.H. Esfe, M. Reza Sarlak, Experimental investigation of switchable behavior of CuO-MWCNT (85%-15%)/10W-40 hybrid nano-lubricants for applications in internal combustion engines, *J. Mol. Liq.* 242 (2017) 326–335.
40. Y. Gan, Y. Syuen Lim, L. Qiao, Combustion of nanofluid fuels with the addition of boron and iron particles at dilute and dense concentrations, *Combust. Flame* 159.4 (2012) 1732–1740.
41. S. Sarvestany, Nasrin, et al., Effects of magnetic nanofluid fuel combustion on the performance and emission characteristics, *J. Dispersion Sci. Technol.* 35.12 (2014) 1745–1750.
42. S. Kumar, S.K. Prasad, J. Banerjee, Analysis of flow and thermal field in nanofluid using a single phase thermal dispersion model, *App. Math. Model.* 34.3 (2010) 573–592.
43. D. Liu, L. Yu, Single-phase thermal transport of nanofluids in a minichannel J. Heat Transfer 133.3 (2011) 031009. https://doi.org/10.1115/1.4002462.
44. A. Behzadmehr, M. Saffar-Avval, N. Galanis, Prediction of turbulent forced convection of a nanofluid in a tube with uniform heat flux using a two phase approach, *Int. J. Heat Fluid Flow* 28.2 (2007) 211–219.
45. M. Kalteh, et al., Eulerian-Eulerian two-phase numerical simulation of nanofluid laminar forced convection in a microchannel, *Int. J. Heat Fluid Flow* 32.1 (2011) 107–116.
46. R.L. Hamilton, O.K. Crosser, Thermal conductivity of heterogeneous two-component systems, *Ind. Eng. Chem. Fundam.* 1.3 (1962) 187–191.
47. Z. Hashin, S. Shtrikman, A variational approach to the theory of the effective magnetic permeability of multiphase materials, *J Appl. Phys.* 33.10 (1962) 3125–3131.
48. D.J. Jeffrey, Conduction through a random suspension of spheres, *Proc. R. Soc. London. A. Math. Phys. Sci.* 335.1602 (1973) 355–367.
49. R.M. Turian, D.-J. Sung, F.-L. Hsu, Thermal conductivity of granular coals, coal-water mixtures and multi-solid/liquid suspensions, *Fuel* 70.10 (1991) 1157–1172.
50. S. Aberoumand, et al., Experimental study on the rheological behavior of silver-heat transfer oil nanofluid and suggesting two empirical based correlations for thermal conductivity and viscosity of oil based nanofluids, *App. Therm. Eng.* 101 (2016) 362–372.
51. M.F. Pakdaman, M.A. Akhavan-Behabadi, P. Razi, An experimental investigation on thermo-physical properties and overall performance of MWCNT/heat transfer oil nanofluid flow inside vertical helically coiled tubes, *Exp. Therm. Fluid Sci.* 40 (2012) 103–111.
52. N. Ahammed, et al., Measurement of thermal conductivity of graphene-water nanofluid at below and above ambient temperatures, *Int. Commun. Heat Mass Transfer* 70 (2016) 66–74.
53. H.E. Patel, T. Sundararajan, S.K. Das, An experimental investigation into the thermal conductivity enhancement in oxide and metallic nanofluids, *J. Nanopart. Res.* 12 (2010) 1015–1031.
54. C.H. Li, G.P. Peterson, Experimental investigation of temperature and volume fraction variations on the effective thermal conductivity of nanoparticle suspensions (nanofluids), *J. App. Phys.* 99.8 (2006) 084314. https://doi.org/10.1063/1.2191571.

55. M.H. Esfe, et al., An experimental study on thermal conductivity of MgO nanoparticles suspended in a binary mixture of water and ethylene glycol, *Int. Commun. Heat Mass Transfer* 67 (2015) 173–175.
56. M.H. Esfe, et al., Experimental study on thermal conductivity of ethylene glycol based nanofluids containing Al_2O_3 nanoparticles, *Int. J. Heat Mass Transfer* 88 (2015) 728–734.
57. M.H. Esfe, et al., Thermal conductivity of Cu/TiO_2-water/EG hybrid nanofluid: Experimental data and modeling using artificial neural network and correlation, *Int. Commun. Heat Mass Transfer* 66 (2015) 100–104.
58. M. Hemmat Esfe, et al., Study on thermal conductivity of water-based nanofluids with hybrid suspensions of CNTs/Al_2O_3 nanoparticles, *J. Therm. Anal. Calorim.* 124 (2016): 455–460.
59. S.S. Harandi, et al., An experimental study on thermal conductivity of F-MWCNTs-Fe_3O_4/EG hybrid nanofluid: Effects of temperature and concentration, *Int. Commun. Heat Mass Transfer* 76 (2016) 171–177.
60. M. Zadkhast, D. Toghraie, A. Karimipour, Developing a new correlation to estimate the thermal conductivity of MWCNT-CuO/water hybrid nanofluid via an experimental investigation, *J. Therm. Anal. Calorim.* 129 (2017) 859–867.
61. A. Kakavandi, M. Akbari, Experimental investigation of thermal conductivity of nanofluids containing of hybrid nanoparticles suspended in binary base fluids and propose a new correlation, *Int. J. Heat Mass Transfer* 124 (2018) 742–751.
62. A. Karimi, et al., Experimental investigation on thermal conductivity of MFe_2O_4 (M= Fe and Co) magnetic nanofluids under influence of magnetic field, *Thermochim. Acta* 598 (2014) 59–67.
63. A. Karimipour, et al., A new correlation for estimating the thermal conductivity and dynamic viscosity of CuO/liquid paraffin nanofluid using neural network method, *Int. Commun. Heat Mass Transfer* 92 (2018) 90–99.
64. R. Ranjbarzadeh, et al., An experimental study on stability and thermal conductivity of water/silica nanofluid: Eco-friendly production of nanoparticles, *J. Cleaner Prod.* 206 (2019) 1089–1100.
65. M. Keyvani, et al., An experimental study on the thermal conductivity of cerium oxide/ethylene glycol nanofluid: developing a new correlation, *J. Mol. Liq.* 266 (2018): 211–217.
66. M. Afrand, D. Toghraie, N. Sina, Experimental study on thermal conductivity of water-based Fe_3O_4 nanofluid: Development of a new correlation and modeled by artificial neural network, *Int. Commun. Heat Mass Transfer* 75 (2016) 262–269.
67. M. Molana, R. Ghasemiasl, T. Armaghani, A different look at the effect of temperature on the nanofluids thermal conductivity: Focus on the experimental-based models, *J. Therm. Anal. Calorim.* 147 (2022) 4553–4577. https://doi.org/10.1007/s10973-021-10836-w

APPENDIX

A.1 NONLINEAR REGRESSION

Nonlinear regression is a form of regression analysis in which data is fit to a model and then expressed as a mathematical function.

Simple linear regression relates two variables (X and Y) with a straight line ($y=mx+b$), while nonlinear regression relates the two variables in a nonlinear (curved) relationship.

The goal of the model is to make the sum of the squares as small as possible. The sum of squares is a measure that tracks how far the Y observations vary from the nonlinear (curved) function that is used to predict Y.

It is computed by first finding the difference between the fitted nonlinear function and every Y point of data in the set. Then, each of those differences is squared. Lastly, all of the squared figures are added together. The smaller the sum of these squared figures, the better the function fits the data points in the set. Nonlinear regression uses logarithmic functions, trigonometric functions, exponential functions, power functions, Lorenz curves, Gaussian functions, and other fitting methods.

For a general summary of the concept, it can be noted that:

- Both linear and nonlinear regression predict Y responses from an X variable (or variables).
- Nonlinear regression is a curved function of an X variable (or variables) that is used to predict a Y variable.
- Nonlinear regression can show a prediction of population growth over time.
- Nonlinear regression is a mathematical function that uses a generated line – typically a curve – to fit an equation to some data.
- The sum of squares is used to determine the fitness of a regression model, which is computed by calculating the difference between the mean and every point of data.
- Nonlinear regression models are used because of their ability to accommodate different mean functions.

Thus, nonlinear regression modeling is similar to linear regression modeling in that both seek to track a particular response from a set of variables graphically. Additionally, nonlinear models are more complicated than linear models to develop because the function is created through a series of approximations (iterations) that may stem from trial-and-error. Mathematicians use several established methods, such as the Gauss-Newton method and the Levenberg–Marquardt method. Therefore, Linear regression relates two variables with a straight line; nonlinear regression relates the variables using a curve.

Nonlinear regression is used for two purposes: (1) to simply fit a smooth curve in order to interpolate values from the curve, or (2) to draw a graph with a smooth curve. If this is the goal, it can be assessed purely by looking at the graph of data and curve.

A.2 NONLINEAR REGRESSION FORMULA

Nonlinear regression uses nonlinear regression equations, which take the form:

$$Y = f(X, \beta) + \varepsilon \tag{A.1}$$

- X = a vector of p predictors,
- β = a vector of k parameters,
- $f(\text{-})$ = a known regression function,
- ε = an error term.

A.3 DIFFERENT TYPES OF NONLINEAR REGRESSION

1. Transformable nonlinear models: models involving a single predictor variable in which transforming Y, X, or both results in a linear relationship between the transformed variables.
2. Polynomial models: models involving one or more predictor variables that include higher-order terms such as $B_{1,1}X_1^2$ or $B_{1,2}X_1X_2$.

The formal definition is that if your regression equation looks like Eq. (A.1), it is nonlinear regression. However, this is more difficult than it sounds. Consider the following nonlinear regression equations:

- The Michaelis-Menten model: $f(x, \beta)=(\beta_1 x)/(\beta_2+x)$.

$$Y = \beta_0 + (0.4 - \beta_0) e^{-\beta_1} (x_i - 5) + \varepsilon_i.$$

These both models meet the requirement of fitting the form $Y=f(X, \beta)+\varepsilon$, but that is not immediately obvious without some in-depth knowledge of algebra and regression analysis.

However, there is a much simpler, more intuitive definition of nonlinear regression: If the given model uses an equation in the form $Y=a_0+b_1X_1$, it is a linear regression model. If not, it is nonlinear. It is much easier to spot a linear regression equation, as it is always going to take the form $Y=a_0+b_1X_1$*.

Note [68]: there is an exception. It is true that if your model has an equation in the form $Y=a+bx$, then it is linear. However, there are a few cases where a nonlinear equation can be transformed to mimic a linear equation. If this happens, the nonlinear equation is called "intrinsically linear." i.e., the nonlinear $Y=B_0X/(B_1+X)$ can be transformed with a little algebra to become intrinsically linear as follows:

$$1/Y = 1/\beta_0 + (\beta_1/\beta_0) * 1/X = \theta_0 + \theta_1 * 1/X.$$

Generally, in order to evaluate nonlinear regression four major steps should be taken:

Step 1: Determine whether the regression line fits your data.
Step 2: Examine the relationship between the predictors and the response.
Step 3: Determine how well the model fits your data.
Step 4: Determine whether your model meets the assumptions of the analysis.

A.4 AN EXAMPLE OF NONLINEAR REGRESSION

One example of how nonlinear regression can be used is to predict population growth over time. A scatterplot of changing population data over time shows that there seems to be a relationship between time and population growth, but that it is a nonlinear relationship, requiring the use of a nonlinear regression model. A logistic population growth model can provide estimates of the population for periods that were not measured, and predictions of future population growth.

Independent and dependent variables used in nonlinear regression should be quantitative. Categorical variables, like region of residence or religion, should be coded as binary variables or other types of quantitative variables.

In order to obtain accurate results from the nonlinear regression model, ensure the function, which is specified, describes the relationship between the independent and dependent variables accurately. Proper starting values are also necessary. Poor starting values may result in a model that fails to converge or a solution that is only optimal locally, rather than globally, even if you have specified the right functional form for the model.

5 Viscosity

The direct association between nanofluids' dynamic viscosity (μ_{nf}) and pressure drop, pumping power, and friction coefficient makes researchers give it a high priority as a factor in their research on the thermophysical property. Simultaneously, it is also a critical parameter in heat exchangers using nanofluids as working fluid. Hence, many researchers have addressed the study of dynamic viscosity, focusing on its dependence on different parameters, such as nanoparticles' size, shape, aggregation, concentration, and temperature. The popular approach to determining these parameters' values is experimentation; Procedural limitations and uncontrollable conditions lead to inaccuracy in experimental results, making simulation approaches, such as MD, a necessary means to study the influence of different parameters on μ_{nf}. MD is technically a simulation method to study the atomistic scale's behavior, structure, and physical mechanisms. This section gives a complete summary of the latest findings on the dynamic viscosity of both conventional and hybrid nanofluids. This section covers the evolution of classical models to advanced models, experimental techniques, and MD simulations to provide insight into the effects of various parameters on nanofluids' dynamic viscosity. Table 5.1 presents an overall description of the theoretical models used to predict the dynamic viscosity of nanofluids.

5.1 CLASSICAL MODELS

Different studies propose some classical theoretical formulas to expound the dynamic viscosity of nanofluids. The first effort was made by Einstein [1], who proposed the most classical equation.

Most of the other derivations are established based on Eq. (5.1). This part will summarize the existing classical models in which they are presented and evolve into classical models. As the efforts continue, the classical model has been continuously improved. Some advanced models have been proposed to predict nanofluids' dynamic viscosity.

5.1.1 Evolution of Classical Models

Einstein is commonly known as the initiator of the attempts to propose a theoretical model for nanofluid's dynamic viscosity [1]. The model is only suitable for spherical nanoparticles dispersed in nanofluids with a volume fraction lower than 2%. This model is written as follows:

$$\mu_{nf} = \left(1 + 2.5\Phi_p\right)\mu_{bf} \tag{5.1}$$

DOI: 10.1201/9781032664118-5

TABLE 5.1
Summary of Theoretical Models for the Prediction of the Dynamic Viscosity of Nanofluids (μ_{nf})

Model	Year	Equation	Features	Ref.
Einstein	1906	$\mu_{nf} = (1+2.5\Phi_p)\mu_{bf}$	Spherical nanoparticles, deficient concentration	[1]
Saito	1950	$\mu_{nf} = \left(1+\frac{1.25\Phi_p}{1-\Phi_p/0.87}\right)\mu_{bf}$	Spherical rigid nanoparticles, Brownian motion, small nanoparticles	[4]
Brinkman	1952	$\mu_{nf} = \frac{\mu_{bf}}{(1-\Phi_p)^{2.5}}$	Spherical nanoparticles, high-concentration Continuous medium	[2]
Frankel and Acrivos	1967	$\mu_{nf} = \frac{9}{8}\left[\frac{(\Phi_p/\Phi_{max})^{1/3}}{1-(\Phi_p/\Phi_{max})^{1/3}}\right]\mu_{bf}$	Evenly distributed nanoparticles, maximum concentration limit	[6]
Batchelor	1977	$\mu_{nf} = \mu_{bf}(1+2.5\Phi_p+6.2\Phi_p^2)$	Brownian motion	[3]
Graham	1981	$\mu_{nf} = (1+2.5\Phi_p)\mu_{bf} + \left[\frac{4.5}{(2w/d_p)\cdot(2+2w/d_p)\cdot(1+2w/d_p)^2}\right]\mu_{bf}$	Maximum concentration limit	[5]
Masoumi et al.	2009	$\mu_{nf} = \mu_{bf} + \frac{\rho_p}{72a}\left(\sqrt{\frac{18k_BT}{\pi\rho_p r_p}}\right)\left(\sqrt[3]{\frac{6\Phi_p}{\pi}}\right)$	Temperature, size, and density of the nanoparticles Brownian motion	[7]
Chevalier	2009	$\frac{\mu_{nf}}{\mu_{bf}} = \left[1-\frac{\varsigma}{\varsigma_m}\left(\frac{r_a}{r}\right)^{3-D_f}\right]^{-2}$	Agglomerate size	[10]
Hosseini	2010	$\frac{\mu_{nf}}{\mu_{bf}} = \exp\left[a+c_1\left(\frac{T}{T_0}\right)+c_2(\Phi_p)+c_3\left(\frac{r_p}{1+d_{nl}}\right)\right]$	Capping layer (nanolayer)	[9]
Chandrasekar	2010	$\mu_{eff} = 1+m\left(\frac{\Phi_p}{1-\Phi_p}\right)^n$	A specific area, density, and sphericity of the nanoparticles	[11]

This model excludes particle–particle interaction and thus is only fit for fluids with low concentrations. Brinkman [2] introduced a new model by adding the nanoparticles to the solution to obtain a continuous medium with finite concentration to resolve this challenge. Their model was designed based on Eq. (5.1) and involved high-concentration nanofluids.

$$\mu_{nf} = \frac{\mu_{bf}}{(1-\Phi_p)^{2.5}} \tag{5.2}$$

Eqs. (5.1 and 5.2) are valid only for spherical nanoparticles and ignore nanoparticle motion. The nanoparticles' Brownian motion was included in the model proposed by Batchelor [3]:

$$\mu_{nf} = \left(1 + A\Phi_p + K_H\Phi_p^2\right)\mu_{bf} = \mu_{bf}\left(1 + 2.5\Phi_p + 6.2\Phi_p^2\right) \tag{5.3}$$

where K_H denotes the Huggins coefficient referring to the Brownian motion's direct effect on dynamic fluid viscosity because of intermolecular interactions, and A is the coefficient from the Einstein dynamic viscosity model [1]. Saito [4] developed the model to include the Brownian motion of rigid spherical particles so that the model be applicable for ultra-small nanoparticles having robust Brownian motion. The model is:

$$\mu_{nf} = \left(1 + \frac{1.25\Phi_p}{1 - \Phi_p/0.87}\right)\mu_{bf} \tag{5.4}$$

It was derived under the condition that the solution concentration did not reach the limit. For a case in which the concentration of spherical nanoparticles approached the maximum limit, Graham introduced cell theory to investigate the relation between the μ and the concentration [5]. The relation is written as:

$$\mu_{nf} = \left(1 + 2.5\Phi_p\right)\mu_{bf} + \left[\frac{4.5}{\left(2w/d_p\right)\cdot\left(2 + 2w/d_p\right)\cdot\left(1 + 2w/d_p\right)^2}\right]\mu_{bf} \tag{5.5}$$

where w represents the distance between the nanoparticles, the cell theory considers a nanoparticle in a solution as a structure surrounded by other nanoparticles. Eq. (5.1) correlates with this model under infinite dilution. Similarly, Frankel and Acrivos [6] introduced a model for nanofluids with evenly distributed nanoparticles using an asymptotic technique when the concentration approximates the maximum value, which is expressed as:

$$\mu_{nf} = \frac{9}{8}\left[\frac{\left(\Phi_p/\Phi_{max}\right)^{1/3}}{1 - \left(\Phi_p/\Phi_{max}\right)^{1/3}}\right]\mu_{bf} \tag{5.6}$$

This model reflects the dependence of effective dynamic viscosity on fluid concentration.

5.2 ADVANCED RECENT MODELS

The development of relevant research has made Einstein's classical model look inaccurate in making predictions about most nanofluids' dynamic viscosity, leading to the emergence of some more modern and realistic models. Masoumi et al. [7] suggested a model considering the nanoparticles' Brownian motion and the related effects on fluid temperature and nanoparticles' size and density to predict μ_{nf}. This model addresses the relative velocity between the base fluid and nanoparticles. The model's critical parameters for obtaining the nanofluid viscosity are the temperature, the mean nanoparticle diameter, the nanoparticle volume fraction, the nanoparticle density, and the base fluid physical properties. During the model's development, a correction factor was proposed to deal with the applied simplifications on the boundary condition. This correction factor was calculated using minimal experimental data for nanofluids consisting of 13 and 28 nm of Al_2O_3 and Al_2O_3 nanoparticles and water as the base fluid. The results obtained from the model were compared with the current data collected from other studies on different nanofluids; the results displayed a significant correlation. This model has an overall higher accuracy compared with other theoretical models. The developed model is as follows:

$$\mu_{\text{eff}} = \mu_{\text{bf}} + \frac{\rho_{\text{p}} v_{\text{BF}} r_{\text{p}}^2}{72aw} \tag{5.7}$$

$$a = \mu_{\text{bf}}^{-1}\left[\left(c_1 r_{\text{p}} + c_2\right)\Phi_{\text{p}} + \left(c_3 r_{\text{p}} + c_4\right)\right] \tag{5.8}$$

In Eq. (5.7), μ_{bf} is the base fluid's dynamic viscosity, and the second term is the apparent viscosity. The Brownian motion velocity v_{BF} and the correction factor a are both functions of temperature. c_1, c_2, c_3, and c_4 are empirical constants determined by experiments. Besides, ρ_p, r_p, Φ_p, and w represent the nanoparticles' density, diameter, volume fraction, and distance between the nanoparticles, respectively. The results indicate that the theoretical predictions are significantly in line with the experimental results for CuO/H_2O, CuO/EG, TiO_2/EG, CuO/EG–water, and Al_2O_3/H_2O nanofluids suggesting better accuracy compared to other models [1,3,8].

Hosseini et al. [9] suggested a dimensionless expression to predict the μ_{nf}:

$$\frac{\mu_{\text{nf}}}{\mu_{\text{bf}}} = \exp\left[a + c_1\left(\frac{T}{T_0}\right) + c_2\left(\Phi_{\text{p}}\right) + c_3\left(\frac{r_{\text{p}}}{1 + d_{\text{nl}}}\right)\right] \tag{5.9}$$

in which a is the factor determined by the nanofluid's properties, c_1, c_2, and c_3 are constants from the experimental data, T_0 and T are the references and actual temperatures of the nanofluid, respectively, and d_{nl} denotes the thickness of the capping layer (nanolayer). For the Al_2O_3/H_2O nanofluids, the values obtained from Eq. (5.9) are in line with the experimental results, suggesting a nonlinear relationship between the dynamic viscosity of nanofluids and the volume fraction of nanoparticles.

Chevalier et al. [10] relied on the fractal theory to investigate the dynamic viscosity of an aggregated suspension at high shear rates using the following equation:

$$\frac{\mu_{\text{nf}}}{\mu_{\text{bf}}} = \left[1 - \frac{\varsigma}{\varsigma_{\text{m}}}\left(\frac{r_{\text{a}}}{r}\right)^{3-D_{\text{f}}}\right]^{-2} \tag{5.10}$$

where r_a, ς_m, and D_f, respectively, denote an average diameter of aggregates, crowding, and fractal dimensions. The fractal dimension is a function of aggregation type, the nanoparticles' shape, and the shear flow. For aggregated nanoparticles, the value of 1.8 is considered for the fractal dimension. This model correlates with nanoparticles' dynamic viscosity experimental results, covering a wide range of volume fractions and nanoparticle sizes.

A theoretical model was proposed by Chandrasekar et al. [11] for predicting the dynamic viscosity of Al_2O_3/H_2O nanofluids:

$$\mu_{eff} = 1 + m\left(\frac{\Phi_p}{1-\Phi_p}\right)^n \quad (5.11)$$

where m and n are constants obtained by least squares regression, the fluid phase's composition and properties, the specific area, density, and sphericity of the nanoparticles affect the constant m. The results obtained from the model mentioned above correlated with Al_2O_3/H_2O nanofluids' experimental measurements reported by Nguyen et al. [12]. There is a linear relationship between dynamic viscosity and nanofluid concentration at low concentrations; however, this relationship becomes nonlinear as the concentration exceeds 2%. The reason for this behavior is attributed to the particle–particle interactions at high concentrations.

The models mentioned fail to predict precisely applicable for a wide range of dispersed nanomaterial types. Hence, these models are only suitable for predicting metal oxide-based nanofluids' dynamic viscosity at different temperatures and concentrations. Many efforts have focused on measuring dynamic viscosity through experimental approaches. The following section gives a full review of the frequent techniques for measuring nanofluid dynamic viscosity.

5.3 EXPERIMENTAL MEASUREMENT TECHNIQUES

Newton's assumption considers an infinite plane to measure a fluid's dynamic viscosity; however, in reality, there is no viscometer made of an infinite place to overcome which some stable and straightforward structures and geometrics have been proposed, including the most popular capillary, concentric cylinders, cone, and plates (see Figure 5.1). This part gives a brief overview of the measurement principle to give better insight into the issue besides discussing the advantages of the three mentioned experimental measurement techniques.

5.3.1 Capillary Viscometer

Capillary viscometers are widely used, cost-effective, and straightforward devices to conveniently measure nanofluids' dynamic viscosity (see Figure 5.1a) [13–20] Concisely, axial pressure drop and the volume flow rate of nanofluid flowing through a nanofluid flow through a capillary with a circular cross-section are measured. These parameters are then used in Hagen–Poiseuille's law to determine the dynamic viscosity. This technique has critical advantages, including straightforward operation and temperature control, low fluid volume requirement, and relatively high shear rates.

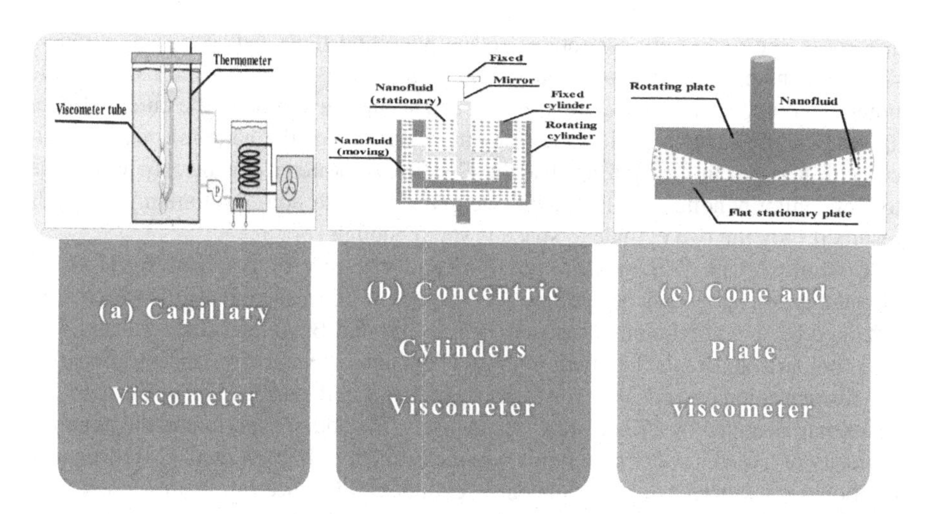

FIGURE 5.1 Schematic representation of the most prevalent measurement methods for μ_{nf}. (a) Capillary viscometer [13], (b) concentric cylinders' viscometer, (c) cone and plate viscometer.

Yiamsawas et al. [13] utilized this technique to measure the dynamic viscosity of TiO_2 and Al_2O_3 nanoparticles suspended in an EG–water mixture. Their experimental setup was composed of a transparent water bath, a high-precision stopwatch, a thermometer (with a precision of 0.1°C), and a viscometer. The viscometer was immersed in the water bath, through which the nanofluid flows. Then, the flow rate and the temperature were measured by stopwatch and thermometer, respectively. Every test for volume fractions of 1%, 2%, 3%, and 4% was repeated three times to ensure measurement accuracy. The results revealed a meaningful difference between the experimental results. They predicted values, indicating the theoretical model's inaccuracy in predicting nanofluids' dynamic viscosity.

The measurement of the dynamic viscosity of nanofluids using a capillary micro viscometers technique was proposed by Chevalier et al. [15]. This method is also applicable for measuring the local pressure drop inside a microchannel. In another attempt, Jarahnejad et al. [16] used the capillary viscometer to measure the dynamic viscosity of Al_2O_3/H_2O and TiO_2/H_2O nanofluids for temperatures ranging from 20°C to 50°C. They used a thermostat bath set up to provide constant temperature during the measurement; all experiments were conducted under atmospheric pressure. The base fluid's dynamic viscosity, that is, water, was also measured to verify the setup's accuracy. The uncertainty analysis was performed by averaging the samples' experimental results, exhibiting good reliability with a maximum standard deviation lower than 1%. Moreover, some other works used the capillary viscometer approach to explore nanofluids' dynamic viscosity [17–22]. It can be concluded from the experimental results mentioned above that the viscometer shows good reliability and accuracy.

5.3.2 Concentric Cylinders

Figure 5.1b shows the Couette concentric cylinder viscometer, the most popular type of rotational viscometers (concentric cylinder, cone, and plate viscometers). In this apparatus, the outer cylinder rotates so the nanofluid flows between two cylinders while the inner cylinder is stationary. Using parameters including the radius of the inner and outer cylinders, the rotational speed ω, and the torque of the outer cylinder applied on the inner cylinder, the dynamic viscosity of the nanofluid can be calculated. It is worth mentioning that the Couette viscometer is generally applicable to Newtonian nanofluids. This technique gives accurate non-Newtonian nanofluids results if only the two cylinders' space is small enough [14].

Contreras et al. [23] measure the dynamic viscosity of graphene-Ag/H_2O-EG nanofluids. To verify their results, they compared the experimental results with results obtained by the correlations proposed by Einstein and Brinkman (Eqs. 5.1 and 5.2). The comparison reveals that there is an excellent agreement at low concentrations. As the concentration exceeds 0.01 vol%, the accuracy of both correlations of Einstein and Brinkman declined by 8% and 12%, respectively. In another experiment, the dynamic viscosity of CuO/DI water nanofluids was measured by Minakov et al. [24] by using a rotational viscometer based on concentric cylinders. The uncertainty analysis of their experiment showed that the uncertainty is approximately 1%. Their results revealed a strong relationship between the dynamic viscosity of nanofluid and the shear rate. It should be noted that for Newtonian nanofluids with a concentration of 0.25 vol%, the dynamic viscosity coefficient is no longer dependent on the shear rate.

5.3.3 Cone and Plate Approach

A cone and plate viscometer, composed of a cone with a wide angle located on a flat stationary plate (see Figure 5.1c), provides researchers with an instrument to measure fluid viscosity. In this instrument, the nanofluid is placed in the gap between the two plates. Since the gap angle is usually small (<4°), the edge effect is negligible, the rotating plate keeps rotating at an angular velocity of ω_0, and the torque can be calculated. Four parameters, including angular velocity, torque, gap angle, and cone radius, are used to calculate the nanofluid's dynamic viscosity. One advantage of the cone and plate viscometer over the other instruments mentioned above is that this approach is applicable for both Newtonian and non-Newtonian fluids because the gap between the two plates is minimal. Therefore, the cone and plate viscometer have extensive application in non-Newtonian fluids [14]. The definition of the shear rate could explain why the cone–plate viscometer is applicable for measuring nanofluids' dynamic viscosity. It is defined as the decreasing flow velocity's relationship and the laminar flow sample's thickness. In the cone–plate viscometer, both thickness and flow velocity are varied. The changes are from the middle (the minor position) to the sample's edge. Since the rotor is spinning at a constant angular velocity, the flow velocity is a function of the distance from the axis of rotation (the smallest one in the middle). As a result, with the appropriate selection of rotor angle and gap, ignoring the measurement error, obtaining a constant shear rate is attainable. By using such a system, the inspection of viscoelastic structures through oscillatory measurements is possible. Observing and studying the viscoelastic structure is of great importance when working with nanofluids exhibiting complex non-Newtonian behavior [25–27].

The theoretical and experimental investigation of the dynamic viscosity of Al_2O_3/H_2O nanofluid was carried out by Chandrasekar et al. [11]. For the experimental part, they utilized a Brookfield cone and plate viscometer to measure the dynamic viscosity. This instrument's outstanding feature, provided by Brookfield Engineering Laboratory, US, was a minimal gap between two plates (0.8°), which is an outstanding feature. The gap provided by the device between the cone and plate was 0.013 mm through which the nanofluid flows. The measurements were done at room temperature, and the results indicate an accuracy of 5%. Besides, the dynamic viscosity of the Ag-MgO/H_2O hybrid nanofluid was also measured by a Brookfield cone and plate viscometer [28]. By an electronic device, the study applied the necessary modifications to the gap to reach the excellent value, that is, 0.013 mm. The measurement of the nanofluid's viscous drag to the spindle was obtained by calibrating the spring's deflection. It was revealed from the results that there is a significant difference between the experimental values of dynamic viscosity and those predicted by theoretical models. This difference can be attributed to factors, including the nanoparticles' concentrations, size, and shape. Asadi et al. [29] prepared a stable suspension of MWCNTs and ZnO dispersed in engine oil. They used a cone and plate viscometer to measure the dynamic viscosity of the MWCNTs–ZnO/engine oil hybrid nanofluid at varying temperatures and concentrations. In their experiment, a device was used to quickly set the nanofluids' temperature to ensure that they remained stable during the measurement. Żyła et al. investigated the non-Newtonian behavior of AlN/EG [27], Si_3N_4/EG [30] as nanofluids, and EG as base fluid in which the graphite and nanodiamonds were dispersed at constant temperature 298.15 K [31].

5.3.4 Other Measurement Methods

In addition to the discussed methods, (i) the Rheo-nuclear magnetic resonance (NMR) method and (ii) the RheoScope method can also be used, although they are less common. The former technique uses a magnetic field, enabling the real-time inspection of the fluid using NMR imaging. This method is superior to other methods in that it can detect a slippery effect between the sample and the device by comparing the linear velocity of the rotor surface and the laminar flow inside the nanofluid [32]. The Rheo-NMR method's advantages include determining whether the chemical structure changes during the shearing of the sample and could be employed in this study. The latter is an instrument to detect and measure fluid viscosity using a rotational rheometer coupled with the optical microscope. With the use of this instrument, the stationary plate can be observed through the measurement. However, in this method, limited magnification is possible [32].

Thus, to investigate the reason behind the shape of dynamic viscosity curves of the $MgAl_2O_4$/DG nanofluids found in their previous study, Żyła et al. [32] utilized both methods to obtain those materials' rheological profile [33]. The Rheo-NMR method can eradicate the chemical changes and slippage effects between the sample and the rotor. Several experiments using this method revealed that particle agglomeration and sedimentation could be the reason behind the unexpected behavior.

5.4 EXPERIMENTAL STUDIES

This section aims to give a full review of experimental advances concerning the factors affecting nanofluid dynamic viscosity and investigate and comprehensively report the influences associated with temperature, concentration, size, and shape of the NPs, pH, and ultrasonication. Figure 5.2 shows the intrinsic mechanism of several factors on the dynamic viscosity of nanofluids. This section provides valuable information to elucidate that a combination of different factors constitutes nanofluid's dynamic viscosity. This information is valuable and of great importance for future engineering applications of nanofluids.

5.4.1 Effect of Temperature

The earliest studies on nanofluids commonly posit that temperature considerably affects nanofluids' dynamic viscosity. Hence, the effect of temperature on nanofluids' dynamic viscosity has been the topic of many studies [34–44]. For instance, Esfe et al. [36] designed different temperature conditions (5°C–65°C) to study μ_{nf} with Al_2O_3 nanoparticles dispersed in engine oil, finding that μ_{nf} quickly ascended in response to temperature increase, a reaction that confirms the increased temperature's ability to reduce the μ_{nf} considerably. In a similar experimental setting, Toghraie et al. [38] tried to corroborate the association between μ of Fe_3O_4/H_2O nanofluids and temperature fluctuations as μ displayed a consistent decreasing trend following an increase from 20°C to 55°C, a reaction visible in different volume concentrations of

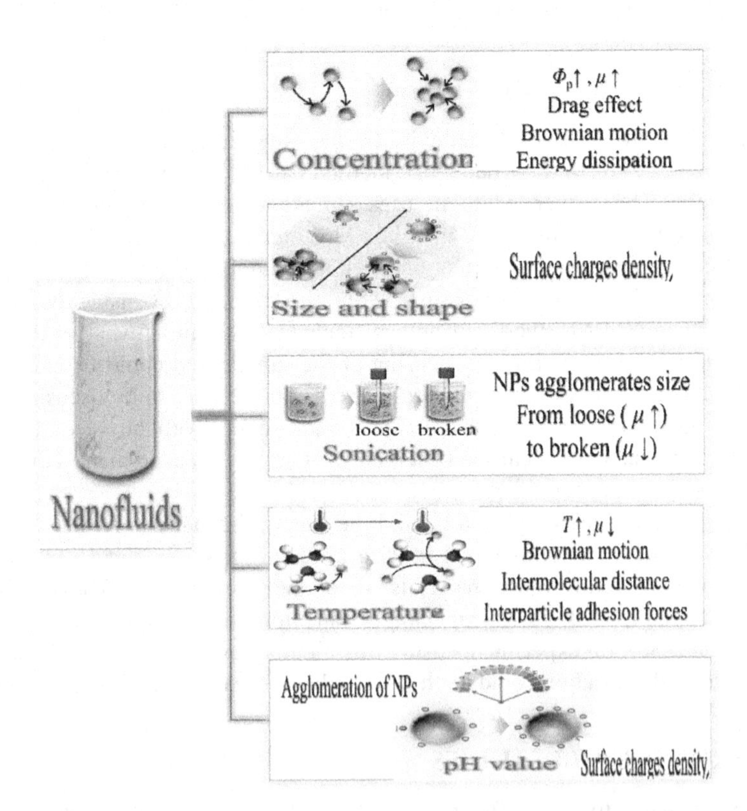

FIGURE 5.2 Schematic diagram of the intrinsic mechanism of various factors on the μ_{nf}.

nanofluids. Similarly, researchers in a relevant study investigated the same association in CuO/H_2O nanofluids under (10°C–50°C) and (1–10 wt%) (Pastoriza-Gallego et al. [35]), reporting that as the temperature increased, μ decreased and under 50°C and %7 wt CuO/H_2O nanofluid, the μ slightly equals the base fluid's μ value at 25°C. Yang et al. [39] investigated the μ of viscoelastic-fluid-based nanofluids with homogeneously dispersed Cu nanoparticles, consisting of an aqueous solution of CTAC and NaSal. They also reported that μ of viscoelastic nanofluids displayed a decreasing trend in reaction to rising temperature regardless of any shear rate, which is associable with more considerable intermolecular distance and less strong interparticle adhesion forces as a result of increased temperature, leading to lower μ_{nf} values.

Besides, several studies have dealt with the developmental trend of relative dynamic viscosity (ratio of nanofluid to base dynamic fluid viscosity, μ_{nf}/μ_{bf}) over varied temperature ranges. For instance, Esfe et al. [38] and Sharifpur et al. [40] agreed on the consistently improving effect of temperature on relative dynamic viscosity. More specifically, on the significantly quicker decreasing rate of μ_{bf} compared to the nanofluid during the heating process. In contrast, the results of many studies suggested otherwise; [41] reported the effect of increasing temperature on relative dynamic viscosity as non-significant, further confirming the temperature's indistinguishable effect on the nanofluid and the base fluid and the close decline rate. The experiments of Baratpour et al. [42] under 30°C–60°C and a volume concentration of below 0.025% revealed the temperature's insignificant impact on the relative dynamic viscosity of SWCNT/EG nanofluids. Despite the reports mentioned so far, the results of many studies have also referred to the nonlinear increasing trend in relative dynamic viscosity, given the observable relative reduction in reaction to increasing temperature [38,45]. Studying the relationship between temperature and the relative dynamic viscosity in a semiconductor SiO_2/silicone oil nanofluid, Murshed et al. [43] reported it highly decreased in reaction to increased temperature.

Current experimental data and relevant literature suggest that temperature on μ_{nf} is a function of temperature range and that temperature similarly affects the nanofluid and the base fluid. Hence, it is critical to design experiments with different temperatures and nanofluid types at different conditions to explore and corroborate the effects of temperature on μ_{nf} in future research.

5.4.2 Effect of Nanoparticle Concentration

Quite expectedly, consistently adding nanoparticles to the base liquid consistently enhances μ_{nf}. The studies mentioned thus far have suggested the direct positive relationship between the highest volume fractions of nanoparticles and the values of μ_{nf}, to confirm which many experiments have been conducted: Chen et al. [46] proposed a new method to prepare a surfactant-free CNT/EG nanofluid with high dispersion of CNT in the base fluid via mechanochemical reaction. Under below 0.4 vol% CNTs, the μ of the CNT/EG the base fluid displayed a higher nanofluid associable with the nanoparticles' lubrication. The relative dynamic viscosity also increased in reaction to the increased nanoparticle loading, an observation curiously more noticeable under over 55°C. They continued to investigate the μ changes of MWCNTs dispersed in different base fluids, including water, EG, glycerin, and silicone oils [47], reporting the

Newtonian behavior of MWCNT/silicone oil and MWCNT/glycerol nanofluids. The MWCNTs lubricate MWCNT/H_2O nanofluids, leading to lower μ_{nf} than the base fluid under lower volume fractions of MWCNTs. Żyła and Fal [48] studied the μ of SiO_2/EG nanofluids at a constant temperature of 298.15 K, finding that μ_{nf} has a positive, linear relationship with the volume fraction. Their results concerning the improvement level outperformed the Einstein model's predictions significantly.

Sharif et al. [49] investigated the impact of nanoparticle volume concentration on μ in Al_2O_3/polyalkylene glycol nanofluids. They examined μ in the range of 0.05–1 vol%, finding that the relative dynamic viscosity varied linearly with concentration at 313.15 K. Moreover, Lee et al. [50] studied the association between μ and SiC/DI water nanofluids concentration in terms of the concentration-dependence of μ. They reported the direct relationship between the μ of SiC/DI water nanofluids and nanoparticle volume fraction as increased volume fraction increased μ. They commonly estimate that the higher the increase rate of μ, the lower the potential application of nanofluids. Corcione et al. [51] examined the influence of many factors on μ_{nf} in multigroup experimental data in the current related literature, mainly focusing on the nanoparticle concentration's impact on μ and finding the increasing linear trend of μ relative to concentration. Suganthi et al. [52] examined the rheological properties of ZnO $C_3H_8O_2$ (PG) nanofluid to explore the combined effects of temperature and concentration on μ_{nf}. They realized that high temperatures impaired the hydrogen bond in the PE, further reducing the μ_{nf}. ZnO nanoparticle loading impaired hydrogen bonds even more. The decreasing trend completely exceeds the increase in μ as a result of adding nanoparticles leads to a decrease in the μ_{nf}.

The Brownian motion's influence generally creates a noticeable drag effect on individual nanoparticles and increases μ_{nf} in reaction to the elevation of nanoparticle volume fraction dispersed in the base fluid, which, in turn, elevates the total drag effect in the nanofluid and the occurrence probability of aggregation, causing the nanofluid to generate severe energy dissipation not conducive to the practical application of the nanofluid [53] and potentially increase μ_{nf}. Thus, a well-dispersed nanofluid suspension generally displays a lower μ_{nf}.

5.4.3 Effect of Nanoparticle Size and Shape

Nanoparticle's size and shape are also essential, indispensable factors when researching μ_{nf}, but different studies' conclusions are not consistent [7,15,16,54–59]. He et al. [54] investigated the μ of TiO_2 nanofluids with nanoparticle sizes of 95, 145, and 210 nm, revealing the direct relationship between size and μ_{nf} improvement. When the nanoparticle size is 210 nm, the μ_{nf} increase is more than 7% at a concentration of 0.6%. Nguyen et al. [56] examined Al_2O_3/H_2O nanofluid and found that under low concentrations (4%), the effect of nanoparticle size on μ_{nf} is limited. However, at high concentrations, a larger nanoparticle size (47 nm) increases μ_{nf} considerably higher than smaller-size nanoparticles (36 nm). This difference was attributed to the molecular structure of nanofluids. Furthermore, Jarahnejad et al. [16] reported a higher μ in 200-nm nanoparticles than 250-nm particles in the study nanofluid, corroborating the aforementioned findings. In contrast, 250-nm nanoparticles give the nanofluid a higher μ_{nf} than that of a nanofluid with 300-nm nanoparticles, which, although sounding contradictory, can be attributed to the inhomogeneity of the nanofluid.

Additionally, many studies refer to the decreasing effect of larger nanoparticle size on μ_{nf}. For instance, the study of Al_2O_3/H_2O by Anoope et al. [57] confirmed that in Al_2O_3 45 nm nanoparticles μ_{nf} is higher than when the size is 150 nm, correlating with Einstein's theoretical prediction [1]. Kwek et al. [58] investigated the same phenomenon in the size range of 10–150 nm and observed that as the nanoparticle size increased, the relative μ_{nf} decreased from 1.9 at a nanoparticle size of 10 nm to 1.5 at 150 nm, associable with higher aggregation probability of the smaller nanoparticles due to weaker surface charges. This observation correlates with the findings of Masoumi et al. [7] and Lu et al. [55]. Studies on SiO_2 nanofluids report similar results. Chevalier's research [15] for three nanoparticle size (35, 94, and 190 nm) in SiO_2/ethanol nanofluids also reported a decreasing trend of μ in responsc to increased nanoparticle size. Namburu et al. [59] examined the values of μ in three nanoparticle sizes, including 20, 50, and 100 nm, in SiO_2/H_2O nanofluids. The largest μ_{nf} was observed in the size of 20 nm, followed by the 50 nm-sized nanofluids. In contrast, increasing the temperature diminished the difference in μ_{nf} caused by nanoparticle size, suggesting the insignificant impact of nanoparticle size on μ_{nf} compared to other factors.

A few studies address the relationship between μ_{nf} and nanoparticle shape, from which some rules can be elucidated. According to a study by Timofeeva et al. [45,60] on the μ of Al_2O_3/EG and Al_2O_3/H_2O nanofluids, slender or flat nanoparticles with lower sphericity generally achieve greater μ_{nf} values than ideal spherical fluids at the same volume fraction, which is associable with the inherent constraints related to the structures of nanoparticle rotation in Brownian motion. Nanoparticles with lower aspect ratios are preferred to obtain a low μ nanofluid.

5.4.4 Effect of pH

The pH affects the nanofluids' stability and dispersion, thus, indirectly affecting μ_{nf}. Hence, the pH value is a critical factor for thoroughly analyzing μ_{nf}. Zhao et al. [61] examined the μ of SiO_2/H_2O nanofluids with different particle sizes and pH values, considering aggregation as a factor. Under the pH condition (5–7), and the nanoparticle diameter of below 20 nm, μ_{nf} was the highest after which the effect of pH value on μ lost its significance as the nanoparticles' diameter increased, remaining slightly constant as the pH value changed, which can be associated with an electric double layer (EDL) over the nanoparticles and the change in the aggregates' fractal dimension.

Timofeeva et al. [45] studied the μ_{nf} of several Al_2O_3 nanoparticle shapes dispersed in a base fluid composed of an equal volume of EG and water. They realized that modifying the suspension's pH value probably influences the surface charge over the nanoparticles. They observed that the μ of the Al_2O_3/H_2O nanofluid was successfully reduced by 31% without affecting k_{nf}. Note that the nanoparticles' surface charge is another critical parameter influencing μ_{nf}, and moderating the suspension's pH value is a determining means of modulating the nanoparticles' surface charge.

Besides, Wankam et al. [62] studied the impacts of pH on μ of heat transfer nanofluids. The rheological curves of 0.1 wt% ZrO_2 nanoparticles dispersed in water at different pH values (4, 6, 8, 10) were obtained, showing that the μ of ZrO_2/H_2O nanofluid was reduced by 20% for a pH=6 or 8 compared to pH=4 or 10. Wang et al. [63] attained a powerful repulsive force and reduced aggregation of nanoparticles by adjusting the suspension's pH to gain a well-dispersed suspension of low μ_{nf}. Several studies reported similar increasing trends. There was an optimized pH value (pH_o) for the lowest μ_{nf} of many nanofluids, for example, $pH_o = 8$ for Al_2O_3/H_2O nanofluid, $pH_o = 9.5$ for Cu/H_2O nanofluids.

The papers discussed briefly thus far stress that a well-adjusted pH can significantly reduce μ_{nf}. Since pH and temperature are both controlling factors of the surface charge density, nanoparticle aggregation and influences on μ generally originate from a decrease in surface charge density.

5.4.5 Sonication Time Effect

Ultrasonication is a popular method of modifying μ_{nf}, destroying aggregation, and promoting nanoparticles' uniform dispersion in the base fluid [64]. Many studies have dealt with the relationship between nanofluids and ultrasonication duration, reporting that μ_{nf} generally displays a decreasing trend under the effect of ultrasonication duration. Nevertheless, no criterion is available to determine an optimum ultrasonication period for nanofluids. The nanofluid type plays a determining role.

Ruan et al. [65] investigated the influence of ultrasonication time on μ of MWCNT/EG nanofluids at a concentration of 0.5 wt%, reporting that prolonged ultrasonication increased μ_{nf} and μ_{nf} reached a maximum value of 1.44×10^5 kJ/m^3 after 40 min, followed by a decrease further ultrasonication until getting close to the μ of pure EG after 1355 min. Furthermore, experiments at different shear rates yielded the same results since well-dispersed MWCNTs have a vast specific surface area and therefore have a higher μ_{nf} while reducing the MWCNTs' aspect ratio and the destruction of the 3-D network results in a lower μ_{nf}. Garg et al. [66] also studied the difference of μ_{nf} of CNT/H_2O nanofluids with ultrasonication time. They observed that μ_{nf} increases to the maximum value and then decreases with ultrasonication time. Mahbubul et al. [67] analyzed the μ_{nf} of Al_2O_3/H_2O nanofluids' ultrasonication effect at a concentration of 0.5 wt%. Their experimental results resembled the findings of Ruan et al. and Garg et al. They reported that 40 min was an optimal ultrasonication time, after which μ_{nf} continues to decrease until it approaches μ_{bf} at 120 min. This phenomenon could be attributed to the nanoparticles' aggregation gradually loosening before 40 min, followed by the destruction of nanoparticles aspect ratio by prolonged ultrasonication, reducing μ_{nf} [68].

The μ_{nf} also displayed the same trend with the ultrasonication duration at different temperatures. Menbari et al. [69] investigated the μ change of Al_2O_3-CuO/H_2O nanofluids, reporting that as ultrasonic time increased, μ_{nf} initially increased then decreased, reaching a maximum after 100–120 min. Silambarasan et al. [70] studied the impact of ultrasonication time on μ of TiO_2/H_2O nanofluids, reporting that μ_{nf} decreased with increasing ultrasonication time associated with decreased nanoparticle agglomeration and size. The findings reported thus far suggest that μ_{nf}'s significant dependence on the ultrasound duration can pave the way for the effortless production of high-performance and high stability nanofluids.

This book examined the changes in μ_{nf} under the effect of many factors. Notice that most of the reported studies have been conducted in a relatively low-temperature range (5°C–60°C), and there is still a lack of exploration at higher temperatures.

Therefore, an elevated temperature favors an increase in the molecular spacing and thus results in the reduction of μ_{nf}, but has an equivocal effect on the relative dynamic viscosity. The nanoparticle concentration mechanism on μ_{nf} is easy to understand, and their contribution to the reduction of μ_{nf} is unfavorable. As the sonication time increases, μ_{nf} usually increases first and then decreases. It is noteworthy that the studies on the effect of nanoparticle size on μ_{nf} have not yet reached a consensus. In particular, the effect of nanoparticle shape on μ_{nf} is still unclear.

Moreover, a well-adjusted pH value can have the same reductive influence on μ_{nf}. As presented in Figure 5.3, with increased loading of nanoparticles, the μ of all types of nanofluids, such as Al_2O_3 [12,13,55,71,72], TiO_2 [13,72–74], CuO [75], Fe_2O_3 [76], MgO [77], SiO_2 [78,79], and SiC [50] represents significant rising trends. Clearly, the volume fraction of nanoparticles in the research on the nanofluids based on graphene [30,80] and CNT [46,47,81] is between 0% and 1% since a trace amount of CNT or graphene significantly improves k_{nf} and even more downgrades μ. In other words, when a trace amount of CNT or graphene is loaded in the base fluid, an apparent increase in μ_{nf} emerges. This is also one of the critical obstacles for graphene and CNT nanofluids used as a new generation of heat transfer fluid at this stage.

5.5 HYBRID NANOFLUIDS

The aim of developing the proposed hybrid nanofluids is a trade-off between the advantages and disadvantages of single-type nanoparticle suspension, aiming to gain a medium with more significant heat transfer characteristics for practical engineering applications. Many papers have addressed the hybrid nanofluids' better heat transfer performance, despite the current gap for further research to resolve the challenges of stability and dispersion for practical applications. Despite all that, μ remains an indispensable problem in hybrid nanofluids. Theoretical models generally fail to predict the (μ_{hnf}) of hybrid nanofluids precisely. Hence, as presented in Table 5.2, researchers have proposed new correlations to estimate μ_{hnf} accurately.

5.5.1 THEORY

As stated earlier, Batchelor [3] introduced Eq. (5.3) as a correlation to predict μ of single-type nanoparticle suspensions with Brownian motion, and Ho et al. [82] represented the accuracy of Eq. (5.3) in μ_{hnf} prediction.

Afrand et al. [83] suggested SiO_2-MWCNT/engine oil hybrid nanofluids as heat engine coolants and lubricants and studied the relationship between temperature and solid volume fraction and μ. They reported a stark difference between μ and the existing well-known model, proposing a correlation that confirmedly predicts the μ_{hnf} accurately. Its general form is as follows:

$$\frac{\mu_{nf}}{\mu_{bf}} = c_0 + c_1\Phi_p + c_2\Phi_p^2 + c_3\Phi_p^3 + c_4\Phi_p^4 \tag{5.12}$$

where μ_{nf} and μ_{bf} represent the nanofluid and base fluid's dynamic viscosity, respectively, and c_0, c_1, c_2, c_3, and c_4 are coefficients for each temperature, with given values

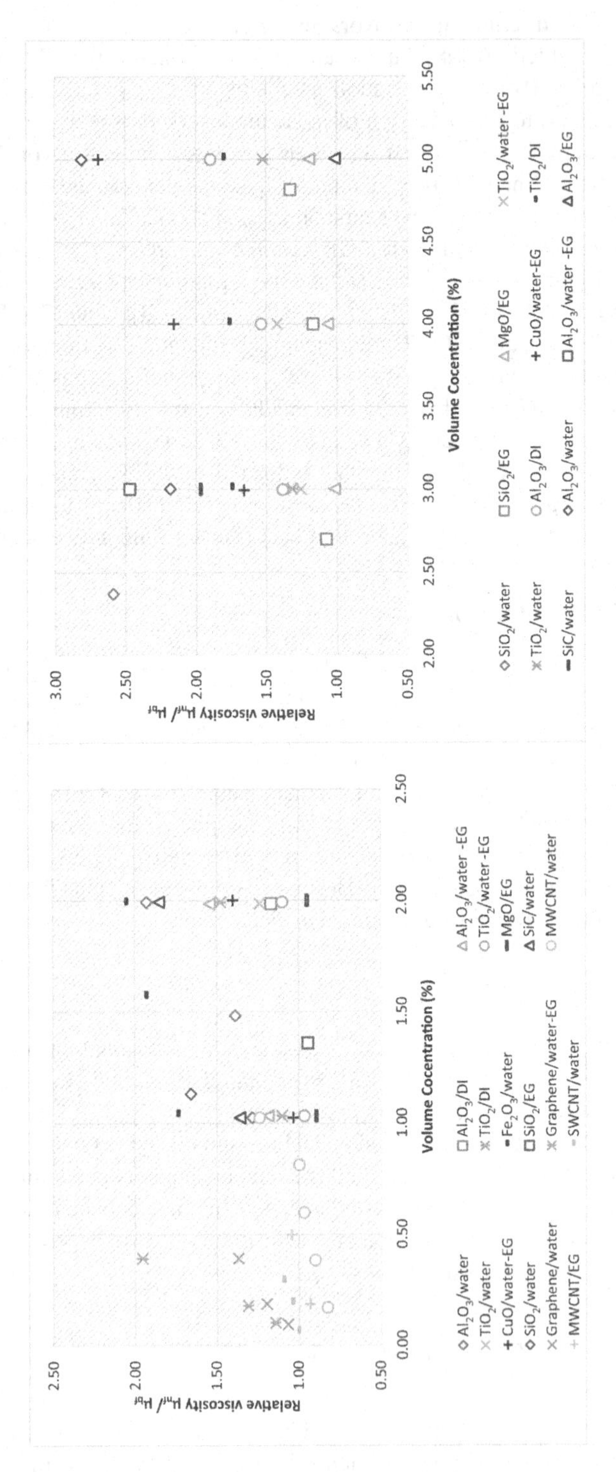

FIGURE 5.3 Summary and comparison of the relative viscosity in different literature.

TABLE 5.2
Summary of Theoretical Models for the Prediction of the Dynamic Viscosity of Hybrid Nanofluids (μ_{hnf})

Model	Year	Equation	Features	Ref.
Afrand et al.	2016	$\frac{\mu_{nf}}{\mu_{bf}} = c_0 + c_1\Phi_p + c_2\Phi_p^2 + c_3\Phi_p^3 + c_4\Phi_p^4$	Valid only in temperatures ranging from 25°C to 60°C and concentrations ranging from 0.05 to 1 vol%	[83]
Esfe et al.	2017	$\mu_{nf} = \left(1 + 32.795\Phi_p - 7214\Phi_p^2 + 714600\Phi_p^3 - 0.1941\times10^8\Phi_p^4\right)\mu_{bf}$	Valid only in a concentration ranging from 0 to 0.02 vol%	[84]
Dardan et al.	2016	$\frac{\mu_{nf}}{\mu_{bf}} = 1.123 + 0.3251\Phi_p - 0.08994T + 0.002552T^2 - 0.00002386T^3 + 0.9695\left(\frac{T}{\Phi_p}\right)^{0.01719}$	Valid only in a concentration ranging from 0.0625 to 1 vol% and temperature from 25°C to 50°C.	[85]
Asadi et al.	2016	$\mu_{nf} = 796.8 + 76.26\Phi_p + 12.88T + 0.7695\Phi_p T + \frac{-196.9T - 16.53\Phi_p T}{\sqrt{T}}$	Valid only in temperature ranging from 5°C to 55°C and concentration ranging from 0.125 to 1 vol%.	[29]

in Ref. [83]. Eq. (5.12) is valid only in the 25°C–60°C temperature range and solid concentration ranging from 0.05 to 1 vol%.

Consequently, Esfe et al. [84] also achieved a similar correlation to Eq. (5.12) based on research on the rheological behavior of MWCNT–ZnO/engine oil hybrid nanofluids. Moreover, the group suggested another model that predicts the μ of Ag-MgO/H_2O hybrid nanofluids by curve-fitting the experimental data. The model is as follows:

$$\mu_{nf} = \left(1 + 32.795\Phi_p - 7214\Phi_p^2 + 714600\Phi_p^3 - 0.1941\times10^8\Phi_p^4\right)\mu_{bf} \quad (5.13)$$

where $0 \le \Phi_p \le 0.02$. They also reported the condition specificity of the model for accurately predicting μ_{nf} [28].

Dardan et al. [85] suggested a new correlation to predict μ of Al_2O_3–MWCNTs/engine oil hybrid nanofluids, a function based on nanoparticle volume concentration and temperature as follows:

$$\frac{\mu_{nf}}{\mu_{bf}} = 1.123 + 0.3251\Phi_p - 0.08994T + 0.002552T^2 - 0.00002386T^3 + 0.9695\left(\frac{T}{\Phi_p}\right)^{0.01719} \quad (5.14)$$

Comparing the results indicated that most of the values obtained in experiments were close to the predicted values, with a maximum deviation is only two and an optimal consistency between the measured data and predicted values, which made the suggested correlation applicable in engineering. Eq. (5.14) is valid for nanofluids with concentrations ranging from 0.0625 to 1 vol% and temperatures from 25°C to 50°C.

Asadi et al. [29] conducted a recent experimental study on MWCNT–ZnO/engine oil hybrid nanofluids' dynamic viscosity, proposing a new correlation to estimate μ_{nf} by fitting the experimental data. Likewise, it is also a function based on the nanofluid's solid volume concentration and temperature, which is described more accurately below:

$$\mu_{nf} = 796.8 + 76.26\Phi_p + 12.88T + 0.7695\Phi_p T + \frac{-196.9T - 16.53\Phi_p T}{\sqrt{T}} \quad (5.15)$$

The comparison indicates that the correlation is significantly consistent with the experimental data. The correlation is only fit under 5°C–55°C and solid concentrations ranging from 0.125 to 1 vol%.

5.5.2 Experimental Studies

Dynamic viscosity is an imperative factor in resisting the relative motion of nanofluids. Hybrid nanofluids can be more effective than mono nanofluids in enhancing heat transfer performance.

Simultaneously, improved heat transfer enhanced μ_{hnf}. That is to say, Afrand et al. [83] revealed that the temperature increase positively affects μ_{hnf}, a phenomenon associated with the increase in the temperature and the impairment of the interactions between molecules. They observed that increased solid volume fraction increased the likelihood of nanoparticle agglomeration, significantly increasing the μ_{hnf}, highlighting the impact of temperature on μ_{hnf}. Many relevant studies have corroborated this effect. Moreover, the association between temperature and solid volume fraction and relative viscosity indicates the direct relationship between these components, as an increase in any of the two increases relative viscosity. However, the effect of solid volume fraction is more pronounced. Soltani et al. [86] experimented with MgO-MWCNT/EG hybrid nanofluids to investigate the impacts of temperature. Particle concentration on μ_{hnf}, reporting findings consistent with the research discussed earlier: μ_{hnf} increased with nanoparticle concentration and decreased with an increase in temperature. At the same time, it gradually improved at a low volume fraction (<0.4%) and rapidly improved at a high volume fraction (0.8% and 1%). Soltani et al. [86] also attributed these associations between the objects of their experiments to the MWCNT chains formed by the interconnection of MWCNTs.

Similarly, Akilu et al. [87] studied the thermophysical properties study of SiO_2–CuO–C/EG–glycerol hybrid nanofluid, showing that μ_{hnf} displayed an increasing trend following increased loading of nanoparticles because of the more vital intermolecular interaction between the base fluid and nanoparticles. In the end, they compared the SiO_2–CuO–C/EG–glycerol hybrid nanofluid's relative viscosity with the SiO_2/EG–glycerol nanofluid, finding that the μ_{hnf}/μ_{bf} was considerably lower than that of the μ_{nf}/μ_{bf} with a difference of up to three times maximum. The underlying reason behind this phenomenon is that since the ρ of SiO_2 is significantly smaller than the ρ of the CuO–C hybrid nanoparticles, fewer particles at the same weight favor the nanoparticles' flow and reduce the flow resistance, consequently reducing μ_{hnf}.

The studies discussed above indicate a general agreement. However, this section has concluded that despite a large body of research exploring the effects of temperature and concentration on μ_{hnf}, few reports deal with the effects of other critical factors. As a modern, high-performance heat transfer fluid, hybrid nanofluids should receive accurate and complete analysis and discussion. Thus, analyzing the impacts of the nanoparticles' size and shape, ultrasonication time, and the base fluid's pH is essential. It is unclear if this effect is the same as the nanofluids' effects, and future research should address this.

5.6 MD SIMULATION

Despite the vast body of experimental research on μ_{nf}, many still endeavor to identify new accurate, and reliable methods to calculate more tangible results. Section 3.16 discussed the application of the MD simulation methods in calculating k_{nf} and studying μ_{nf}. Some pioneering researchers have used this method to investigate μ_{nf} and have reported novel results. In MD simulation, EMD and NEMD apply to calculate μ_{nf}. Specifically, the EMD simulation principle is based on the Green–Kubo formula, while NEMD offers three specific μ_{nf} calculation methods. The first method can be concisely described as μ_{nf} calculated by the SLLOD equation of motion, the second method creates shear stress between the fluid layers by moving the walls. The last method is the standard reverse non-equilibrium method.

Vakili-Nezhaad et al. [88] investigated the relationship between μ and different graphene nanosheet sizes (length: 10×10 nm and 20×20 nm) in graphene/H_2O nanofluids. They used L–J potential to describe all the water molecules' interactions and Coulomb's law to model electrostatic interactions. A combination of L–J potential and Coulomb potential was employed to describe the atomic interactions between graphene atoms. In contrast, they used Lorentz–Berthelot's rule to address the interaction between the molecules and the graphene nanosheets. They specifically designated μ_{nf} via the Green–Kubo method based on the fluctuation–dissipation theorem and linear response theory, showing that all the higher μ values in all nanofluids are compared to water. Also, μ displayed a decreasing trend as the nanoparticles dispersed in the base fluid increased; hence, a negative correlation between the graphene nanosheets' size and μ_{nf} was observed mainly due to the more robust intermolecular interaction between H_2O and graphene. Their findings were in line with similar findings in other studies: the smallest nanoparticles had the highest μ_{nf} values [59].

Besides, Rudyak et al. [89] also investigated the μ_{nf} of different-sized nanoparticles dispersed in a base fluid using MD. The nanofluids were formed by dispersing Al and Li nanoparticles in Ar. Similarly, the Green–Kubo method was used to calculate the shear viscosity. L–J potential, RK potential, and specially devised potentials describe the interactions between molecules in the carrier medium, the molecules, the nanoparticles, and between the nanoparticles, respectively. MD simulation results unambiguously demonstrate the dependence of μ_{nf} on nanoparticle size. At constant volume concentrations, the finer the nanoparticles and the higher μ_{nf}. The increase in μ_{nf} is mainly due to nanoparticle–molecule interactions and the correlations between molecules–molecules and nanoparticle–molecule interactions. Lou et al. [90] also corroborated the hypothesis that smaller nanoparticle sizes generate higher μ_{nf}, justifiable by stronger interactions between water and Al_2O_3 nanoparticles. The observations of a relevant study on influential factors on μ in SWCNT/ethanol nanofluids with SWCNT diameters of 0.54–1.08 nm indicated a lower μ_{nf} than μ_{bf} at all temperatures with μ_{nf} decreasing in reaction to increasing SWCNT diameter. The chirality of SWCNT's relative effect on the shear viscosity [91] was also confirmed.

MD simulation allows for investigating the impact of nanoparticle size on μ_{nf} and identifying the influence of temperature, volume fraction, and aggregation. In a study with similar objectives, Jabbari et al. [92] studied atomic and molecular-level physical phenomena on SWCNT/H_2O nanofluids. The Green–Kubo relation results indicate that a higher volume fraction directly improves μ_{nf}, attributable to the more substantial effect of CNTs' higher volume fraction on H_2O molecules generating more vital van der Waals forces between them. Increased temperature makes reaction H_2O molecules display lower reaction tendency and impairs the bonds between the nanoparticles, reducing μ_{nf}. Simultaneously, they reported a direct association between the nanoparticle volume fraction and μ_{nf} for CNT/H_2O nanofluids [93]. That is to say, increasing nanoparticle volume fraction influences a more significant number of H_2O molecules, and the interactions between them impair the momentum transfer between the nanofluid layers, and thus, μ_{nf} improves.

Contrariwise, the nanoparticles and the base liquid molecules display lower interaction in reaction to raised temperature, thus increasing the intermolecular spacing and, hence better dispersion. Additionally, Zeroual et al. [94] used the MD method to investigate μ of Cu/Ar nanofluids, ascribing the influence of volume fraction on μ_{nf} to a dense nanolayer at the interface between the nanoparticles and the base fluid. In the MD simulation for μ of Al_2O_3/H_2O nanofluids, Lou et al. [90] contend that controlling the concentration and temperature turns the nanoparticles' interaction with water molecules into a strong μ_{nf} determinant.

It can be said that all MD simulations reviewed thus far reported similar conclusions, confirming several researchers' experimental observations that displayed MD simulation can fulfill the thermophysical results observed experimentally and give a more tangible insight into the atomic and molecular-level behaviors of nanofluids. There is substantial research potential in this field, although MD studies on nanofluids have increased.

Ultimately, it is imperative to highlight the significant impact of viscosity on the single-phase modeling of nanofluids in free convection heat transfer. This consideration arises from conflicting reports within the literature, where instances of both decreased [95] and increased [96–98] heat transfer following the addition of nanoparticles have been documented. Chapter 8 of this book will provide readers with a detailed exploration of this dual behavior and the extent to which numerical results are influenced by the selection of viscosity models.

REFERENCES

1. A. Einstein, Eine neue bestimmung der molekuldimensionen, *Ann. Phys.* 19 (1906) 289–306. https://doi.org/10.1002/andp.19063240204.
2. H.C. Brinkman, The viscosity of concentrated suspensions and solutions, *J. Chem. Phys.* 20 (1952) 571. https://doi.org/10.1063/1.1700493.
3. G.K. Batchelor, The effect of Brownian motion on the bulk stress in a suspension of spherical particles, *J. Fluid Mech.* 83 (1977) 97–117. https://doi.org/10.1017/S0022112077001062.
4. N. Saito, Concentration dependence of the viscosity of high polymer solutions. I, *J. Phys. Soc. Japan* 5 (1950) 4–8. https://doi.org/10.1143/JPSJ.5.4.
5. A.L. Graham, On the viscosity of suspensions of solid spheres, *Chem. Eng.* 37 (1981) 275–286. https://doi.org/10.1007/bf00951252.
6. N.A. Frankel, A. Acrivos, On the viscosity of a concentrated suspension of solid spheres, *Chem. Eng. Sci.* 22 (1967) 847–853. https://doi.org/10.1016/0009-2509(67)80149-0.
7. N. Masoumi, N. Sohrabi, A. Behzadmehr, A new model for calculating the effective viscosity of nanofluids, *J. Phys. D. Appl. Phys.* 42 (2009) 055501. https://doi.org/10.1088/0022-3727/42/5/055501.
8. S.D. Pandey, V.K. Nema, Experimental analysis of heat transfer and friction factor of nanofluid as a coolant in a corrugated plate heat exchanger, *Exp. Therm. Fluid Sci.* 38 (2012) 248–256. https://doi.org/10.1016/j.expthermflusci.2011.12.013.
9. S.M. Hosseini, A.R. Moghadassi, D.E. Henneke, A new dimensionless group model for determining the viscosity of nanofluids, *J. Therm. Anal. Calorim.* 100 (2010) 873–877. https://doi.org/10.1007/s10973-010-0721-0.
10. J. Chevalier, O. Tillement, F. Ayela, Structure and rheology of SiO_2 nanoparticle suspensions under very high shear rates, *Phys. Rev. E* 80 (2009) 051403. https://doi.org/10.1103/PhysRevE.80.051403.
11. M. Chandrasekar, S. Suresh, A. Chandra Bose, Experimental investigations and theoretical determination of thermal conductivity and viscosity of Al_2O_3/water nanofluid, *Exp. Therm. Fluid Sci.* 34 (2010) 210–216. https://doi.org/10.1016/j.expthermflusci.2009.10.022.

12. C.T. Nguyen, F. Desgranges, N. Galanis, G. Roy, T. Maré, S. Boucher, H.A. Mintsa, Viscosity data for Al_2O_3-water nanofluid-hysteresis: is heat transfer enhancement using nanofluids reliable?, *Int. J. Therm. Sci.* 47 (2008) 103–111. https://doi.org/10.1016/j.ijthermalsci.2007.01.033.
13. T. Yiamsawas, O. Mahian, A.S. Dalkilic, S. Kaewnai, S. Wongwises, Experimental studies on the viscosity of TiO_2 and Al_2O_3 nanoparticles suspended in a mixture of ethylene glycol and water for high temperature applications, *Appl. Energy* 111 (2013) 40–45. https://doi.org/10.1016/j.apenergy.2013.04.068.
14. K. Walters, W.M. Jones, Measurement of viscosity, in: W. Boyes (Ed.), *Instrumentation Reference*, Fourth Ed., Elsevier, 2010: pp. 69–75.
15. J. Chevalier, O. Tillement, F. Ayela, Rheological properties of nanofluids flowing through microchannels, *Appl. Phys. Lett.* 91 (2007) 233103. https://doi.org/10.1063/1.2821117.
16. M. Jarahnejad, E.B. Haghighi, M. Saleemi, N. Nikkam, R. Khodabandeh, B. Palm, M.S. Toprak, M. Muhammed, Experimental investigation on viscosity of water-based Al_2O_3 and TiO_2 nanofluids, *Rheol. Acta.* 54 (2015) 411–422. https://doi.org/10.1007/s00397-015-0838-y.
17. J. Chen, J. Jia, Experimental study of TiO_2 nanofluid coolant for automobile cooling applications, *Mater. Res. In*nov. 21 (2017) 177–181. https://doi.org/10.1080/14328917.2016.1198549.
18. A.S. Dalkilic, B.O. Küçükyıldırım, A. Akdogan Eker, A. Çebi, S. Tapan, C. Jumpholkul, S. Wongwises, Experimental investigation on the viscosity of water-CNT and antifreeze-CNT nanofluids, *Int. Commun. Heat Mass Transf.* 80 (2017) 47–59. https://doi.org/10.1016/j.icheatmasstransfer.2016.11.011.
19. E.V. Timofeeva, A.N. Gavrilov, J.M. McCloskey, Y.V. Tolmachev, S. Sprunt, L.M. Lopatina, J.V. Selinger, Thermal conductivity and particle agglomeration in alumina nanofluids: Experiment and theory, *Phys. Rev. E* 76 (2007) 28–39. https://doi.org/10.1103/PhysRevE.76.061203.
20. M.M. Heyhat, F. Kowsary, A.M. Rashidi, S. Alem Varzane Esfehani, A. Amrollahi, Experimental investigation of turbulent flow and convective heat transfer characteristics of alumina water nanofluids in fully developed flow regime, *Int. Commun. Heat Mass Transf.* 39 (2012) 1272–1278. https://doi.org/10.1016/j.icheatmasstransfer.2012.06.024.
21. J. Jeong, C. Li, Y. Kwon, J. Lee, S.H. Kim, R. Yun, Particle shape effect on the viscosity and thermal conductivity of ZnO nanofluids, *Int. J. Refrig.* 36 (2013) 2233–2241. https://doi.org/10.1016/j.ijrefrig.2013.07.024.
22. S. Ferrouillat, A. Bontemps, O. Poncelet, O. Soriano, J.A. Gruss, Influence of nanoparticle shape factor on convective heat transfer and energetic performance of water-based SiO2 and ZnO nanofluids, *Appl. Therm. Eng.* 51 (2013) 839–851. https://doi.org/10.1016/j.applthermaleng.2012.10.020.
23. E.M.C. Contreras, G.A. Oliveira, E.P. Bandarra Filho, Experimental analysis of the thermohydraulic performance of graphene and silver nanofluids in automotive cooling systems, *Int. J. Heat Mass Transf.* 132 (2019) 375–387. https://doi.org/10.1016/j.ijheatmasstransfer.2018.12.014.
24. A.V. Minakov, V.Y. Rudyak, D.V. Guzei, A.S. Lobasov, Measurement of the heat transfer coefficient of a nanofluid based on water and copper oxide particles in a cylindrical channel, *High Temp.* 53 (2015) 246–253. https://doi.org/10.1134/S0018151X15020169.
25. G. Żyła, J. Fal, P. Estellé, Thermophysical and dielectric profiles of ethylene glycol based titanium nitride (TiN-EG) nanofluids with various size of particles, *Int. J. Heat Mass Transf.* 113 (2017) 1189–1199. https://doi.org/10.1016/j.ijheatmasstransfer.2017.06.032.
26. M.J. Pastoriza-Gallego, L. Lugo, J.L. Legido, M.M. Piñeiro, Rheological non-Newtonian behaviour of ethylene glycol-based Fe_2O_3 nanofluids, *Nanoscale Res. Lett.* 6 (2011) 560. https://doi.org/10.1186/1556-276X-6-560.

27. G. Żyła, J. Fal, Experimental studies on viscosity, thermal and electrical conductivity of aluminum nitride-ethylene glycol (AlN-EG) nanofluids, *Thermochim. Acta.* 637 (2016) 11–16. https://doi.org/10.1016/j.tca.2016.05.006.
28. M.H. Esfe, A.A.A. Arani, M. Rezaie, W.M. Yan, A. Karimipour, Experimental determination of thermal conductivity and dynamic viscosity of Ag-MgO/water hybrid nanofluid, *Int. Commun. Heat Mass Transf.* 66 (2015) 189–195. https://doi.org/10.1016/j.icheatmasstransfer.2015.06.003.
29. M. Asadi, A. Asadi, Dynamic viscosity of MWCNT/ZnO-engine oil hybrid nanofluid: An experimental investigation and new correlation in different temperatures and solid concentrations, *Int. Commun. Heat Mass Transf.* 76 (2016) 41–45. https://doi.org/10.1016/j.icheatmasstransfer.2016.05.019.
30. G. Żyła, J. Fal, S. Bikić, M. Wanic, Ethylene glycol based silicon nitride nanofluids: An experimental study on their thermophysical, electrical and optical properties, *Phys. E Low-Dimensional Syst. Nanostructures.* 104 (2018) 82–90. https://doi.org/10.1016/j.physe.2018.07.023.
31. G. Żyła, J. Fal, P. Estellé, The influence of ash content on thermophysical properties of ethylene glycol based graphite/diamonds mixture nanofluids, *Diam. Relat. Mater.* 74 (2017) 81–89. https://doi.org/10.1016/j.diamond.2017.02.008.
32. G. Żyła, M. Cholewa, On unexpected behavior of viscosity of diethylene glycol-based $MgAl_2O_4$ nanofluids, *RSC Adv.* 4 (2014) 26057–26062. https://doi.org/10.1039/c4ra03143a.
33. G. Żyła, M. Cholewa, A. Witekb, Rheological properties of diethylene glycol-based $MgAl_2O_4$ nanofluid, *RSC Adv.* 3 (2013) 6429–6434. https://doi.org/10.1039/c3ra40187a.
34. M.M. Ghosh, S. Ghosh, S.K. Pabi, Effects of particle shape and fluid temperature on heat-transfer characteristics of nanofluids, *J. Mater. Eng. Perform.* 22 (2013) 1525–1529. https://doi.org/10.1007/s11665-012-0441-7.
35. M.J. Pastoriza-Gallego, C. Casanova, J.L. Legido, M.M. Piñeiro, CuO in water nanofluid: Influence of particle size and polydispersity on volumetric behaviour and viscosity, *Fluid Phase Equilib.* 300 (2011) 188–196. https://doi.org/10.1016/j.fluid.2010.10.015.
36. T. Yiamsawas, A.S. Dalkilic, O. Mahian, S. Wongwises, Measurement and correlation of the viscosity of water-based Al_2O_3 and TiO_2 nanofluids in high temperatures and comparisons with literature reports, *J. Dispers. Sci. Technol.* 34 (2013) 1697–1703. https://doi.org/10.1080/01932691.2013.764483.
37. M.H. Esfe, M. Afrand, S. Gharehkhani, H. Rostamian, D. Toghraie, M. Dahari, An experimental study on viscosity of alumina-engine oil: Effects of temperature and nanoparticles concentration, *Int. Commun. Heat Mass Transf.* 76 (2016) 202–208. https://doi.org/10.1016/j.icheatmasstransfer.2016.05.013.
38. D. Toghraie, S.M. Alempour, M. Afrand, Experimental determination of viscosity of water based magnetite nanofluid for application in heating and cooling systems, *J. Magn. Magn. Mater.* 417 (2016) 243–248. https://doi.org/10.1016/j.jmmm.2016.05.092.
39. J. Yang, F. Li, W. Zhou, Y. He, B. Jiang, Experimental investigation on the thermal conductivity and shear viscosity of viscoelastic-fluid-based nanofluids, *Int. J. Heat Mass Transf.* 55 (2012) 3160–3166. https://doi.org/10.1016/j.ijheatmasstransfer.2012.02.052.
40. M. Sharifpur, S.A. Adio, J.P. Meyer, Experimental investigation and model development for effective viscosity of Al_2O_3-glycerol nanofluids by using dimensional analysis and GMDH-NN methods, *Int. Commun. Heat Mass Transf.* 68 (2015) 208–219. https://doi.org/10.1016/j.icheatmasstransfer.2015.09.002.
41. N. Ahammed, L.G. Asirvatham, S. Wongwises, Effect of volume concentration and temperature on viscosity and surface tension of graphene-water nanofluid for heat transfer applications, *J. Therm. Anal. Calorim.* 123 (2016) 1399–1409. https://doi.org/10.1007/s10973-015-5034-x.

42. M. Baratpour, A. Karimipour, M. Afrand, S. Wongwises, Effects of temperature and concentration on the viscosity of nanofluids made of single-wall carbon nanotubes in ethylene glycol, *Int. Commun. Heat Mass Transf.* 74 (2016) 108–113. https://doi.org/10.1016/j.icheatmasstransfer.2016.02.008.
43. S.M.S. Murshed, F.J.V. Santos, C.A. de Castro, Investigations of viscosity of silicone oil-based semiconductor nanofluids, *J. Nanofluids* 2 (2013) 261–266. https://doi.org/10.1166/jon.2013.1062.
44. K. Bashirnezhad, S. Bazri, M.R. Safaei, M. Goodarzi, M. Dahari, O. Mahian, A.S. Dalkiliça, S. Wongwises, Viscosity of nanofluids: A review of recent experimental studies, *Int. Commun. Heat Mass Transf.* 73 (2016) 114–123. https://doi.org/10.1016/j.icheatmasstransfer.2016.02.005.
45. E.V. Timofeeva, J.L. Routbort, D. Singh, Particle shape effects on thermophysical properties of alumina nanofluids, *J. Appl. Phys.* 106 (2009) 014304. https://doi.org/10.1063/1.3155999.
46. L. Chen, H. Xie, Y. Li, W. Yu, Nanofluids containing carbon nanotubes treated by mechanochemical reaction, *Thermochim. Acta.* 477 (2008) 21–24. https://doi.org/10.1016/j.tca.2008.08.001.
47. L. Chen, H. Xie, W. Yu, Y. Li, Rheological behaviors of nanofluids containing multi-walled carbon nanotube, *J. Dispers. Sci. Technol.* 32 (2011) 550–554. https://doi.org/10.1080/01932691003757223.
48. G. Żyła, J. Fal, Viscosity, thermal and electrical conductivity of silicon dioxide-ethylene glycol transparent nanofluids: An experimental studies, *Thermochim. Acta.* 650 (2017) 106–113. https://doi.org/10.1016/j.tca.2017.02.001.
49. M.Z. Sharif, W.H. Azmi, A.A.M. Redhwan, R. Mamat, Investigation of thermal conductivity and viscosity of Al_2O_3/PAG nanolubricant for application in automotive air conditioning system, *Int. J. Refrig.* 70 (2016) 93–102. https://doi.org/10.1016/j.ijrefrig.2016.06.025.
50. S.W. Lee, S.D. Park, S. Kang, I.C. Bang, J.H. Kim, Investigation of viscosity and thermal conductivity of SiC nanofluids for heat transfer applications, *Int. J. Heat Mass Transf.* 54 (2011) 433–438. https://doi.org/10.1016/j.ijheatmasstransfer.2010.09.026.
51. M. Corcione, Empirical correlating equations for predicting the effective thermal conductivity and dynamic viscosity of nanofluids, *Energy Convers. Manag.* 52 (2011) 789–793. https://doi.org/10.1016/j.enconman.2010.06.072.
52. K.S. Suganthi, N. Anusha, K.S. Rajan, Low viscous ZnO-propylene glycol nanofluid: A potential coolant candidate, *J. Nanoparticle Res.* 15 (2013) 1986. https://doi.org/10.1007/s11051-013-1986-6.
53. J.P. Meyer, S.A. Adio, M. Sharifpur, P.N. Nwosu, The viscosity of nanofluids: A review of the theoretical, empirical, and numerical models, *Heat Transf. Eng.* 37 (2016) 387–421. https://doi.org/10.1080/01457632.2015.1057447.
54. Y. He, Y. Jin, H. Chen, Y. Ding, D. Cang, H. Lu, Heat transfer and flow behaviour of aqueous suspensions of TiO_2 nanoparticles (nanofluids) flowing upward through a vertical pipe, *Int. J. Heat Mass Transf.* 50 (2007) 2272–2281. https://doi.org/10.1016/j.ijheatmasstransfer.2006.10.024.
55. W. Lu, Q. Fan, Study for the particle's scale effect on some thermophysical properties of nanofluids by a simplified molecular dynamics method, *Eng. Anal. Bound. Elem.* 32 (2008) 282–289. https://doi.org/10.1016/j.enganabound.2007.10.006.
56. C.T. Nguyen, F. Desgranges, G. Roy, N. Galanis, T. Maré, S. Boucher, H.A. Mintsa, Temperature and particle-size dependent viscosity data for water-based nanofluids-Hysteresis phenomenon, *Int. J. Heat Fluid Flow.* 28 (2007) 1492–1506. https://doi.org/10.1016/j.ijheatfluidflow.2007.02.004.

57. K.B. Anoop, T. Sundararajan, S.K. Das, Effect of particle size on the convective heat transfer in nanofluid in the developing region, *Int. J. Heat Mass Transf.* 52 (2009) 2189–2195. https://doi.org/10.1016/j.ijheatmasstransfer.2007.11.063.
58. D. Kwek, A. Crivoi, F. Duan, Effects of temperature and particle size on the thermal property measurements of Al_2O_3–water nanofluids, *J. Chem. Eng.* 55 (2010) 5690–5695. https://doi.org/10.1021/je1006407.
59. P.K. Namburu, D.P. Kulkarni, A. Dandekar, D.K. Das, Experimental investigation of viscosity and specific heat of silicon dioxide nanofluids, *Micro Nano Lett.* 2 (2007) 27–71. https://doi.org/10.1049/mnl:20070037.
60. E.V. Timofeeva, W. Yu, D.M. France, D. Singh, J.L. Routbort, Nanofluids for heat transfer: An engineering approach, *Nanoscale Res. Lett.* 6 (2011) 182. https://doi.org/10.1186/1556-276X-6-182.
61. J. Zhao, Z. Luo, M. Ni, K. Cen, Dependence of nanofluid viscosity on particle size and pH value, *Chinese Phys. Lett.* 26 (2009) 066202. https://doi.org/10.1088/0256-307X/26/6/066202.
62. C.T. Wamkam, M.K. Opoku, H. Hong, P. Smith, Effects of pH on heat transfer nanofluids containing ZrO_2 and TiO_2 nanoparticles, *J. Appl. Phys.* 109 (2011) 024305. https://doi.org/10.1063/1.3532003.
63. X. Wang, X. Li, Influence of pH on nanofluids' viscosity and thermal conductivity, *Chinese Phys. Lett.* 26 (2009) 056601. https://doi.org/10.1088/0256-307X/26/5/056601.
64. Babita, S.K. Sharma, S.M. Gupta, Preparation and evaluation of stable nanofluids for heat transfer application: A review, *Exp. Therm. Fluid Sci.* 79 (2016) 202–212. https://doi.org/10.1016/j.expthermflusci.2016.06.029.
65. B. Ruan, A.M. Jacobi, Ultrasonication effects on thermal and rheological properties of carbon nanotube suspensions, *Nanoscale Res. Lett.* 7 (2012) 127. https://doi.org/10.1186/1556-276X-7-127.
66. P. Garg, J.L. Alvarado, C. Marsh, T.A. Carlson, D.A. Kessler, K. Annamalai, An experimental study on the effect of ultrasonication on viscosity and heat transfer performance of multi-wall carbon nanotube-based aqueous nanofluids, *Int. J. Heat Mass Transf.* 52 (2009) 5090–5101. https://doi.org/10.1016/j.ijheatmasstransfer.2009.04.029.
67. I.M. Mahbubul, T.H. Chong, S.S. Khaleduzzaman, I.M. Shahrul, R. Saidur, B.D. Long, M.A. Amalina, Effect of ultrasonication duration on colloidal structure and viscosity of alumina-water nanofluid, *Ind. Eng. Chem. Res.* 53 (2014) 6677–6684. https://doi.org/10.1021/ie500705j.
68. Y. Yang, E.A. Grulke, Z.G. Zhang, G. Wu, Thermal and rheological properties of carbon nanotube-in-oil dispersions, *J. Appl. Phys.* 99 (2006) 114307. https://doi.org/10.1063/1.2193161.
69. A. Menbari, A.A. Alemrajabi, Y. Ghayeb, Investigation on the stability, viscosity and extinction coefficient of CuO-Al_2O_3/water binary mixture nanofluid, *Exp. Therm. Fluid Sci.* 74 (2016) 122–129. https://doi.org/10.1016/j.expthermflusci.2015.11.025.
70. M. Silambarasan, S. Manikandan, K.S. Rajan, Viscosity and thermal conductivity of dispersions of sub-micron TiO_2 particles in water prepared by stirred bead milling and ultrasonication, *Int. J. Heat Mass Transf.* 55 (2012) 7991–8002. https://doi.org/10.1016/j.ijheatmasstransfer.2012.08.030.
71. W. Williams, J. Buongiorno, L.W. Hu, Experimental investigation of turbulent convective heat transfer and pressure loss of Alumina/water and Zirconia/water nanoparticle colloids (nanofluids) in horizontal tubes, *J. Heat Transfer.* 130 (2008) 042412. https://doi.org/10.1115/1.2818775.
72. S. Murshed, K. Leong, C. Yang, Thermophysical properties of nanofluids, in: K. D. Sattler (Ed.), *Handbook of Nanophysics Nanoparticles Quantum Qots*, CRC Press, Boca Raton, 2010: pp. 32(1–14).

73. S.M.S. Murshed, K.C. Leong, C. Yang, Investigations of thermal conductivity and viscosity of nanofluids, *Int. J. Therm. Sci.* 47 (2008) 560–568. https://doi.org/10.1016/j.ijthermalsci.2007.05.004.
74. A. Turgut, I. Tavman, M. Chirtoc, H.P. Schuchmann, C. Sauter, S. Tavman, Thermal conductivity and viscosity measurements of water-based TiO_2 nanofluids, *Int. J. Thermophys.* 30 (2009) 1213–1226. https://doi.org/10.1007/s10765-009-0594-2.
75. P.K. Namburu, D.P. Kulkarni, D. Misra, D.K. Das, Viscosity of copper oxide nanoparticles dispersed in ethylene glycol and water mixture, *Exp. Therm. Fluid Sci.* 32 (2007) 397–402. https://doi.org/10.1016/j.expthermflusci.2007.05.001.
76. L.S. Sundar, M.K. Singh, A.C.M. Sousa, Investigation of thermal conductivity and viscosity of Fe_3O_4 nanofluid for heat transfer applications, *Int. Commun. Heat Mass Transf.* 44 (2013) 7–14. https://doi.org/10.1016/j.icheatmasstransfer.2013.02.014.
77. H. Xie, W. Yu, W. Chen, MgO nanofluids: Higher thermal conductivity and lower viscosity among ethylene glycol-based nanofluids containing oxide nanoparticles, *J. Exp. Nanosci.* 5 (2010) 463–472. https://doi.org/10.1080/17458081003628949.
78. H. Masuda, A. Ebata, K. Teramae, N. Hishinuma, Alteration of thermal conductivity and viscosity of liquid by dispersing ultra-fine particles. Dispersion of Al_2O_3, SiO_2 and TiO_2 ultra-fine particles, *Netsu Bussei.* 7 (1993) 227–233. https://doi.org/10.2963/jjtp.7.227.
79. V.Y. Rudyak, S.V. Dimov, V.V. Kuznetsov, S.P. Bardakhanov, Measurement of the viscosity coefficient of an ethylene glycol-based nanofluid with silicon-dioxide particles, *Dokl. Phys.* 58 (2013) 173–176. https://doi.org/10.1134/s1028335813050042.
80. M. Kole, T.K. Dey, Investigation of thermal conductivity, viscosity, and electrical conductivity of graphene based nanofluids, *J. Appl. Phys.* 113 (2013) 084307. https://doi.org/10.1063/1.4793581.
81. S. Harish, K. Ishikawa, E. Einarsson, S. Aikawa, T. Inoue, P. Zhao, M. Watanabe, S. Chiashi, J. Shiomi, S. Maruyama, Temperature dependent thermal conductivity increase of aqueous nanofluid with single walled carbonnanotube inclusion, *Mater. Express.* 2 (2012) 213–223. https://doi.org/10.1166/mex.2012.1074.
82. C.J. Ho, J.B. Huang, P.S. Tsai, Y.M. Yang, Preparation and properties of hybrid water-based suspension of Al_2O_3 nanoparticles and MEPCM particles as functional forced convection fluid, *Int. Commun. Heat Mass Transf.* 37 (2010) 490–494. https://doi.org/10.1016/j.icheatmasstransfer.2009.12.007.
83. M. Afrand, K.N. Najafabadi, M. Akbari, Effects of temperature and solid volume fraction on viscosity of SiO_2-MWCNTs/SAE40 hybrid nanofluid as a coolant and lubricant in heat engines, *Appl. Therm. Eng.* 102 (2016) 45–54. https://doi.org/10.1016/j.applthermaleng.2016.04.002.
84. M.H. Esfe, M. Afrand, S.H. Rostamian, D. Toghraie, Examination of rheological behavior of MWCNTs/ZnO-SAE40 hybrid nano-lubricants under various temperatures and solid volume fractions, *Exp. Therm. Fluid Sci.* 80 (2017) 384–390. https://doi.org/10.1016/j.expthermflusci.2016.07.011.
85. E. Dardan, M. Afrand, A.H. Meghdadi Isfahani, Effect of suspending hybrid nano-additives on rheological behavior of engine oil and pumping power, *Appl. Therm. Eng.* 109 (2016) 524–534. https://doi.org/10.1016/j.applthermaleng.2016.08.103.
86. O. Soltani, M. Akbari, Effects of temperature and particles concentration on the dynamic viscosity of MgO-MWCNT/ethylene glycol hybrid nanofluid: Experimental study, *Phys. E Low-Dimensional Syst. Nanostructures* 84 (2016) 564–570. https://doi.org/10.1016/j.physe.2016.06.015.
87. S. Akilu, A.T. Baheta, M.A.M. Said, A.A. Minea, K.V. Sharma, Properties of glycerol and ethylene glycol mixture based SiO_2-CuO/C hybrid nanofluid for enhanced solar energy transport, *Sol. Energy Mater. Sol. Cells* 179 (2018) 118–128. https://doi.org/10.1016/j.solmat.2017.10.027.

88. G.R. Vakili-Nezhaad, M. Mohammadi, A.M. Gujarathi, M. Al-Wadhahi, R. Al-Maamari, Molecular dynamics simulation of water-graphene nanofluid, *SN Appl. Sci.* 1 (2019) 214. https://doi.org/10.1007/s42452-019-0224-y.
89. V.Y. Rudyak, S.L. Krasnolutskii, Simulation of the nanofluid viscosity coefficient by the molecular dynamics method, *Tech. Phys.* 60 (2015) 798–804. https://doi.org/10.1134/S1063784215060237.
90. Z. Lou, M. Yang, Molecular dynamics simulations on the shear viscosity of Al_2O_3 nanofluids, *Comput. Fluids.* 117 (2015) 17–23. https://doi.org/10.1016/j.compfluid.2015.05.006.
91. G.R. Vakili-Nezhaad, M. Al-Wadhahi, A.M. Gujrathi, R. Al-Maamari, M. Mohammadi, Effect of temperature and diameter of narrow single-walled carbon nanotubes on the viscosity of nanofluid: A molecular dynamics study, *Fluid Phase Equilib.* 434 (2017) 193–199. https://doi.org/10.1016/j.fluid.2016.11.032.
92. F. Jabbari, A. Rajabpour, S. Saedodin, Viscosity of carbon nanotube/water nanofluid, *J. Therm. Anal. Calorim.* 135 (2019) 1787–1796. https://doi.org/10.1007/s10973-018-7458-6.
93. F. Jabbari, S. Saedodin, A. Rajabpour, Experimental investigation and molecular dynamics simulations of viscosity of CNT-water nanofluid at different temperatures and volume fractions of nanoparticles, *J. Chem. Eng. Data.* 64 (2019) 262–272. https://doi.org/10.1021/acs.jced.8b00783.
94. S. Zeroual, H. Loulijat, E. Achehal, P. Estellé, A. Hasnaoui, S. Ouaskit, Viscosity of Ar-Cu nanofluids by molecular dynamics simulations: Effects of nanoparticle content, temperature and potential interaction, *J. Mol. Liq.* 268 (2018) 490–496. https://doi.org/10.1016/j.molliq.2018.07.090.
95. A.M. Rashad, et al., Entropy generation and MHD natural convection of a nanofluid in an inclined square porous cavity: Effects of a heat sink and source size and location, *Chinese Journal of Physics* 56.1 (2018) 193–211.
96. A. Chamkha, et al., Entropy generation and natural convection of CuO-water nanofluid in C-shaped cavity under magnetic field, *Entropy* 18.2 (2016) 50.
97. M.A. Ismael, T. Armaghani, A.J. Chamkha, Conjugate heat transfer and entropy generation in a cavity filled with a nanofluid-saturated porous media and heated by a triangular solid, *Journal of the Taiwan Institute of Chemical Engineers* 59 (2016) 138–151.
98. T. Armaghani, M.A. Ismael, A.J. Chamkha, Analysis of entropy generation and natural convection in an inclined partially porous layered cavity filled with a nanofluid, *Canadian Journal of Physics* 95.3 (2017) 238–252.

6 Statistical Study and Overview of Nanofluid Viscosity Correlations

A. Barkhordar

6.1 INTRODUCTION

Nanofluids are considered the top candidates to replace surface cooling systems, making it essential to study the effect of nanoparticles on the thermophysical properties of the base fluid when it is added. Viscosity is a crucial factor in heat transfer, especially convection heat transfer.

In most of the studies published, the correlations obtained from experiments were performed without examining statistical tests, and the effect of different parameters, including temperature, volume (mass) fraction, and so on, on the viscosity of nanofluid in the proposed correlations was not specified. Moreover, in some correlations, it was shown that the elimination of one of the parameters did not affect the response of that correlation.

In this section, for statistical analysis, analysis of variance and sensitivity analysis were used to determine the relationship of the correlation with its variable parameters.

One of the perspectives for solving conservation equations for nanofluids is the single-phase method. In this method, the thermophysical properties of the nanofluid replace the thermophysical properties of the base fluid [1–3]. The single-phase model is used in some investigations [4–7]. This shows the impact of the thermophysical properties of the nanofluid, especially its viscosity, on heat transfer [8–10].

Among the thermophysical properties of nanofluids, viscosity indicates fluid resistance. Therefore, viscosity has determined the performance of energy and heating systems [11–13].

In industrial equipment and scientific research where heat transfer is in the forms of forced convection and natural convection, the viscosity of nanofluids plays a crucial role in determining the flow regime, pumping power, pressure drop, and workability of systems [14–17].

The first viscosity model for suspensions containing metal particles was introduced by Einstein in 1906 [18]. Later on, viscosity models proposed by Brinkman [19], Batchelor [20], and other equations started to be used to model the heat transfer of nanofluids. However, these correlations each have weaknesses, including the inability to estimate the viscosity of nanofluids in a wide range of temperatures and concentrations used in heat transfer.

DOI: 10.1201/9781032664118-6

In the experiments conducted by Duangthongsuk and Wongwises [21] on the behavior of TiO_2 and water nanofluid, they presented a correlation by applying the effects of base fluid viscosity and volume fraction variables on the viscosity model. In their model, nanofluid's temperatures and volume fraction ranged between 15°C and 35°C and 0.2% and 2%, respectively.

In an experimental test performed by Esfe and Saedodin [22] on the viscosity of ZnO nanofluid with ethylene glycol-based fluid at a temperature between 25°C and 50°C and a volume fraction of 0.25% to 5%, the viscosity model with the variables of temperature and volume fraction and the viscosity of the base fluid presented that the ratio of mean variation of the model data compared to other the experimental values was less than 2%.

Sharifpur et al. [23] also introduced a viscosity model based on the data derived from experiments using D-glycerin nanofluid with an accuracy of 0.9495. In their viscosity model, in addition to the variables of temperature and volume fraction and the viscosity of the base fluid, the effect of the thickness of the nanoparticles is also taken into account. It is used for nanofluids in the temperature range of 20°C–70°C volume fraction of 0%–5%, and diameter of nanoparticles about 19–160 nm. But Aberoumand et al. [24] presented a viscosity model for oil-silver nanofluid that depends only on the variables of volume fraction and viscosity of the base fluid and is valid for nanofluids at temperatures between 25°C and 60°C and volume fraction of 0%–2%.

In an experimental study performed by Akbari et al. [25] on SiO_2 and ethylene glycol nanofluids in the temperature range of 30°C–50°C and volume fraction of 0.5%–3%, using temperature and volume fraction components and the viscosity of the base fluid proposed a viscosity model for the nanofluid. Li and Zuo [26] proposed a viscosity model for a nanofluid including TiO_2 nanoparticles and a mixture of water-based fluid and ethylene glycol at a temperature between 20°C and 50°C and a volume fraction of 0.25%–1%.

Yu et al. [27] also proposed a new viscosity model based on the data derived from experiments using multi-walled carbon nanotubes (MWCNT) and water nanofluids. In their model, in addition to the role of temperature, mass fraction, and base fluid viscosity, the effect of shear rate on viscosity variations is also considered. This model is valid in a temperature range of 275–283 K, and mass fraction range of 0.1%–0.6%, and a shear rate of 10–1000/s. According to the experiment performed by Yan et al. [28] on a hybrid nanofluid with MWCNT nanoparticles and TiO_2 with a base fluid of ethylene glycol at 25°C–55°C and a volume fraction of 0.05%–1%, the viscosity model with volume fraction and non-dimensional temperature components have been presented with an accuracy of 0.995.

Figure 6.1 has been plotted to know the year of publication of the evaluated correlations in the present study. Therefore, it can be concluded that researchers have considered the study and presentation of models for the viscosity of nanofluids in recent years.

Figure 6.2 has been plotted to indicate the temperature range at which the viscosity models are valid. The correlations are separated according to the temperature range they cover in different temperature ranges that differ by 10°. (Temperature difference of less than 5°C in the classification has been neglected.)

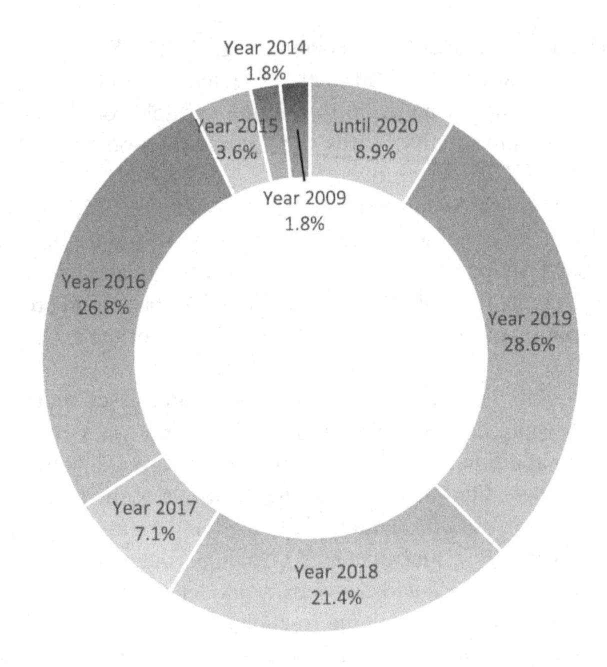

FIGURE 6.1 Year of publication of evaluated correlations in articles.

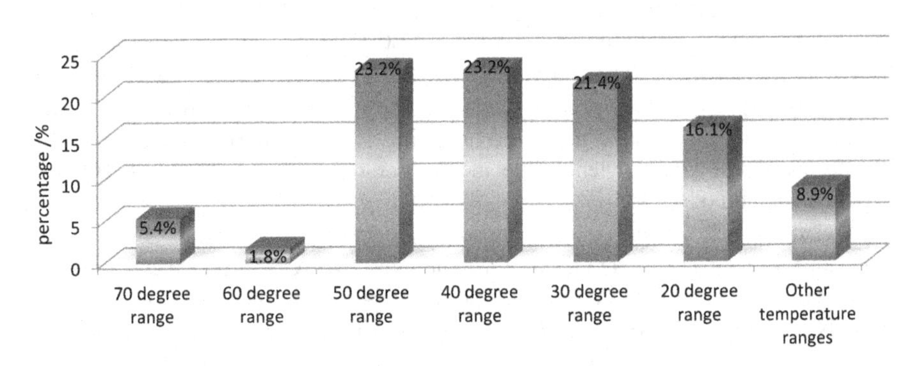

FIGURE 6.2 The temperature ranges are covered by the viscosity models.

According to Figure 6.2, 21.4% and 23.2% of the temperature range covers 30° and 40°, respectively, and only 23.2% of the correlations in the 50° temperature range can estimate the viscosity of the nanofluid.

Similar to the temperature diagram, the viscosity models were separated into 1% by volume (mass) fraction intervals relative to the concentration range in which they are valid, and Figure 6.3 shows that only 19.6% of the correlations can cover the concentration range of 2%.

The authors also worked on the thermal conductivity of nanofluid based on present studies and introduced a new general model named MAG [29].

In this study, the correlations presented for the viscosity of nanofluids were completely reviewed and investigated thoroughly in terms of compliance with the physics

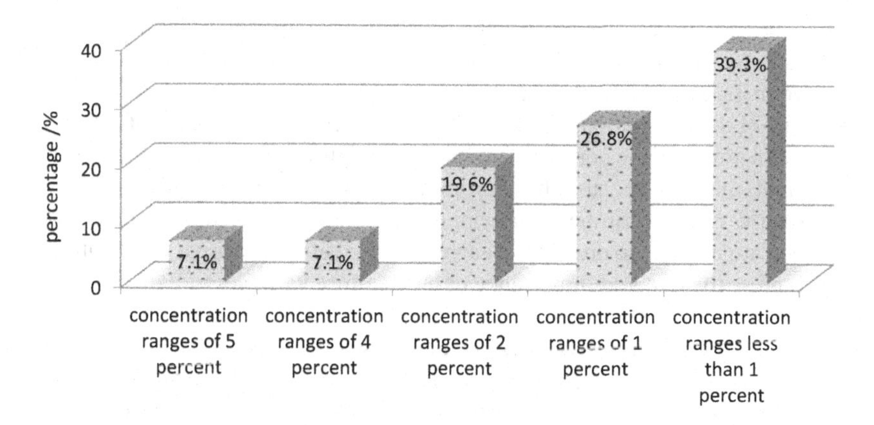

FIGURE 6.3 The concentration ranges are covered by the viscosity models.

and viscosity of nanofluids. In the following, the relationship between viscosity and variables of temperature and volume (mass) fraction of nanofluids was evaluated according to the statistical test of variance. Moreover, all the correlations presented for nanofluid viscosity were investigated with the sensitivity analysis test to identify the variable with the most significant effect on the viscosity model. Finally, two general models for water-based nanofluids and ethylene glycol were presented to predict the viscosity behavior of nanofluids.

6.2 STRATEGY

6.2.1 Analysis of Variance

Statistics is a broad field of mathematics that studies how data collection, summary, and conclusion are studied. Here, the status of variables related to the viscosity of nanofluid is investigated using statistical science based on probability theory and mathematics.

ANOVA test or analysis of variance is a subset of statistical science, which analyzes and compares the means of different statistical groups and determines the effect of independent variables on the dependent variable. This method has been introduced by the famous statistician and geneticist "Ronald Fisher" [30].

This method tries to estimate the differences between several statistical populations. In other words, using the mean index in statistical populations, we will be able to express the characteristics of the population; thus, if the mean of one group is different from other groups in society, we conclude that the statistical populations are not the same. In the one-way analysis of variance, the null hypothesis indicates that the mean of the experimental groups is equal to each other. The opposite assumption is that at least one of the means is different from the others; if the null hypothesis is confirmed, it will be accepted that there is no difference between the means of the groups, and the variable has no role in the correlation. Therefore, to better understand the correspondence of variables on nanofluid viscosity, it is necessary to perform variance analysis [31].

Thus, by groups that will be created in terms of temperature and concentration variables for each equation and with the help of a one-way ANOVA test, the results presented in Table 6.1 are obtained.

In some of the correlations proposed by researchers due to the lack of attention to the response power of the correlation in the temperature range affecting the viscosity of nanofluids and also because the appropriate relationship is not used in correlations to express the relationship between temperature and viscosity of nanofluid and the correlation are not able to predict the nanofluid viscosity at sensitive temperature and do not express the role and importance of temperature variables in nanofluid viscosity models.

One of the issues that researchers had not considered in presenting correlations is the effect of terms on the viscosity model. For example, Dalkılıç et al. [32] presented Eq. (6.1) for nanofluid viscosity by experimental investigation of the viscosity of a hybrid nanofluid containing graphene and SiO_2 nanoparticles in a water-based nanofluid in the temperature range of 15°C–60°C and a volume fraction of 0.001%–0.02%.

$$\mu_{nf} = \left[1.00527 \times \left(T^{0.00035} \right) \times (1+\varphi)^{9.36265} \times \left(\frac{\varphi_{w,\,G}}{\varphi_{w,\,SiO_2}} \right)^{-0.028935} \right] \mu_{bf} \tag{6.1}$$

In the case where $\varphi = 1$ and other components are a constant value, if the lower- and upper-temperature limits are set in Eq. (6.1), the range of viscosity changes will be less than 0.04%. Therefore, Eq. (6.1) is not dependent on temperature, but the researcher had given in the equation, and this is not considered significant by the researcher.

In addition to the temperature variable, the concentration variable also has an undeniable role in the viscosity models of nanofluids. So with the increase in the concentration of nanofluid, the viscosity of nanofluid increases significantly; therefore, in most viscosity models, the prominent role of the concentration variable was considered by researchers. However, in some correlations recently published by researchers, the correlations have been measured in the range of inappropriate volume (mass) fractions and only at low concentrations. On the other hand, due to the use of irrational relationships to express the relationship between the viscosity of nanofluid and the concentration of nanoparticles, viscosity models cannot respond commensurately with the expected concentration range in heat transfer.

By conducting experimental tests on nanofluids of Al_2O_3 and ethylene glycol, Li et al. [33] presented the new viscosity model in compliance with the trend of temperature variation from 25°C to 60°C and a mass fraction of 0%–2% in Eq. (6.2).

$$\mu_{nf} = -334.9\varphi_w^{4.044} \left(\frac{1}{T} \right)^{10.03} + 296.8 \left(\frac{1}{T} \right)^{0.7795} - 6.841 \tag{6.2}$$

Concentration variable in term $-334.9\varphi_w^{4.044} \left(\frac{1}{T} \right)^{10.03}$ less than 0.01% affects the nanofluid viscosity. Therefore, the viscosity model of Li et al. does not have the expected dependence on the concentration variable.

TABLE 6.1
Experimental correlation of nanofluid viscosity and their analysis

No.	Author	Material Nanoparticle (Base Fluid)	Correlation	Range of Temperature Nanoparticle Diameter Concentration	Analysis: ANOVA[1] [Variances] P-Value for T	Analysis: ANOVA[1] [Variances] P-Value for φ	Analysis: Statistical Status[5]	Analysis: Data[2] Term[3] Trend[4]	Overall Conclusion
1	Li et al. [40]	SiO_2 (liquid paraffin)	$\mu_{nf} = \begin{bmatrix} 6.8376 + 15.2522\varphi_w \\ +0.038779\varphi_w T - 2.63029\varphi_w^2 \end{bmatrix} \mu_{bf}$	$25°C < T_{nf} < 70°C$ 20–45 nm $0.005 < \varphi < 5$ Wt.	1.000	0.000	Rejected	✓ ✓ ✓	According to the results of the statistical test of variance, the correlation is not dependent on the temperature variable.
2	Li et al. [33]	Al_2O_3 (EG)	$\mu_{nf} = -334.9\varphi_w^{4.044}\left(\frac{1}{T}\right)^{10.03} + 296.8\left(\frac{1}{T}\right)^{0.7795} - 6.841$	$25°C < T_{nf} < 60°C$ 20 nm $0 < \varphi < 2$ Wt.	0.000	1.000	Rejected	✓ ✓ ✓	According to the results of the statistical test of variance, the correlation is not dependent on the concentration variable.
3	Sahoo and Kumar [41]	Al_2O_3,SiO_2,CuO (Water)	$\mu_{nf} = \begin{bmatrix} 0.955 - 0.00271 \times T + 1.858 \times \frac{\varphi}{100} \\ +\left(705 \times \frac{\varphi}{100}\right)^{1.223} \end{bmatrix} \mu_{bf}$	$35°C < T_{nf} < 50°C$ 30–50 nm $0.01 < \varphi < 0.1$ Vol.	1.000	0.000	Rejected	✓ ✓ ✓	According to the results of the statistical test of variance, the correlation is not dependent on the temperature variable.

(Continued)

TABLE 6.1 (*Continued*)
Experimental correlation of nanofluid viscosity and their analysis

No.	Author	Material Nanoparticle (Base Fluid)	Correlation	Range of Temperature Nanoparticle Diameter Concentration	Analysis: ANOVA[1] [Variances] P-Value for T	Analysis: ANOVA[1] [Variances] P-Value for φ	Analysis: Statistical Status[5]	Analysis: Data[2] Term[3] Trend[4]	Overall Conclusion
4	Yan et al. [28]	$MWCNT/TiO_2$ (80%-20%) Hybrid (EG)	$\mu_{nf} = [0.90463 + 280.20104\varphi + 0.257340\theta + 368.052390\theta\varphi - 28643.68399\varphi^2 - 0.0120510\theta^2 - 1968.73612\varphi^2\theta - 235.047290\theta^2\varphi + 2.09629 \times 10^6\varphi^3 - 0.0996940\theta^3]\ \mu_{bf}$	$25°C < T_{nf} < 55°C$ 30 nm 0.05 < <1 Vol.	0.000	0.983	Rejected	✓ × ✓	According to the results of the statistical test of variance, the correlation is not dependent on the concentration variable. The correlation has multiple terms. (θ is the dimensionless temperature)
5	Tian et al. [42]	Cu/MWCNT (water/EG) [70 – 30]	$\mu_{nf} = \begin{bmatrix} 0.50013 + 0.019722T \\ +4.23827\varphi - 0.099694T\varphi \end{bmatrix} \mu_{bf}$	$20°C < T_{nf} < 60°C$ - $0.025 < \varphi < 0.25$ Vol.	0.000	0.000	Acceptable	✓ ✓ ✓	
6	Li et al. [43]	SWCNT (water)	$\mu_{nf} = 0.001732 - 0.000046T + 4.832 \times 10^{-7}T^2 + 0.0259\varphi - 1.262\varphi^2 - 0.000305T\varphi$	$5°C < T_{nf} < 50°C$ 20 nm $0 < \varphi < 0.5$ Vol.	0.978	0.682	Rejected	✓ ✓ ✓	According to the results of the statistical test of variance, the correlation is not dependent on the temperature variable. The correlation has multiple terms.

(*Continued*)

TABLE 6.1 (*Continued*)
Experimental correlation of nanofluid viscosity and their analysis

No.	Author	Material Nanoparticle (Base Fluid)	Correlation	Range of Temperature Nanoparticle Diameter Concentration	Analysis: ANOVA[1] [Variances] P-Value for T	Analysis: ANOVA[1] [Variances] P-Value for φ	Analysis: Statistical Status[5]	Analysis: Data[2] Term[3] Trend[4]	Overall Conclusion
7	Esfe and Esfandeh [39]	MWCNT/ZnO [20–80] Hybrid (5W30)	$\mu_{nf} = 458.77 - 23.13T - 0.01\dot{\gamma} - 3.16\varphi T + 5.82 \times 10^{-4} T\dot{\gamma} + 0.44T^2 + 2.48\times10^{-7}\dot{\gamma}^2 + 9.6\times10^{-5}\varphi T\dot{\gamma} + 0.02\varphi T^2 - 8.37\times10^{-6}T^2\dot{\gamma} + 71.26\varphi^3 - 2.81\times10^{-3}T^3$	$5°C < T_{nf} < 55°C$ - $0.05 < \varphi < 1$ Vol.	0.000	0.000	Acceptable	✓ × ×	The variation trend of the viscosity model is not consistent with the increase in the volume fraction of the nanofluid. The correlation has multiple terms.
8	Ganesh Kumar et al. [44]	MWCNTs (water/solar glycol) [70 – 30]	$\mu_{nf} = \left[0.8834\varphi_w^2 + 0.3903\varphi_w + 1.0077\right]\mu_{bf}$	$30°C < T_{nf} < 50°C$ 30–50 nm $0.15 < \varphi < 0.45$ Wt.	-	1.000	Rejected	✓ ✓ ✓	According to the results of the statistical test of variance, the correlation is not dependent on the concentration variable.
9	Asadi and Pourfattah [45]	ZnO/MgO Hybrid (Oil)	$\mu_{nf} = a + b\varphi$	$15°C < T_{nf} < 55°C$ 35–45 nm ZnO 40 nm MgO $0.125 < \varphi < 1.5$ Vol.	0.000	0.000	Acceptable	✓ ✓ ✓	(a, b are dependent on temperature)

(Continued)

TABLE 6.1 (*Continued*)
Experimental correlation of nanofluid viscosity and their analysis

No.	Author	Material Nanoparticle (Base Fluid)	Correlation	Range of Temperature Nanoparticle Diameter Concentration	Analysis: ANOVA[1] [Variances] P-Value for T	Analysis: ANOVA[1] [Variances] P-Value for φ	Analysis: Statistical Status[5]	Analysis: Data[2] Term[3] Trend[4]	Overall Conclusion
10	Shahsavar et al. [46]	Fe_3O_4 (Paraffin)	$\mu_{nf} = a_{04} + a_{14}(b_{2233}) + a_{24}(b_{1311}) + a_{34}(b_{2233})^2$ $+ a_{44}(b_{1311})^2 + a_{54}(b_{2233})(b_{1311})$	$20°C < T_{nf} < 90°C$ 20 nm $0.5 < \varphi < 3$ Vol.	0.999	1.000	Rejected	✓ ✓ ✓	According to the statistical variance analysis, the viscosity model does not have the expected dependence on the variables of temperature and concentration. (Constants are dependent on temperature)
11	Esfe et al. [38]	MWCNT/Al_2O_3 (30%-70%) (5W50)	$\mu_{nf} = 688.46 + 347.09\varphi - 33.12T - 0.04\dot{\gamma}$ $-7.36\varphi T - 0.0087\varphi\dot{\gamma}$ $+0.0014T\dot{\gamma} - 305.24\varphi^2 + 0.61T^2$ $+1.49\times10^{-6}\dot{\gamma}^2$ $+0.0001\varphi T\dot{\gamma} + 0.46\varphi^2 T + 0.0014\varphi^2\dot{\gamma}$ $+0.065\varphi T^2$ $+1.87\times10^{-7}\varphi\dot{\gamma}^2 - 7.25\times10^{-6}T^2\dot{\gamma}$ $-7.33\times10^{-8}T\dot{\gamma}^2$ $+169.62\varphi^3 - 0.0043T^3 + 0.00000000011\dot{\gamma}^3$	$5°C < T_{nf} < 55°C$ 5–15 nm $0.05 < \varphi < 1$ Vol.	0.000	0.000	Acceptable	✓ × ✓	The correlation has multiple terms.

(Continued)

TABLE 6.1 (*Continued*)
Experimental correlation of nanofluid viscosity and their analysis

No.	Author	Material Nanoparticle (Base Fluid)	Correlation	Range of Temperature Nanoparticle Diameter Concentration	Analysis: ANOVA[1] [Variances] P-Value for T	Analysis: ANOVA[1] [Variances] P-Value for φ	Analysis: Statistical Status[5]	Analysis: Data[2] Term[3] Trend[4]	Overall Conclusion
12	Alarifi et al. [34]	$MWCNT/TiO_2$ [80%–20%] (Oil)	$\mu_{nf} = 2.936T + \frac{2e^4}{1.68 + T - 1.68\varphi} - 448.8 - \tan((1.68\varphi) - 1.68)$	$25°C < T_{nf} < 50°C$ - $0.25 < \varphi < 2$ Vol.	0.000	0.000	Acceptable	✓ × ✓	The correlation has an irrelevant function.
13	Li et al. [47]	Cu (EG)	$\mu_{nf} = \left[1.045 + 2.105\varphi_w - 5.015\varphi_w^2\right]\mu_{bf}$	$20°C < T_{nf} < 60°C$ 50 nm $1 < \varphi < 3.8$ Wt.	-	1.000	Rejected	✓ ✓ ✓	According to the results of the statistical test of variance, the correlation is not dependent on the concentration variable.
14	Ruhani et al. [35]	SiO_2 (EG/Water) [30–70]	$\mu_{nf} = [2.030 - \begin{pmatrix} 931.616 \times \varphi^{0.9305} \\ \times 5.4597 \times T^{-3.4574} \end{pmatrix} - \exp\left(-0.0028 \times \varphi^{2.1421} \times T^{1.0133}\right)^2]\ \mu_{bf}$	$25°C < T_{nf} < 50°C$ 20–30 nm $0.1 < \varphi < 1.5$ Vol.	1.000	0.000	Rejected	✓ × ✓	According to the results of the statistical test of variance, the correlation is not dependent on the temperature variable. The correlation has a complex term.

(*Continued*)

TABLE 6.1 (*Continued*)
Experimental correlation of nanofluid viscosity and their analysis

No.	Author	Material Nanoparticle (Base Fluid)	Correlation	Range of Temperature Nanoparticle Diameter Concentration	Analysis: ANOVA[1] [Variances] P-Value for T	Analysis: ANOVA[1] [Variances] P-Value for φ	Analysis: Statistical Status[5]	Analysis: Data[2] Term[3] Trend[4]	Overall Conclusion
15	Esfe et al. [37]	MWCNT/ZnO [10%–90%] Hybrid (10W40)	$\mu_{nf} = 679.78306 + 259.62463\varphi - 33.64131T - 0.045393\dot{\gamma} - 6.0695 \times \varphi T - 0.00031841\varphi\dot{\gamma} + 0.00129007T\dot{\gamma} - 236.23287\varphi^2 + 0.65796T^2 + 2.31776E - 06\dot{\gamma}^2 + 1.55167\varphi^2 T + 0.049805\varphi T^2 - 9.63258E-8T\dot{\gamma}^2 + 94.57115\varphi^3 - 0.0051586T^3 + 0.0000000001\dot{\gamma}^3$	$5°C < T_{nf} < 55°C$ 5–15 nm $0.05 < \varphi < 1$ Vol.	0.000	1.000	Rejected	✓ ✓ ✓	According to the results of the statistical test of variance, the correlation is not dependent on the concentration variable. The correlation has multiple terms.
16	Esfe et al. [48]	MWCNTs/TiO_2 [20%–80%] Hybrid (Water/EG) [70%–30%]	$\mu_{nf} = 6.35 + 2.56\varphi - 0.24T - 0.068\varphi T + 0.905\varphi^2 + 0.0027T^2$	$10°C < T_{nf} < 50°C$ 5–15 nm $0.05 < \varphi < 0.85$ Vol.	0.000	*0.001*	Acceptable	✓ ✓ ✓	

(*Continued*)

TABLE 6.1 (*Continued*)
Experimental correlation of nanofluid viscosity and their analysis

No.	Author	Material Nanoparticle (Base Fluid)	Correlation	Range of Temperature Nanoparticle Diameter Concentration	Analysis: ANOVA[1] [Variances] *P*-Value for *T*	Analysis: ANOVA[1] [Variances] *P*-Value for φ	Analysis: Statistical Status[5]	Analysis: Data[2] Term[3] Trend[4]	Overall Conclusion
17	Yu et al. [27]	MWCNT (water)	$\mu_{nf}=\begin{bmatrix}1-1634\varphi+8T\varphi+(10T-2190)\\ \varphi\exp\left(-\frac{\dot{\gamma}}{60}\right)\end{bmatrix}\mu_{bf}$	$275K \le T_{nf} \le 283K$ 8–15 nm $0 \le \varphi \le 0.2381$ Vol.	1.000	0.000	Rejected	✓ ✓ ✓	According to the results of the statistical test of variance, the correlation is not dependent on the temperature variable.
18	Soman et al. [49]	BMImBr ionic liquid (water)	$\mu_{nf}=-3.4807+0.0304T+0.4674\varphi_w+0.0005T\varphi_w$ $-0.1233\varphi_w^2-0.000052T^2$	$296K \le T_{nf} \le 336K$ 20–30 nm $0.1 \le \varphi \le 0.6$ Wt.	*1.000*	0.000	Rejected	✓ ✓ ✓	According to the results of the statistical test of variance, the correlation is not dependent on the temperature variable.
19	Ruhani et al. [50]	ZnO/Ag [50%–50%] Hybrid (water)	$\mu_{nf}=\left[a_0+a_1\varphi+a_2\varphi^2+a_3\varphi^3\right]\mu_{bf}$	$25°C < T_{nf} < 50°C$ 10–30 nm ZnO 30–50 nm Ag $0.125 \le \varphi \le 2$ Vol.	0.999	0.000	Rejected	✓ ✓ ✓	According to the results of the statistical test of variance, the correlation is not dependent on the temperature variable. (a_0, a_1, a_2, a_3 are dependent on temperature)

(*Continued*)

TABLE 6.1 (*Continued*)
Experimental correlation of nanofluid viscosity and their analysis

No.	Author	Material Nanoparticle (Base Fluid)	Correlation	Range of Temperature Nanoparticle Diameter Concentration	Analysis: ANOVA[1] [Variances] *P*-Value for *T*	Analysis: ANOVA[1] [Variances] *P*-Value for φ	Analysis: Statistical Status[5]	Analysis: Data[2] Term[3] Trend[4]	Overall Conclusion
20	Huminic et al. [36]	La_2O_3 lanthanum oxide (water)	$\mu_{nf} = a_1 + a_2T + a_3\varphi + a_4T^2 + a_5T\varphi + a_6\varphi^2$ $+a_7T^3 + a_8T^2\varphi + a_9T\varphi^2 + a_{10}\varphi^3 + a_{11}T^3\varphi$ $+a_{12}T^2\varphi^2 + a_{13}T\varphi^3 + a_{14}\varphi^4$	$293K \le T_{nf} \le 323K$ 60–70 nm diameter and 500–700 nm length. $0 < \varphi < 0.03$ Vol.	*1.000*	0.961	Rejected	✓ × ×	According to the statistical variance analysis, the viscosity model does not have the expected dependence on the variables of temperature and concentration. Given the zero concentration in the correlation, the value of the viscosity of the base fluid is not correctly predicted. The correlation has multiple terms.

(*Continued*)

TABLE 6.1 (*Continued*)
Experimental correlation of nanofluid viscosity and their analysis

No.	Author	Material Nanoparticle (Base Fluid)	Correlation	Range of Temperature Nanoparticle Diameter Concentration	Analysis: ANOVA[1] [Variances] *P*-Value for *T*	Analysis: ANOVA[1] [Variances] *P*-Value for φ	Analysis: Statistical Status[5]	Analysis: Data[2] Term[3] Trend[4]	Overall Conclusion
21	Elcioglu et al. [51]	Al_2O_3 (water)	$\mu_{nf}=1.00973-0.0148T+34.2292\varphi+0.0041d_p-0.3458T\varphi$	$20°C<T_{nf}<50°C$ 10–30nm $0.01<\varphi<0.03$ Vol.	1.000	0.999	Rejected	✓ ✓ ✓	According to the statistical variance analysis, the viscosity model does not have the expected dependence on the variables of temperature and concentration. (ANOVA [d_p]: 0.986) (d_p: nanoparticles diameter)
22	Li and Zou [26]	TiO_2 (EG/water) [20–80]	$\mu_{nf}=0.838\varphi^{0.188}T^{0.089}\mu_{bf}^{1.1}$	$20°C<T_{nf}<60°C$ 15nm $0.025<\varphi<0.1$ Vol.	0.220	0.000	Acceptable	✓ ✓ ✓	
23	Ghasemi and Karimipour [52]	CuO (liquid paraffin)	$\mu_{nf}=\left[a_1T^{b_1}+a_2\varphi_w^{b_2}+a_3\varphi_w^{b_3}.T^{b_4}+a_4\right]\mu_{bf}$	$20°C<T_{nf}<60°C$ 15–30nm $0.25<\varphi<6$ Wt.	1.000	0.998	Rejected	✓ ✓ ✓	According to the statistical variance analysis, the viscosity model does not have the expected dependence on the variables of temperature and concentration.

(*Continued*)

TABLE 6.1 (*Continued*)
Experimental correlation of nanofluid viscosity and their analysis

No.	Author	Material Nanoparticle (Base Fluid)	Correlation	Range of Temperature Nanoparticle Diameter Concentration	Analysis: ANOVA[1] [Variances] P-Value for T	Analysis: ANOVA[1] [Variances] P-Value for φ	Analysis: Statistical Status[5]	Analysis: Data[2] Term[3] Trend[4]	Overall Conclusion
24	Dalkılıça et al. [32]	SiO_2/graphite hybrid (water)	$\mu_{nf} = \left[1.00527 \times \left(T^{0.00035} \right) \times (1+\varphi)^{9.36265} \times \left(\frac{\varphi_{w,G}}{\varphi_{w,SiO_2}} \right)^{-0.028935} \right] \mu_{bf}$	$15°C < T_{nf} < 60°C$ SiO_2 7nm Graphite 6–10 nm $0.001 < \varphi < 0.02$ Vol.	0.998	0.000	Rejected	✓ ✓ ✓	According to the results of the statistical test of variance, the correlation is not dependent on the temperature variable.
25	Moldoveanu et al. [53]	Al_2O_3, SiO_2 Hybrid and separately (water)	$Al_2O_3: \mu_{nf} = \left[4135\varphi^2 - 91.72\varphi + 2.06 \right] \mu_{bf}$ $SiO_2: \mu_{nf} = \left[-769\varphi^2 + 42\varphi + 1.1 \right] \mu_{bf}$	$10°C < T_{nf} < 60°C$ Al_2O_3 43 nm SiO_2 20 nm $1 < \varphi < 5$ Al_2O_3 $1 < \varphi < 3$ SiO_2 Vol.	-	0.000	Acceptable	✓ ✓ ✓	
26	Moldoveanu et al. [54]	Al_2O_3, TiO_2 Hybrid and separately (water)	$Al_2O_3: \mu_{nf} = \left[0.6152\varphi^2 - 1.5449\varphi + 2.3792 \right] \mu_{bf}$ $TiO_2: \mu_{nf} = \left[0.2302 - 0.3202\varphi + 1.5056 \right] \mu_{bf}$	25°C Al_2O_3 43 nm TiO_2 30 nm $1 < \varphi < 5$ Al_2O_3 $1 < \varphi < 3$ TiO_2 Vol.	-	0.000	Acceptable	✓ ✓ ✓	
27	Saeedi et al. [55]	CeO_2 (EG)	$\mu_{nf} = \left[781.4 \times T^{-2.117} \times \varphi^{0.2722} + \frac{0.05776}{T^{-0.7819} \times \varphi^{-0.04009}} + 0.511 \times \varphi^2 - 0.1779 \times \varphi^3 \right] \mu_{bf}$	$25°C < T_{nf} < 50°C$ 10–30 nm $0.05 < \varphi < 1.2$ Vol.	0.000	0.000	Acceptable	✓ ✓ ✓	

(*Continued*)

TABLE 6.1 (*Continued*)
Experimental correlation of nanofluid viscosity and their analysis

No.	Author	Material Nanoparticle (Base Fluid)	Correlation	Range of Temperature Nanoparticle Diameter Concentration	Analysis: ANOVA[1] [Variances] P-Value for T	Analysis: ANOVA[1] [Variances] P-Value for φ	Analysis: Statistical Status[5]	Analysis: Data[2] Term[3] Trend[4]	Overall Conclusion
28	Karimipour et al. [56]	CuO (paraffin)	$\mu_{nf} = \left[a_1 + a_2 T.\varphi_w + a_3\varphi_w\right] \mu_{bf}$	$25°C < T_{nf} < 100°C$ 15–30 nm $0.25 < \varphi < 6$ Wt.	0.371	0.000	Acceptable	✓ ✓ ✓	
29	Alrashed et al. [57]	MWCNT/COOH (water)	$\mu_{nf} = 0.00215 - \left(0.00020 \times \begin{bmatrix} \ln(T-1) \\ -\ln(\varphi) \end{bmatrix}\right)$	$20°C < T_{nf} < 50°C$ 9–15 Diamond 3–5 MWCNTs $0 < \varphi < 0.2$ Vol.	0.997	0.000	Rejected	✓ ✓ ✓	According to the results of the statistical test of variance, the correlation is not dependent on the temperature variable.
30	Khodadadi et al. [58]	MgO (water)	$\mu_{nf} = \left[a_1 + a_2 e^{\varphi^{b_1}} + a_3 e^{T^{b_2}} + a_4 T\right] \mu_{bf}$	$25°C < T_{nf} < 60°C$ 20 nm $0.07 < \varphi < 1.25$ Vol.	0.000	*0.000*	Acceptable	✓ × ✓	The correlation has complex terms.
31	Esfe amd Arani [59]	$MWCNT/SiO_2$ (40–60) hybrid (oil) [5W50]	$\mu_{nf} = [1.047 + 0.19\varphi + 0.0011T - 1.51\times10^{-5}\dot{\gamma}$ $-1.88\times10^{-7}T\dot{\gamma} + 9.974\varphi^2 + 1.5\times10^{-9}\dot{\gamma}^2]\ \mu_{bf}$	$5°C < T_{nf} < 55°C$ 5~30 nm $0 < \varphi < 1$ Vol.	1.000	*0.000*	Rejected	✓ ✓ ✓	According to the results of the statistical test of variance, the correlation is not dependent on the temperature variable.

(*Continued*)

TABLE 6.1 (*Continued*)
Experimental correlation of nanofluid viscosity and their analysis

No.	Author	Material Nanoparticle (Base Fluid)	Correlation	Range of Temperature Nanoparticle Diameter Concentration	Analysis: ANOVA[1] [Variances] P-Value for T	Analysis: ANOVA[1] [Variances] P-Value for φ	Analysis: Statistical Status[5]	Analysis: Data[2] Term[3] Trend[4]	Overall Conclusion
32	Esfe et al. [60]	$MWCNT/Al_2O_3$ [10%–90%] hybrid (10w40)	$\mu_{nf} = 697.4317382 + 431.879068\varphi - 33.39840555T - 0.041346927\dot{\gamma} - 10.77912341\varphi T - 0.006913725\varphi\dot{\gamma} + 0.001487159T\dot{\gamma} - 334.024913\varphi^2 + 0.6233416667T^2 + 0.00000133838\dot{\gamma}^2 + 0.000149409\varphi T\dot{\gamma} + 2.908513579\varphi^2 T + 0.076573024\varphi T^2 - 0.00000120931T\dot{\gamma}^2 + 121.11947\varphi^3 - 0.004802059T^3$	$5°C < T_{nf} < 55°C$ - $0.05 < \varphi < 1$ Vol.	0.000	0.000	Acceptable	✓ × ✓	The correlation has multiple terms.
33	Abdul Hamid et al. [61]	TiO_2/SiO_2 hybrid (water/EG) [20–80, 40–60, 50–50, 60–40 80–20]	$\mu_{nf} = \left[1.42(1+R)^{-0.1063}\left(\frac{T}{80}\right)^{0.2321}\right]\mu_{bf}$	$30°C < T_{nf} < 80°C$ 50 nm TiO_2 22 nm SiO_2 1.0 Vol.	0.000	-	Acceptable	✓ ✓ ✓	(R is ratio of TiO_2 to SiO_2)

(*Continued*)

TABLE 6.1 (*Continued*)
Experimental correlation of nanofluid viscosity and their analysis

No.	Author	Material Nanoparticle (Base Fluid)	Correlation	Range of Temperature Nanoparticle Diameter Concentration	Analysis: ANOVA[1] [Variances] P-Value for T	Analysis: ANOVA[1] [Variances] P-Value for φ	Analysis: Statistical Status[5]	Analysis: Data[2] Term[3] Trend[4]	Overall Conclusion
34	Akbari et al. [25]	SiO_2 (EG)	$\mu_{nf} = [-24.81 + 3.23T^{0.08014} \exp\left(1.838\varphi^{0.002334}\right) -0.0006779T^2 + 0.024\varphi^3]\ \mu_{bf}$	$30°C < T_{nf} < 50°C$ 25 nm $1 < \varphi < 3$ Vol.	0.000	0.000	Acceptable	✓ ✓ ✓	The correlation has complex terms.
35	Nabil et al. [62]	TiO_2/SiO_2 hybrid (water/EG) [60–40]	$\mu_{nf} = \left[37\left(0.1 + \frac{\varphi}{100}\right)^{1.59}\left(0.1 + \frac{T}{80}\right)^{0.31}\right]\mu_{bf}$	$30°C < T_{nf} < 80°C$ 30–50 nm TiO_2 22 nm SiO_2 $0.5 < \varphi < 3$ Vol.	0.000	*0.000*	Acceptable	✓ ✓ ✓	
36	Zyła and Fal [63]	SiO_2 (EG)	$\mu_{nf} = \left[1 + 15.39\varphi\right]\mu_{bf}$	298.15 K 7–14 nm $0 < \varphi < 0.26$ Vol.	-	0.999	Rejected	✓ ✓ ✓	According to the results of the statistical test of variance, the correlation is not dependent on the concentration variable.
37	Amani et al. [64]	Fe_2O_4 magnetic field (water)	$\mu_{nf} = \left[a_1\left(1+\varphi\right)^{a_2} T^{a_3} B^{a_4}\right]\mu_{bf}$	$20°C < T_{nf} < 60°C$ 20 nm $0.25 < \varphi < 3$ Vol.	*1.000*	0.000	Rejected	✓ ✓ ✓	According to the results of the statistical test of variance, the correlation is not dependent on the temperature variable. (Constant parameters are dependent on magnetic field)

(*Continued*)

TABLE 6.1 (*Continued*)
Experimental correlation of nanofluid viscosity and their analysis

No.	Author	Material Nanoparticle (Base Fluid)	Correlation	Range of Temperature Nanoparticle Diameter Concentration	Analysis: ANOVA[1] [Variances] P-Value for T	Analysis: ANOVA[1] [Variances] P-Value for φ	Analysis: Statistical Status[5]	Analysis: Data[2] Term[3] Trend[4]	Overall Conclusion
38	Soltani and Akbari [65]	MgO/MWCNT hybrid (EG)	$\mu_{nf} = \left[\begin{array}{l}\left[0.191\varphi + 0.240\left(T^{-0.342}\varphi^{-0.473}\right)\right]\\ \exp\left(1.45T^{0.120}\varphi^{0.158}\right)\end{array}\right]\mu_{bf}$	$30°C < T_{nf} < 60°C$ 5–20 nm MWCNT 40 nm MgO $0 < \varphi < 1$ Vol.	0.516	0.000	Acceptable	✓ × ×	The variation trend of the viscosity model is not consistent with the increase in the volume fraction of the nanofluid. The correlation has complex terms.
39	Aberoumand et al. [24]	Ag (oil)	$\mu_{nf} = \left[\begin{array}{l}1.15 + 1.061\varphi - 0.5442\varphi^2\\ +0.1181\varphi^3\end{array}\right]\mu_{bf}$	$25°C < T_{nf} < 60°C$ 20 nm $0 < \varphi < 2$ Vol.	-	0.000	Acceptable	✓ ✓ ✓	
40	İlhan et al. [66]	Hexagonal boron nitride [hBN] (water/EG)	$\mu_{nf} = \left[15\varphi^2 + 11.47\varphi + 1\right]\mu_{bf}$	25°C 70 nm $0.03 < \varphi < 3$ Vol.	-	0.000	Acceptable	✓ ✓ ✓	
41	Esfe et al. [67]	TiO_2 (water)	$\mu_{nf} = [1.431 - 0.01864T + 0.6073\varphi_w + 0.01334T^2 + 0.02586T\varphi_w + 0.3092\varphi_w^2 + 0.006043T^3 + 0.005644T^2\varphi_w + 0.03323T\varphi_w^2 + 0.08318\varphi_w^3]\ \mu_{bf}$	$280K \le T_{nf} \le 350K$ - $1 < \varphi < 3.5$ Wt.	0.000	0.000	Acceptable	✓ ✓ ✓	The correlation has multiple terms.

(*Continued*)

TABLE 6.1 (*Continued*)
Experimental correlation of nanofluid viscosity and their analysis

No.	Author	Material Nanoparticle (Base Fluid)	Correlation	Range of Temperature Nanoparticle Diameter Concentration	Analysis: ANOVA[1] [Variances] P-Value for T	Analysis: ANOVA[1] [Variances] P-Value for φ	Analysis: Statistical Status[5]	Analysis: Data[2] Term[3] Trend[4]	Overall Conclusion
42	Toghraie et al. [68]	Fe_3O_4 (water(	$\mu_{nf}=\begin{bmatrix}1.01+\left(0.007165T^{1.171}\varphi^{1.509}\right)\\ \times\exp(-0.00719T\varphi)\end{bmatrix}\mu_{bf}$	$20°C<T_{nf}<55°C$ 20–30 nm $0.01<\varphi<3$ Vol.	0.609	0.000	Acceptable	✓ ✓ ✓	
43	Abdolbaqi et al. [69]	SiO_2 (BioGlycol/ water) [20% -80%] [30% -70%]	$\mu_{nf}=\left[0.906\exp\left(10.975\varphi+0.169\frac{T}{80}\right)\right]\mu_{bf}$	$30°C<T_{nf}<80°C$ 22 nm $0.5<\varphi<2$ Vol.	0.819	0.000	Acceptable	✓ ✓ ✓	
44	Syam Sundar et al. [70]	Nanodiamond [ND] (water)	$\mu_{nf}=1.097\mu_{bf}\left[(1+\varphi)^{0.632}\left(\frac{T_{min}}{T_{max}}\right)^{0.056}\right]$	$293K\le T_{nf}\le 333K$ 5–10 nm $0<\varphi<1$ Vol.	1.000	0.970	Rejected	✓ ✓ ✓	According to the statistical variance analysis, the viscosity model does not have the expected dependence on the variables of temperature and concentration.
45	Mostafizur et al. [71]	SiO_2 (methanol)	$\mu_{nf}=0.2861\varphi^2+0.4752\varphi+1.056$	$5°C<T_{nf}<25°C$ 5–15 nm $0.005<\varphi<0.15$ Vol.	-	0.000	Acceptable	✓ ✓ ✓	

(*Continued*)

TABLE 6.1 (*Continued*)
Experimental correlation of nanofluid viscosity and their analysis

No.	Author	Material Nanoparticle (Base Fluid)	Correlation	Range of Temperature Nanoparticle Diameter Concentration	Analysis: ANOVA[1] [Variances] P-Value for T	Analysis: ANOVA[1] [Variances] P-Value for φ	Analysis: Statistical Status[5]	Analysis: Data[2] Term[3] Trend[4]	Overall Conclusion
46	Asadi and Asadi [72]	MWCNT/ZnO (oil)	$\mu_{nf} = 796.8 + 76.26\varphi + 12.88T + 0.7695\varphi T + \frac{-169.9T - 16.53\varphi T}{\sqrt{T}}$	$5°C < T_{nf} < 55°C$ 30 nm $0.125 < \varphi < 1$ Vol.	0.000	1.000	Rejected	✓ ✓ ✓	According to the results of the statistical test of variance, the correlation is not dependent on the concentration variable.
47	Adio et al. [73]	MgO (EG)	$\mu_{nf} = \left[1 + a_o\phi + a_1\left(\frac{T}{T_o}\right)\varphi + a_2\left(\frac{d_p}{h}\right)\varphi + a_3\left(\frac{T}{T_o}\right)\varphi + a_4\left(\left(\frac{d_p}{h}\right)\varphi\right)^2 + a_5\left(\left(\frac{T}{T_o}\right)\varphi\right)^2 + a_6\varphi^2 + a_7\left(\frac{T}{T_o}\right)^2 \varphi^{\frac{1}{3}} \right] \mu_{bf}$	$20°C < T_{nf} < 70°C$ ~21, ~105 and ~125 nm $0 < \varphi < 5$ Vol.	0.000	0.000	Acceptable	✓ ✓ ✓	(a is constant, h is the thickness of the capping layer (nanolayer))

(*Continued*)

TABLE 6.1 (*Continued*)
Experimental correlation of nanofluid viscosity and their analysis

No.	Author	Material Nanoparticle (Base Fluid)	Correlation	Range of Temperature Nanoparticle Diameter Concentration	Analysis: ANOVA[1] [Variances] P-Value for T	Analysis: ANOVA[1] [Variances] P-Value for φ	Analysis: Statistical Status[5]	Analysis: Data[2] Term[3] Trend[4]	Overall Conclusion
48	Esfe et al. [74]	Al_2O_3 (Oil)	$\mu_{nf} = \begin{bmatrix} 1.402 - 0.08407\varphi + 0.2916\varphi^2 \\ -0.05465\varphi^3 \end{bmatrix} \mu_{bf}$ 5°C $\mu_{nf} = \begin{bmatrix} 1.392 - 0.1682\varphi + 0.5401\varphi^2 \\ -0.1553\varphi^3 \end{bmatrix} \mu_{bf}$ 25°C $\mu_{nf} = \begin{bmatrix} 1.438 + 0.1002\varphi + 0.2995\varphi^2 \\ -0.08701\varphi^3 \end{bmatrix} \mu_{bf}$ 45°C $\mu_{nf} = \begin{bmatrix} 1.053 + 0.4609\varphi - 0.2516\varphi^2 \\ +0.1286\varphi^3 \end{bmatrix} \mu_{bf}$ 65°C	$5°C < T_{nf} < 65°C$ 20 nm $0.25 < \varphi < 2$ Vol.	0.000	0.000	Acceptable	✓ ✓ ✓	The correlation has multiple terms.
49	Baratpour et al. [75]	SWCNTs (EG)	$\mu_{nf} = \left[1.089 + \begin{bmatrix} -7.722 \times 10^{-9} \left(T/\varphi \right)^2 \\ +1.19177T^{0.298}\varphi^{0.4777} \end{bmatrix} \times \exp\left(19457T^{-0.453}\varphi^{3.219}\right) \right] \mu_{bf}$	$30°C < T_{nf} < 60°C$ - $0.0125 < \varphi < 0.1$ Vol.	0.220	0.000	Acceptable	✓ ✓ ✓	The correlation has complex terms.

(*Continued*)

TABLE 6.1 (*Continued*)
Experimental correlation of nanofluid viscosity and their analysis

No.	Author	Material Nanoparticle (Base Fluid)	Correlation	Range of Temperature Nanoparticle Diameter Concentration	Analysis: ANOVA[1] [Variances] P-Value for T	Analysis: ANOVA[1] [Variances] P-Value for φ	Analysis: Statistical Status[5]	Analysis: Data[2] Term[3] Trend[4]	Overall Conclusion
50	Afrand et al. [76]	SiO_2/MWCNT Hybrid (oil) [SAE40]	$\mu_{nf} = \left[a_o + a_1\varphi + a_2\varphi^2 + a_3\varphi^3 + a_4\varphi^4\right]\mu_{bf}$	$25°C < T_{nf} < 60°C$ 5–15 nm $0 < \varphi < 1$ Vol.	0.995	0.000	Rejected	✓ ✓ ✓	According to the results of the statistical test of variance, the correlation is not dependent on the temperature variable. (a is dependent on temperature)
51	Abdolbaqi et al. [77]	TiO_2 (BioGlycol/water) [20%–80%] [30%–70%]	$\mu_{nf} = \left[0.918\exp\left(14.696\varphi + 0.161\frac{T}{80}\right)\right]\mu_{bf}$	$30°C < T_{nf} < 80°C$ 15 nm $0.5 < \varphi < 2$ Vol.	0.813	0.000	Acceptable	✓ ✓ ✓	
52	Dalkilic et al. [78]	Graphite (Water)	$\mu_{nf} = 1.1686\mu_{bf} + 1.3764\times10^{-4}\varphi - 1.8027\times10^{-4}$	$20°C < T_{nf} < 60°C$ 6–10 nm $0 < \varphi < 2$ Vol.	-	0.945	Acceptable	✓ ✓ ✓	
53	Li et al. [79]	SiC (EG)	$\mu_{nf} = \begin{bmatrix}1.07879 + 0.45546\varphi \\ +0.4051\varphi^2 - 0.2871\varphi^3\end{bmatrix}\mu_{bf}$	$25°C < T_{nf} < 60°C$ 30 nm $0.2 < \varphi < 1$ Vol.	-	0.000	Acceptable	✓ ✓ ✓	

(*Continued*)

TABLE 6.1 (*Continued*)
Experimental correlation of nanofluid viscosity and their analysis

No.	Author	Material Nanoparticle (Base Fluid)	Correlation	Range of Temperature Nanoparticle Diameter Concentration	Analysis: ANOVA[1] [Variances] *P*-Value for *T*	Analysis: ANOVA[1] [Variances] *P*-Value for φ	Analysis: Statistical Status[5]	Analysis: Data[2] Term[3] Trend[4]	Overall Conclusion
54	Sharifpur et al. [23]	Al_2O_3 (glycerol)	$\mu_{nf} = \left[1 + \Re[\eta].\left[\left(\frac{T}{T_o}\right)^{a_1}.\varphi^{a_2}.\left(\frac{d_p}{h}\right)^{a_3}\right]\right]\mu_{bf}$	$20°C < T_{nf} < 70°C$ 19–160 nm $0 < \varphi < 5$ Vol.	0.997	0.000	Rejected	✓ ✓ ✓	According to the results of the statistical test of variance, the correlation is not dependent on the temperature variable. ($\Re$ is the systemparameter, $[\eta]$ is the intrinsic viscosity, a_1,a_2,a_3 are correlation coefficients)
55	Esfe and Saedodin [22]	ZnO (EG)	$\mu_{nf} = \begin{bmatrix} 0.9118\exp\left(5.94\varphi - 0.00001359T^2\right) \\ +0.0303Ln(T) \end{bmatrix}\mu_{bf}$	$25°C < T_{nf} < 50°C$ 18 nm $0.25 < \varphi < 5$ Vol.	1.000	0.000	Rejected	✓ ✓ ✓	According to the results of the statistical test of variance, the correlation is not dependent on the temperature variable. The correlation has complex terms.
56	Duangthongsuk and Wongwises [21]	TiO_2 (water)	$\mu_{nf} = \left[\left(a_1 + a_2\varphi + a_3\varphi^2\right)\right]\mu_{bf}$	$15°C < T_{nf} < 35°C$ 21 nm $0.2 < \varphi < 2$ Vol.	-	0.345	Acceptable	✓ ✓ ✓	(*a*, *b* and *c* are constant values and dependent on temperature)

(*Continued*)

TABLE 6.1 (*Continued*)
Experimental correlation of nanofluid viscosity and their analysis

No.	Author	Material Nanoparticle (Base Fluid)	Correlation	Range of Temperature Nanoparticle Diameter Concentration	Analysis				Overall Conclusion
					ANOVA[1] [Variances]		Statistical Status[5]	Data[2]	
					P-Value for T	P-Value for φ		Term[3] Trend[4]	

Note:
Variances analysis[1]: Results of variance analysis determine the dependence of the correlation on the variable. ✓: According to the type of physical evaluation, the correlation is reliable in this case.
Data[2]: By considering the concentration of zero in the correlation, the ability of the correlation to estimate the viscosity of the base fluid is investigated. ×: According to the type of physical evaluation, the correlation, in this case is not reliable.
Terms[3]: correlations with multiple or irrational terms are identified.
Trends[4]: According to the physical laws governing the viscosity of nanofluids, the decreasing or increasing trend of viscosity is evaluated according to changes in temperature and concentration.
Statistical[5]: Results of physical evaluations of correlations.

By evaluating the correlations presented for nanofluid viscosity, the situation of temperature and volume (mass) fraction variables in the correlations are determined. The results showed If there is a significant relationship between variables and nanofluid viscosity.

Therefore, it is necessary to consider the effects of variables on the viscosity models of nanofluids in the temperature range and volume (mass) fraction of heat transfer.

6.2.2 Physical Analysis of Correlations

Viscosity models at different dimensions in this part of the research are studied. First, the structure of the proposed correlations to calculate the viscosity has been examined and then evaluated for analyzing complex, heterogeneous, or ambiguous components. Meanwhile, in another section, based on the experimental studies of researchers in the temperature range and volume (mass) fraction, the data extracted from the correlations are evaluated. Also, the accuracy of the experimental model for the case where the concentration of nanofluid is considered zero with the viscosity of base fluid has been investigated. Finally, the correlations have been measured in terms of the laws governing the physics of nanofluids and the changes due to an increase or decrease in the temperature and concentration of nanofluids.

According to Table 6.1 of the Term section, in most of the experimental relations studied for calculating viscosity, it is observed that mathematical expressions and terms do not interfere with the calculation of the viscosity of nanofluids. Thus, these expressions have only caused the complexity and inefficiency of empirical relations, which increases the possibility of errors in the viscosity calculations of nanofluids.

Given the above, Alarifi et al. [34] studied the viscosity of hybrid nanofluids, which are composed of a mixture of MWCNT and TiO_2 nanoparticles in oil, and they presented a new model by Eq. (6.3) for the viscosity of the nanofluid by examining the effects of temperature and concentration on the viscosity of the nanofluid. According to the equation, a trigonometric ratio has been used to express the relationship between concentration and viscosity. Therefore, a disproportionate function in relationships is not necessary and can only cause problems in calculations.

$$\mu_{\text{nf}} = 2.936T + \frac{2e^4}{1.68 + T - (1.68\varphi)} - 448.8 - \tan\left((1.68\varphi) - 1.68\right) \tag{6.3}$$

Based on an experimental test on the viscosity behavior of SiO_2 nanoparticles dispersed in a mixture of water and ethylene glycol, Ruhani et al. [35] proposed a viscosity model in Eq. (6.4), valid in the temperature range of 25°C–50°C and a volume fraction of 0.1%–1.5%. The temperature variable in the viscosity model is both in the position of the power function and is powered by the exponential function in terms of position. Therefore, using such functions one after the other is not justified and causes the calculations to be complex.

$$\begin{aligned}\mu_{\text{nf}} = \Big[&2.030 - \left(931.616 \times \varphi^{0.9305} \times 5.4597 \times T^{-3.4574}\right) \\ &- \exp\left(-0.0028 \times \varphi^{2.1421} \times T^{1.0133}\right)^2\Big]\,\mu_{\text{bf}}\end{aligned} \tag{6.4}$$

Yan et al. [28] have presented Eq. (6.5) for the hybrid nanofluid of MWCNT and TiO_2 in ethylene glycol in the temperature range of 25°C–55°C and at the volume fraction between 0.05% and 1%;

$$\begin{aligned}\mu_{nf} = [&0.90463 + 280.20104\varphi + 0.257340\theta + 368.052390\theta\varphi \\ &-28643.68399\varphi^2 - 0.0120510\theta^2 - 1968.73612\varphi^2\theta \\ &-235.047290\theta^2\varphi + 2.09629\times10^6\varphi^3 - 0.0996940\theta^3]\ \mu_{bf}\end{aligned} \tag{6.5}$$

Equation (6.5) has many terms that entering the equation for subsequent heat transfer calculations may be associated with many errors. On the other hand, by reducing the terms of the equation with increasing the accuracy, the equation becomes easier to use. Therefore, Eq. (6.5) can be presented more simply as Eq. (6.6).

$$\mu_{nf} = \left(0.0029741 + \varphi^{1.07982}\right)\times\left(T^{-1.25573}\right)\times 311157 \tag{6.6}$$

Equation (6.6) has been obtained by the nonlinear regression method from experimental data of Saeedi et al. for hybrid nanofluid viscosity.

According to Figure 6.4, the proposed model in Eq. (6.6), while having higher accuracy than Eq. (6.5), has a simpler form compared to Eq. (6.5). Figure 6.4 plotted at a volume fraction of 0%–1% and a temperature of 30°C for nanofluids.

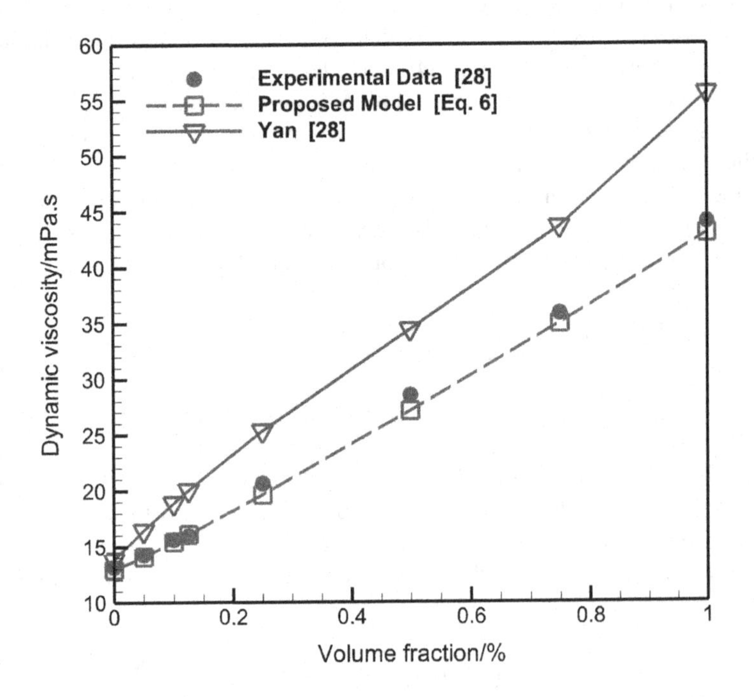

FIGURE 6.4 Comparison of the results of Yan et al.'s correlation [28] and the currently proposed correlation with experimental data.

Huminic et al. [36] proposed Eq. (6.7) for La_2O_3 and water nanofluids in the temperature range of 293–323 K and a volume fraction between 0% and 0.03%.

According to Eq. (6.7), there are many components in the correlation that are similar to Eq. (6.5), and the experimental data provided by Huminic et al. are used, and Eq. (6.8) is presented as follows.

$$\mu_{nf} = a_1 + a_2T + a_3\varphi + a_4T^2 + a_5T\varphi + a_6\varphi^2 + a_7T^3 + a_8T^2\varphi + a_9T\varphi^2 + a_{10}\varphi^3 + a_{11}T^3\varphi + a_{12}T^2\varphi^2 + a_{13}T\varphi^3 + a_{14}\varphi^4 \tag{6.7}$$

$$\mu_{nf} = \left(0.458474 + \varphi^{1.10104}\right) \times \left(\frac{T}{323}\right)^{-0.636006} \times 1.5773 \tag{6.8}$$

According to Figure 6.5, Eq. (6.8), simplicity has an accuracy of more than 2% more than that in Eq. (6.7).

A closer look reveals similar cases in which researchers try to present complex correlations; however, using the same experimental data, simple and sometimes linear equations can be given with much higher accuracy than the desired equations, so Eqs. (6.6 and 6.8) can replace Eqs. (6.5 and 6.7).

While the models proposed for nanofluid viscosity by Esfe et al. [37,38] presented in Eqs. (6.9 and 6.10), there are several terms in the viscosity model; therefore, the presence of multiple terms in the equations is not necessary and makes the equations

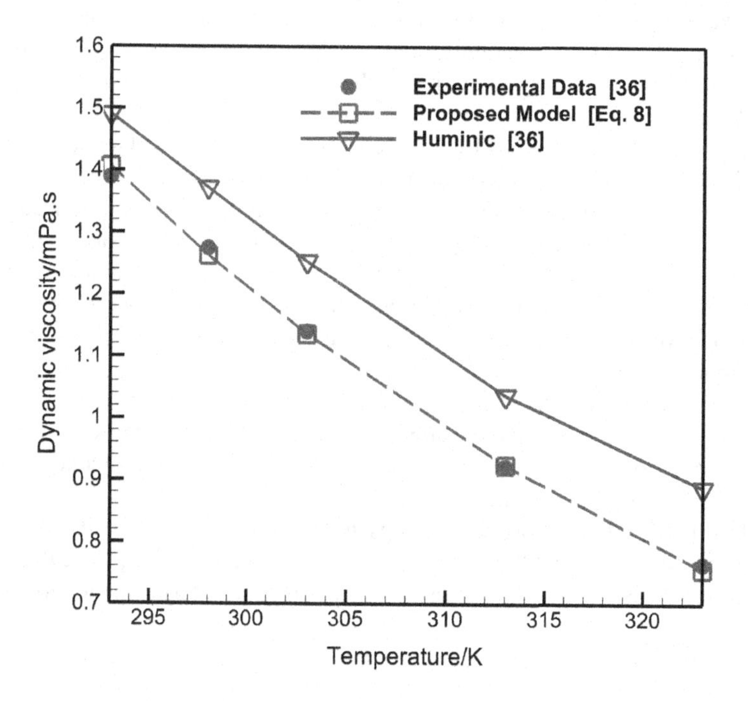

FIGURE 6.5 Comparison of the results of Huminic et al.'s [36] correlation and the proposed correlation with experimental data.

more complex, and these multiple components in viscosity calculation will lead to increased computational error.

$$\mu_{nf} = 679.78306 + 259.62463\varphi - 33.64131T - 0.045393\dot{\gamma} - 6.0695 \times \varphi T$$
$$-0.00031841\varphi\dot{\gamma} + 0.00129007\mathrm{T}\dot{\gamma} - 236.23287\varphi^2 + 0.65796T^2 + 2.31776\mathrm{E}$$
$$-06\dot{\gamma}^{\ 2} + 1.55167\varphi^2 T + 0.049805\varphi T^2 - 9.63258E - 8T\dot{\gamma}^{\ 2} + 94.57115\varphi^3$$
$$-0.0051586T^3 + 0.0000000001\dot{\gamma}^{\ 3} \tag{6.9}$$

$$\mu_{nf} = 688.46 + 347.09\varphi - 33.12T - 0.04\dot{\gamma} - 7.36\varphi T - 0.0087\varphi\dot{\gamma} + 0.0014T\dot{\gamma}$$
$$-305.24\varphi^2 + 0.61T^2 + 1.49 \times 10^{-6}\dot{\gamma}^2 + 0.0001\varphi T\dot{\gamma} + 0.46\varphi^2 T + 0.0014\varphi^2\dot{\gamma}$$
$$+0.065\varphi T^2 + 1.87 \times 10^{-7}\varphi\dot{\gamma}^2 - 7.25 \times 10^{-6} T^2\dot{\gamma} - 7.33 \times 10^{-8} T\dot{\gamma}^2 + 169.62\varphi^3$$
$$-0.0043T^3 + 0.00000000011\dot{\gamma}^3 \tag{6.10}$$

In the continuation of reviewing the results and data extracted from viscosity models, it has been observed that sometimes the equations at zero concentration and in a certain range of temperature and concentration have an unusual response. The results of this analysis have been presented in Table 6.1 of the Data section.

For example, Huminic et al. [36] presented the viscosity model for La_2O_3 and water nanofluids in the temperature range of 293–323 K and at a volume fraction between 0 and 0.03 contrary to the researcher claims, the correlation is not able to respond at zero concentration.

Finally, according to studies by researchers on the viscous behavior of nanofluids, variables of temperature and concentration independently and directly affect the nanofluid viscosity, so that with increasing temperature, the viscosity of nanofluid decreases, and by adding nanoparticles to the base fluid, the viscosity of nanofluid increase. Therefore, at this stage, the variation trend of viscosity values at different temperatures and concentrations according to the physical laws governing nanofluid viscosity is investigated. The results of this analysis have been presented in Table 6.1 of the Trend section.

According to the presented issue, in the viscosity model proposed by Esfe and Esfandeh [39] for the viscosity of nanofluids, including oil and hybrid particles, the variation trend of viscosity values with increasing concentration is contrary to the physical laws governing nanofluid viscosity.

Table 6.1 exhibits the correlations presented for nanofluid viscosity by various researchers from 2009 to 2020, along with statistical analysis and physical analysis. Statistical analysis of the ANOVA test was performed, and terms related to physical examination [Data, Term, Trend] entirely have been presented for each experimental equation.

In addition, the validity range of the equations and the overall conclusion have been presented by the correlation evaluation.

6.3 SENSITIVITY ANALYSIS WITH MONTE CARLO TEST

The sensitivity analysis method has been used to continue the statistical study process of relationships and know the position of variables in correlations. According to the general definition of statistics, sensitivity analysis is the study of the effectiveness of output variables from a set of assumed input variables in a statistical model.

As a result, the researcher can determine how changes in a component affect the model's output. Therefore, in the continuation of the statistical study article, the sensitivity analysis will give a deeper look at viscosity models with the variables of base fluid viscosity, temperature, and concentration [80].

Therefore, to obtain a complete conclusion about the performance of viscosity models and the effectiveness of variables, only the correlations in which the variables have the expected dependence on the viscosity equations in terms of variance test are examined.

In the present study, to analyze viscosity models, also a method known as the Monte Carlo test is used.

Variables with little effect and little change in the equations are displayed as flat lines in the graph. So, the more curved lines show the more dependence of the variable on the equation.

Li and Zou [26] introduced the viscosity model of Eq. (6.11) for nanofluids consisting of Al_2O_3 nanoparticles and water-based nanofluid, and Saeedi et al. [55] proposed the viscosity model of Eq. (6.12) for CeO_2 nanoparticles dispersed in ethylene glycol.

$$\mu_{nf} = \left[781.4 \times T^{-2.117} \times \varphi^{0.2722} + \frac{0.05776}{T^{-0.7819} \times \varphi^{-0.04009}} + 0.511 \times \varphi^2 - 0.1779 \times \varphi^3\right] \mu_{bf} \tag{6.11}$$

$$\mu_{nf} = 0.838\varphi^{0.188} T^{0.089} \mu_{bf}^{1.1} \tag{6.12}$$

The test results in Figures 6.6 and 6.7 show that the curvature of the miobf (μ_{bf}) line is greater than that of the phi (φ) and T lines, and the presence of the variable of the viscosity of the base fluid in correlation is more important than other variables. In addition, concentration and temperature, respectively, have an influential role in the equations.

Equation (6.13) presents the Nabil et al. [62] viscosity model for the hybrid nanofluid of TiO_2 and SiO_2 in a mixture of water and ethylene glycol.

$$\mu_{nf} = \left[37\left(0.1 + \frac{\varphi}{100}\right)^{1.59} \left(0.1 + \frac{T}{80}\right)^{0.31}\right] \mu_{bf} \tag{6.13}$$

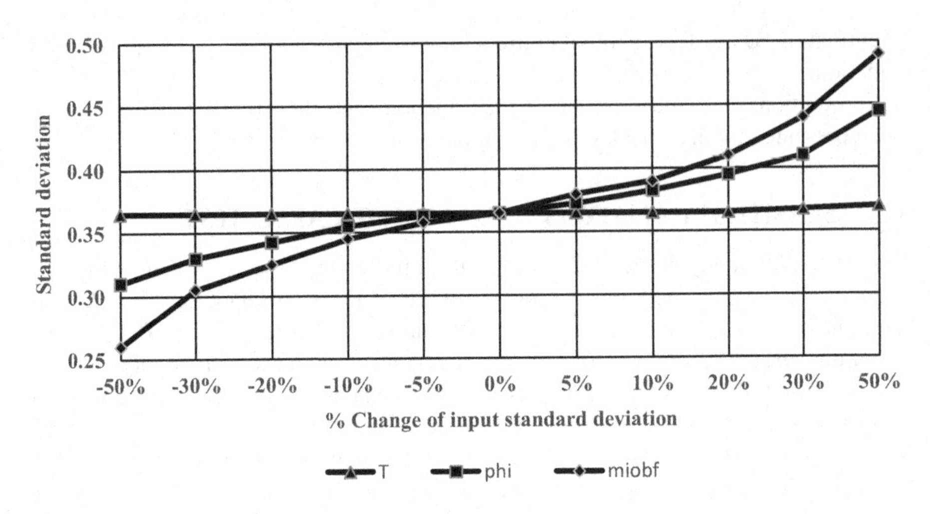

FIGURE 6.6 Result of sensitivity analysis on the correlation proposed by Li and Zou [26]. (Rhombus pattern μ_{bf} (miobf): base fluid viscosity variable/Square pattern φ (phi): volume fraction variable/Triangle pattern (T): temperature variable).

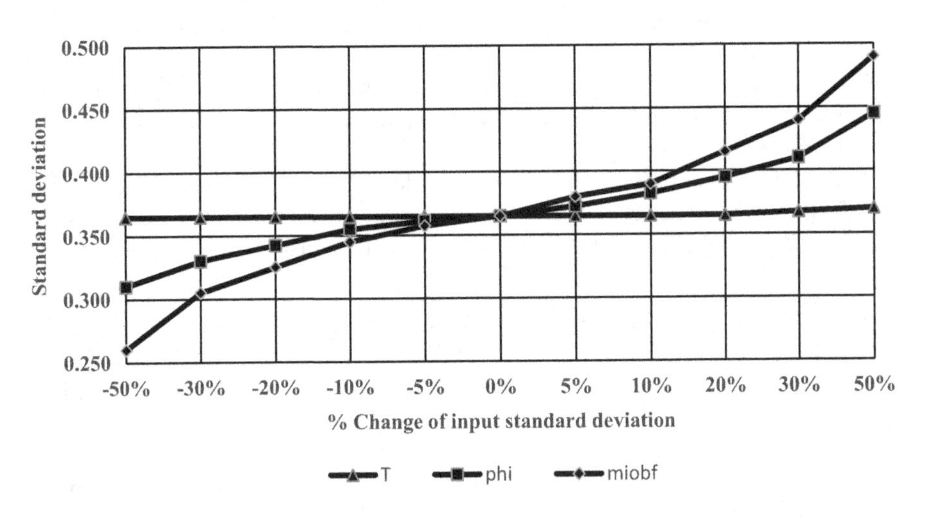

FIGURE 6.7 Result of sensitivity analysis on the correlation proposed by Saeedi et al. [55]. (Rhombus pattern μ_{bf} (miobf): base fluid viscosity variable/Square pattern φ (phi): volume fraction variable/Triangle pattern (T): temperature variable).

In Figure 6.8, the miobf (μ_{bf}) line has higher curvature than the phi (φ) and T lines. Therefore, the sensitivity analysis results showed that the variables of base fluid viscosity, concentration, and temperature significantly affect Eq. (6.13), respectively.

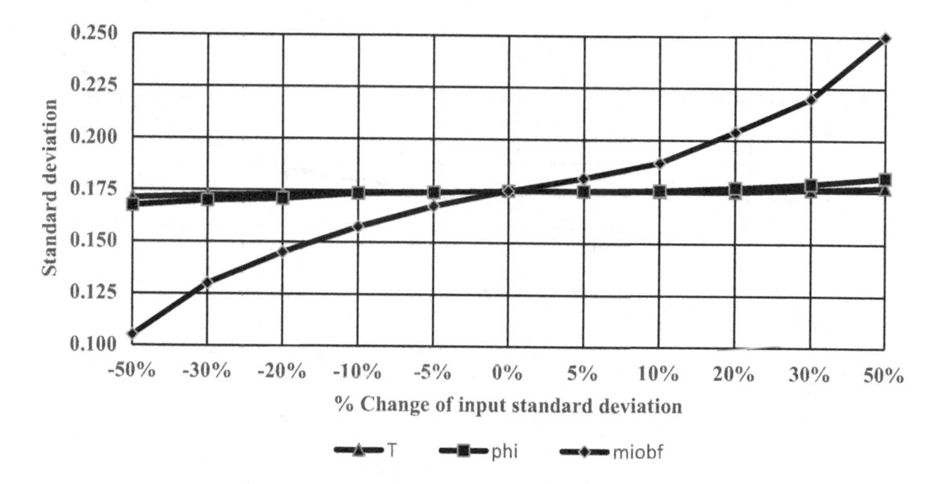

FIGURE 6.8 Result of sensitivity analysis on the correlation proposed by Nabil et al. [62]. (Rhombus pattern μ_{bf} (miobf): base fluid viscosity variable/ Square pattern φ (phi): volume fraction variable/Triangle pattern (T): temperature variable).

The variables of base fluid viscosity, temperature, and concentration are available in the above viscosity equations. The variable of base fluid viscosity applies the value of the base fluid viscosity in proportion to the reference temperature in the viscosity model. Therefore, the dependence of the viscosity equations on the viscosity variable of the base fluid expresses the relationship of the equations on temperature. On the other hand, the sensitivity analysis results show that the variable of the base fluid viscosity has a greater contribution in estimating the viscosity of nanofluid than other variables. Therefore, considering the mentioned conditions, it is concluded that the temperature factor indirectly has a more significant effect on viscosity equations than other variables.

Esfe et al. [48] have presented Eq. (6.14) for the viscosity of hybrid nanoparticles of MWCNT and TiO_2 in a mixture of water-based nanofluid and ethylene glycol;

$$\mu_{nf} = 6.35 + 2.56\varphi - 0.24T - 0.068\varphi T + 0.905\varphi^2 + 0.0027T^2 \tag{6.14}$$

According to the sensitivity analysis results presented in Figure 6.9, the effect of the temperature is greater than the concentration in the equation because the curvature of the T line is more significant.

$$\begin{aligned}\mu_{nf} = {} & 688.46 + 347.09\varphi - 33.12T - 0.04\dot{\gamma} - 7.36\varphi T - 0.0087\varphi\dot{\gamma} + 0.0014T\dot{\gamma} \\ & -305.24\varphi^2 + 0.61T^2 + 1.49\times10^{-6}\dot{\gamma}^2 + 0.0001\varphi T\dot{\gamma} + 0.46\varphi^2 T + 0.0014\varphi^2\dot{\gamma} \\ & +0.065\varphi T^2 + 1.87\times10^{-7}\varphi\dot{\gamma}^2 - 7.25\times10^{-6}T^2\dot{\gamma} - 7.33\times10^{-8}T\dot{\gamma}^2 + 169.62\varphi^3 \\ & -0.0043T^3 + 0.00000000011\dot{\gamma}^3\end{aligned} \tag{6.15}$$

Esfe et al. [38] presented the nanofluid viscosity model for the MWCNT and Al_2O_3 hybrid nanoparticles dispersed in oil in Eq. (6.15).

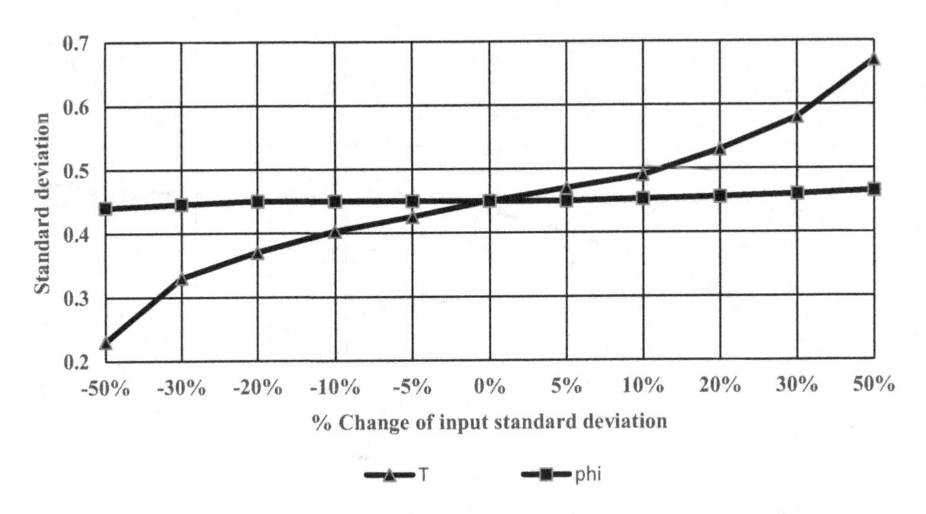

FIGURE 6.9 Result of sensitivity analysis on the correlation proposed by Esfe et al. [48]. (Triangle pattern (T): temperature variable/Square pattern φ (phi): concentration variable).

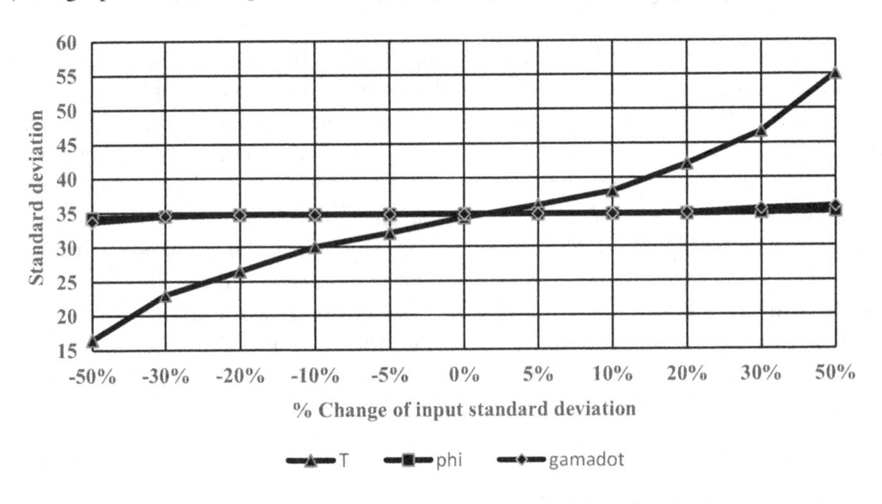

FIGURE 6.10 Result of sensitivity analysis on the correlation proposed by Esfe et al. [37]. (Triangle pattern (T): temperature variable/Square pattern φ (phi): volume fraction variable/ Rhombus pattern $\dot{\gamma}$ (gamadot): variable of shear rate).

The sensitivity analysis results in Figure 6.10 showed that the T line has higher curvature than the $\dot{\gamma}$ (gamadot) and phi (φ) lines, so the effect of the temperature is greater than the concentration in the equation.

According to the results and the role of temperature and concentration in Eq. (6.15), the shear rate of nanofluid is also effective in calculating the viscosity of nanofluid.

The results of sensitivity analysis of the previous two equations show that when the variable of base fluid viscosity is not present in the viscosity equations, conditions are created that the effect of temperature factor is directly applied in the viscosity equations, thus, the temperature variable has a more influential role than other variables.

Equation (6.16) presents Li et al. [79] model for the viscosity of SiC and water nanofluids.

$$\mu_{nf} = \left[1.07879 + 0.45546\varphi + 0.4051\varphi^2 - 0.2871\varphi^3\right]\mu_{bf} \tag{6.16}$$

Based on the sensitivity test results in Figure 6.11 and given the curvature of the miobf (μ_{bf}) line, the viscosity of base fluid has a higher contribution to the concentration variable in calculating the viscosity of the nanofluid.

There is a base fluid viscosity variable μ_{bf} in most of the correlations reviewed here and in the known nanofluid viscosity models, and this factor determines the viscosity value of the base fluid relative to the reference temperature in the viscosity models, so the effects of temperature through the base fluid viscosity variable is considered in viscosity models. Therefore, the high dependence of the viscosity models on the variable of base fluid viscosity is due to the dependence of the viscosity models on temperature.

It is concluded that the effects of temperature are not directly considered in the correlations, and the μ_{bf} component is not an independent variable, which can cause problems.

According to the issues mentioned above and based on the sensitivity analysis results performed on viscosity models, the temperature factor plays a decisive role in viscosity models. Therefore, the results show the inherent dependence of nanofluid viscosity on temperature.

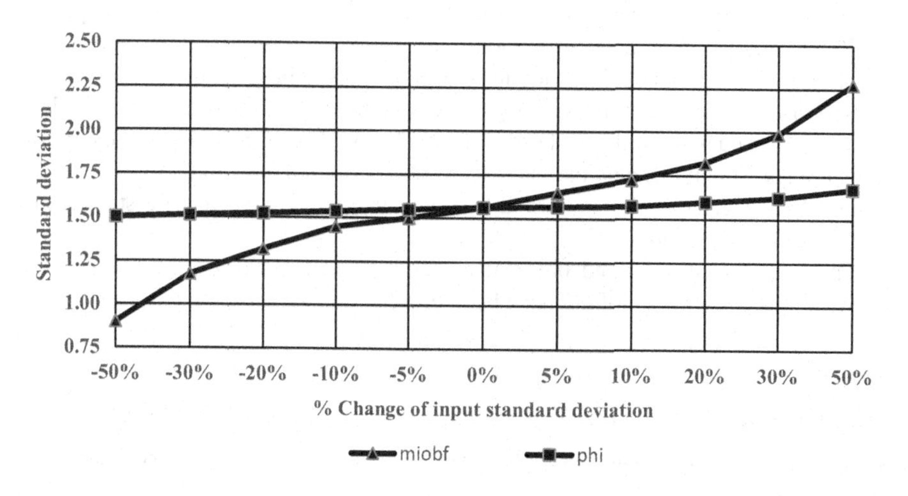

FIGURE 6.11 Result of sensitivity analysis on the correlation proposed by Li et al. [79]. (Rhombus pattern μ_{bf} (miobf): base fluid viscosity variable/Square pattern φ (phi): concentration variable).

6.4 OVERALL ANALYSIS OF EMPIRICAL CORRELATIONS

Table 6.1 has statistically and physically examined the prediction correlations of nanofluid viscosity. In the statistical study, variance analysis for temperature and volume (mass) fraction of nanofluid have been performed, and the results have been presented in the column related to the statistical study. In addition, in physical examination, three factors of Data, Term, and Trend have been considered. The Data column in Table 6.1 examines the experimental correlations' ability to respond at zero concentration to reach the value of the base fluid viscosity. In the Term column of Table 6.1, the results have been mentioned regarding the existence of numerous and irrational terms for estimating the viscosity of the mentioned experimental correlations. The compliance and the role of the variables introduced in the empirical correlations presented in Table 6.1 relative to the physics governing the viscosity of the nanofluid have been evaluated in the Trend column of Table 6.1.

In evaluating viscosity relationships, it was observed that there are relationships that have good conditions in the physical examination but are not statistically similar. Also, Reverse conditions for equations are possible. Therefore, it was decided to report the relationships with good status in two physical and statistical states in the total section.

By examining all the correlations in Table 6.1 and their statistical analysis and examining the physical performance of the correlations and the validity of each equation, it can be concluded that 53.6% of the equations are statistically valid. Also, 73.2% of the equations have accuracy and simplicity in terms of performance; in total, 35.7% of the equations in both physical and statistical states have an acceptable condition (Table 6.2).

The results of the evaluation of the experimental correlations presented in Table 6.1 for nanofluids based on water and ethylene glycol, which are widely used in the field of research, have been developed. Accordingly, in a comprehensive study on the correlations of Table 6.1 for water-based nanofluids, it is statistically and physically determined that statistically, 38.9% of the correlations are acceptable, 83.3% are physically reliable correlations, and a total of 27.7% of the correlations are acceptable. The results of this study have been presented in Table 6.3.

Suppose the examination for water-based nanofluid is performed again for ethylene glycol-based nanofluids, as shown in Table 6.4. In that case, 54.5% of the correlations are statistically acceptable, 54.5% are physically reliable, and a total of 27.2% of the ethylene glycol-based nanofluid equations in both physical and statistical states had an acceptable condition.

Thus, the statistically and physically acceptable correlations for the water and ethylene glycol-based nanofluid have been presented in Tables 6.5 and 6.6.

TABLE 6.2
Status of Statistical and Physical Analysis of all Equations of Table 6.1

Viscosity Correlations of Nanofluids	Acceptable (Reliable) (%)	Rejected (Unreliable) (%)
Statistically status	53.6	46.4
Physical examination	73.2	26.8
Total status	35.7	64.3

TABLE 6.3
Status of Evaluation of Viscosity Equations Based on Water-Based Nanofluid

Viscosity Correlations of Water-Based Nanofluid	Acceptable (Reliable) (%)	Rejected (Unreliable) (%)
Statistically status	38.9	61.1
Physical examination	83.3	16.7
Total status	27.7	72.3

TABLE 6.4
Status of Evaluation of Viscosity Equations Based on Ethylene Glycol-Based Nanofluid

Viscosity Correlations of EG Based-Nanofluid	Acceptable (Reliable) (%)	Rejected (Unreliable) (%)
Statistically status	54.5	45.5
Physical examination	54.5	45.5
Total status	27.2	72.8

TABLE 6.5
Acceptable Correlations for the Viscosity of Water-Based Nanofluid

No.	Authors	Correlation	Material Nanoparticle (Base Fluid)
1	Duangthongsuk and Wongwises 2009 [21]	$\mu_{nf} = \left[\left(a_1 + a_2\varphi + a_3\varphi^2\right)\right]\mu_{bf}$	TiO_2 (water)
2	Moldoveanu et al. 2018 [53]	Al_2O_3: $\mu_{nf} = \left[4135\varphi^2 - 91.72\varphi + 2.06\right]\mu_{bf}$ SiO_2: $\mu_{nf} = \left[-769\varphi^2 + 42\varphi + 1.1\right]\mu_{bf}$	Al_2O_3, SiO_2 Hybrid Separately (Water)
3	Moldoveanu et al. 2018 [54]	Al_2O_3 : $\mu_{nf} = \left[0.6152\varphi^2 - 1.5449\varphi + 2.3792\right]\mu_{bf}$ TiO_2 : $\mu_{nf} = \left[0.2302\varphi^2 - 0.3202\varphi + 1.5056\right]\mu_{bf}$	Al_2O_3, TiO_2 Hybrid Separately (Water)
4	Toghraie et al. 2016 [68]	$\mu_{nf} = \begin{bmatrix} 1.01 + \left(0.007165T^{1.171}\varphi^{1.509}\right) \\ \times\exp(-0.00719T\varphi) \end{bmatrix}\mu_{bf}$	Fe_3O_4 (water)
5	Dalkiliç et al. 2016 [78]	$\mu_{nf} = 1.1686\mu_{bf} + 1.3764\times10^{-4}\varphi - 1.8027\times10^{-4}$	Graphite (Water)

TABLE 6.6
Acceptable Correlations for the Viscosity of Ethylene Glycol-Based Nanofluid

No.	Authors	Correlation	Material Nanoparticle (Base Fluid)
1	Saeedi et al. 2018 [55]	$\mu_{nf} = \begin{bmatrix} 781.4 \times T^{-2.117} \times \varphi^{0.2722} + \dfrac{0.05776}{T^{-0.7819} \times \varphi^{-0.04009}} \\ +0.511 \times \varphi^2 - 0.1779 \times \varphi^3 \end{bmatrix} \mu_{bf}$	CeO_2 (EG)
2	Adio et al. 2016 [73]	$\mu_{nf} = \begin{bmatrix} 1 + a_0\varphi + a_1\left(\dfrac{T}{T_0}\right)\varphi + a_2\left(\dfrac{d_p}{h}\right)\varphi + a_3\left(\dfrac{T}{T_0}\right)\varphi \\ +a_4\left(\left(\dfrac{d_p}{h}\right)\varphi\right)^2 + a_5\left(\left(\dfrac{T}{T_0}\right)\varphi\right)^2 \end{bmatrix} \mu_{bf}$	MgO (EG)
5	Li et al. 2015 [79]	$\mu_{nf} = \begin{bmatrix} 1.07879 + 0.45546\varphi \\ +0.4051\varphi^2 - 0.2871\varphi^3 \end{bmatrix} \mu_{bf}$	SiC (EG)

6.5 PROPOSING A VISCOSITY MODEL AND ITS VALIDATION

6.5.1 Preliminary Analysis

By studying the correlations proposed by the researchers, the relationship of variables with the viscosity of nanofluid was determined. Therefore, it was confirmed by the analysis of variance that temperature and concentration variables play a decisive role in the relationship between the viscosity of nanofluids. In addition, the results of the statistical tools of sensitivity analysis showed that the viscosity of nanofluid is directly dependent on the temperature factor.

Given that the volume fraction variable of nanoparticles affects nanofluid's viscosity, the rate of viscosity changes relative to the volume fraction of nanoparticles depends on the type of the base fluid.

On the other hand, it is crucial for the viscosity of the nanofluid with water-based nanofluid and ethylene glycol, which does not have limited use in terms of temperature, volume fraction, and especially particle material. Therefore, in the present study, two models with very high accuracy for nanofluids with water-based nanofluid in a wide range of volume fractions and temperature have been presented, and this important has been done for ethylene glycol-based nanofluid.

Evaluating the experimental studies on the viscosity of nanofluids in proportion to temperature and concentration, approximately the physical conditions of more than 1200 experimental data were examined, and dispersed viscosity values were observed for the experimental data under the same physical conditions. Therefore, to increase the accuracy of the proposed correlation, a group of articles has been removed, and articles with appropriate and centralized experimental data have been selected to provide the correlation.

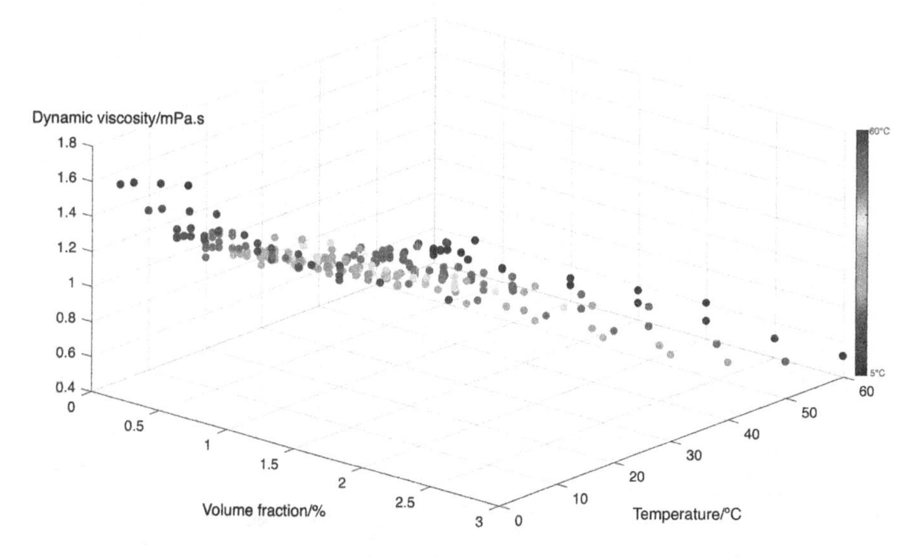

FIGURE 6.12 Three-dimensional representation of the viscosity dispersion of nanofluids with water-based nanofluid [21,43,64,78,81–84].

Figure 6.12 is plotted in terms of temperature and volume fraction of the water-based nanofluid to know the physical condition of the experimental data used [21,43,64,78,81–84].

Figure 6.12 shows that the congestion of experimental data for the viscosity of water-based nanofluid at concentrations less than 1%, and the temperature range of 30°C to 40°C is higher.

Also, the dispersion of experimental data to present the viscosity model is shown in Figure 6.13 in terms of temperature and volume fraction of ethylene glycol-based nanofluid [55,85–88].

Also, according to Figure 6.13, at low concentrations and the temperature range of 30°C–50°C, the viscosity of ethylene glycol-based nanofluid has higher congestion.

6.5.2 Proposed Correlation

Experimental viscosity correlations proposed by previous researchers cover a limited temperature range and volume fraction. Most experimental correlations proposed for nanofluid viscosity are unable to estimate the base fluid viscosity. On the other hand, in the analysis of variance, it was found that most of the mentioned correlations do not depend on the independent variables of those correlations. The sensitivity analysis results also showed that the factor of temperature directly affects the viscosity of nanofluid, and the variable of temperature has a more significant contribution in estimating the viscosity of nanofluid than other variables. In such conditions, to eliminate these shortcomings, a model has been presented for estimating the viscosity of water and ethylene glycol-based nanofluid entitled BAG, Barkhordar–Armaghani–Ghasemiasl. The summary of the statistical and physical studies of the BAG model

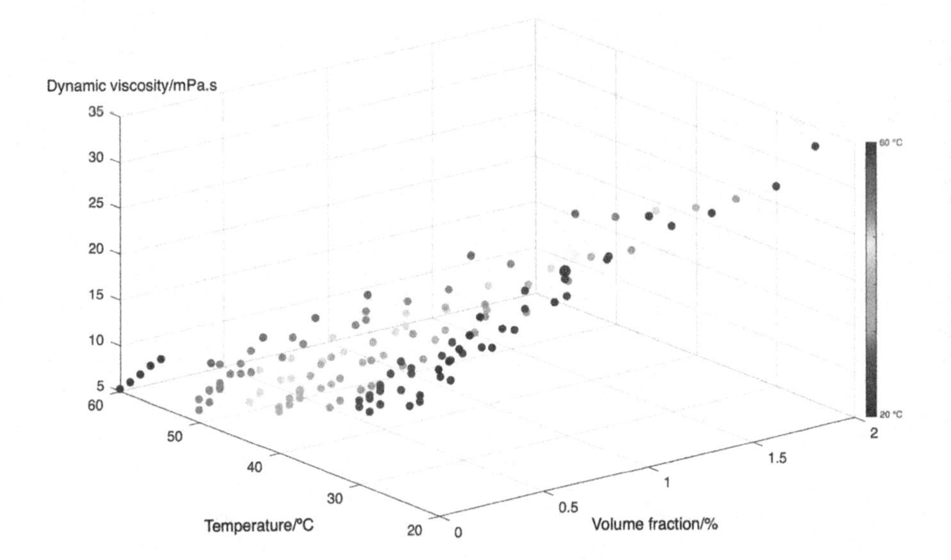

FIGURE 6.13 Three-dimensional representation of the viscosity dispersion of nanofluids with ethylene glycol-based nanofluid [55,85–88].

is given in Table 6.7. According to the variance analysis, the temperature variable and the volume fraction have the appropriate *P*-values for the BAG model.

The results of estimating the viscosity of water and ethylene glycol-based nanofluid based on the BAG model in Table 6.7 compared to the experimental data in Figures 6.12 and 6.13 based on R^2 of water-based nanofluid is 97.01% and for ethylene glycol-based nanofluid is 96.08%.

6.5.3 Evaluation of BAG Viscosity Model

For assessing the validity of the accuracy of the presented correlations, it is necessary to compare the obtained results with the conventional and selected correlations. For this purpose, some conventional correlations in articles are introduced as follows. Einstein [1] was the first to introduce a microfluidic viscosity model for suspensions containing metal particles in 1906. This correlation applies to the viscosity of microfluid with spherical particles at a volume fraction of less than 5%. This model has been given in Eq. (6.17).

$$\mu_{\text{nf}} = (1 + 2.5\varphi)\mu_{\text{bf}} \tag{6.17}$$

Brinkman [19] proposed a new model in 1952 according to Einstein's model. This correlation is suitable for suspensions with a volume fraction of less than 4%. This correlation has been given in Eq. (6.18), used in most studies by researchers.

$$\mu_{\text{nf}} = \left(\frac{1}{(1-\varphi)^{2.5}}\right)\mu_{\text{bf}} \tag{6.18}$$

TABLE 6.7
BAG Models for Nanofluids Viscosity

No.	Symbol	Base Fluid	Correlation	Range of Temperature Concentration	Analysis: ANOVA [Variances] P-Value for T	Analysis: ANOVA [Variances] P-Value for φ	Analysis: Data Term Trend	Status: Statistical	Status: Physical
1	BAG I	Water	$\mu_{water}=\left(1.89841+\left(\frac{\varphi}{0.03}\right)^{3.3616}\right)\times\left(-0.27265-\left(\frac{T}{60}\right)^{0.701322}\right)+\left(\frac{T}{60}\right)^{-0.213772}+0.248161-\left(\frac{\varphi}{0.03}\right)^{1.65192}$	$5°C<T<60°C$ $0<\varphi<3$ vol.%	0.000	0.000	✓ ✓ ✓	Acceptable	Reliable
2	BAG II	Ethylene glycol	$\mu_{EG}=\left(0.85614+\left(\frac{\varphi}{0.02}\right)^{0.81593}\right)\times\left(6.30106-\left(\frac{T}{60}\right)^{-1.10622}\right)+\left(\frac{T}{60}\right)^{-0.23249}+\left(\frac{\varphi}{0.02}\right)^{0.24584}$	$20°C<T<60°C$ $0<\varphi<2$ vol.%	0.000	0.000	✓ ✓ ✓	Acceptable	Reliable

In 1977, Batchelor [20] proposed a viscosity model for single-phase suspensions based on the Brownian motion of particles. Moreover, Eq. (6.19) has been derived according to the Einstein equation and the existence of spherical particles.

$$\mu_{nf} = \left(1 + 2.5\varphi + 6.2\varphi^2\right)\mu_{bf} \tag{6.19}$$

In recent studies, Wang et al. [89] performed experiments on Al_2O_3 nanofluids separately for water and ethylene glycol-based nanofluid. Equations (6.20 and 6.21) were obtained for water-based Al_2O_3 nanofluid and ethylene glycol-based Al_2O_3 nanofluid.

$$\mu_{nf} = \left(1 + 7.3\varphi + 123\varphi^2\right)\mu_{bf} \tag{6.20}$$

$$\mu_{nf} = \left(1 + 4.6\varphi + 6.7\varphi^2\right)\mu_{bf} \tag{6.21}$$

In another study, Chen et al. [90] presented Eq. (6.22) for nanofluid viscosity. This correlation has been used in numerous previous articles.

$$\mu_{nf} = \left(1 + 10.6\varphi + (10.6\varphi)^2\right)\mu_{bf} \tag{6.22}$$

Ho et al. [91] then performed an experimental experiment based on convection heat transfer and examined the variation trend of viscosity with increasing nanofluid concentration and presented Eq. (6.23).

$$\mu_{nf} = \left(1 + 4.93\varphi + 222.4\varphi^2\right)\mu_{bf} \tag{6.23}$$

Then, the results of estimating the viscosity of water-based nanofluid based on the BAG I model in row 1 of Table 6.7 have been evaluated with the conventional and selected correlations in Eqs. (6.17–6.23) according to the experimental data in Figure 6.12.

In Figure 6.14, parts (a) and (b), the results for the nanofluid viscosity have been plotted at the temperature of 20°C and 50°C and a variable volume fraction of 0% to 3%, respectively. In the results, where scattered data are available, the values estimated by the BAG I model are more accurate than the points where the data are most concentrated.

Given that the present study uses a variety of experimental data, the experimental data used in this study have been extracted from several sources; also, the existence of data scatter in constant physical conditions seems reasonable. Therefore, in such cases, it is expected that BAG models can predict the values in which the data are more focused.

According to Figure 6.14, parts (c) and (d), the results for the nanofluid viscosity have been plotted at volume fractions of 0.1% and 1% and a variable temperature of 20°C–60°C, respectively. The Results indicated the BAG I model accurately predicts nanofluid viscosity according to the trend of temperature changes.

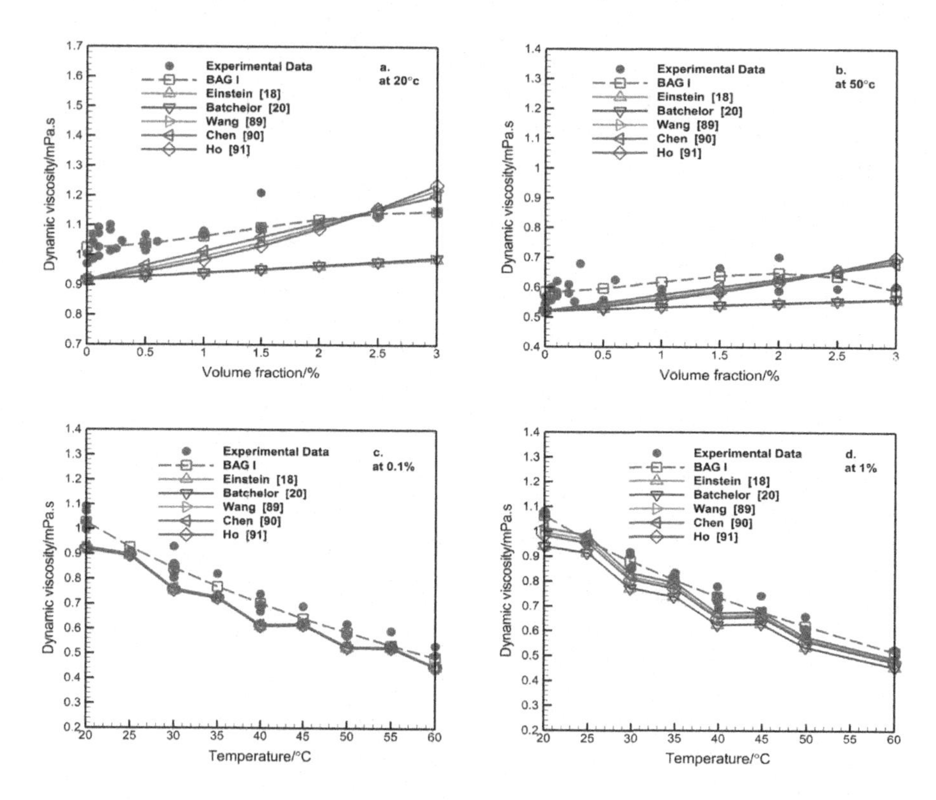

FIGURE 6.14 Comparison of BAG I model with conventional correlations in Eqs. (6.17–6.23). (a) Dynamic viscosity by volume fraction at $T=20°C$ and $0<\varphi<3\%$. (b) Dynamic viscosity by volume fraction at $T=50°C$ and $0<\varphi<3\%$. (c) Dynamic viscosity by temperature °C at $\varphi=0.1\%$ and $20°C<T<60°C$. (d) Dynamic viscosity by temperature °C at $\varphi=1\%$ and $20°C<T<60°C$.

In another analysis, the BAG I model with the acceptable correlations in Table 6.1 for water-based nanofluid has been examined based on the experimental data in Figure 6.12. However, the accuracy range of the acceptable correlations in Table 6.1 is not the same as the range of experimental data in Figure 6.12, but to quantitatively express the estimation of the considered correlations relative to the BAG I model, which can estimate over a wide range of temperature 5°C–60°C and volume fraction 0%–3%. The results in the same physical conditions in Figure 6.14 have also been shown in Figure 6.15; as can be seen, the results obtained from the BAG I model relative to the acceptable correlations in Table 6.1 are in good agreement with the experimental data.

RMSE measures the error rate of two datasets. This parameter compares the predicted values and the experiment's values with each other, and the lower value leads to the lower error of the model. Thus, RMSE is an appropriate tool to compare correlations.

Based on Eqs. (6.24 and6.25), to determine the error of the equations in predicting the experimental viscosity values, the "root mean square error," or RMSE index, has

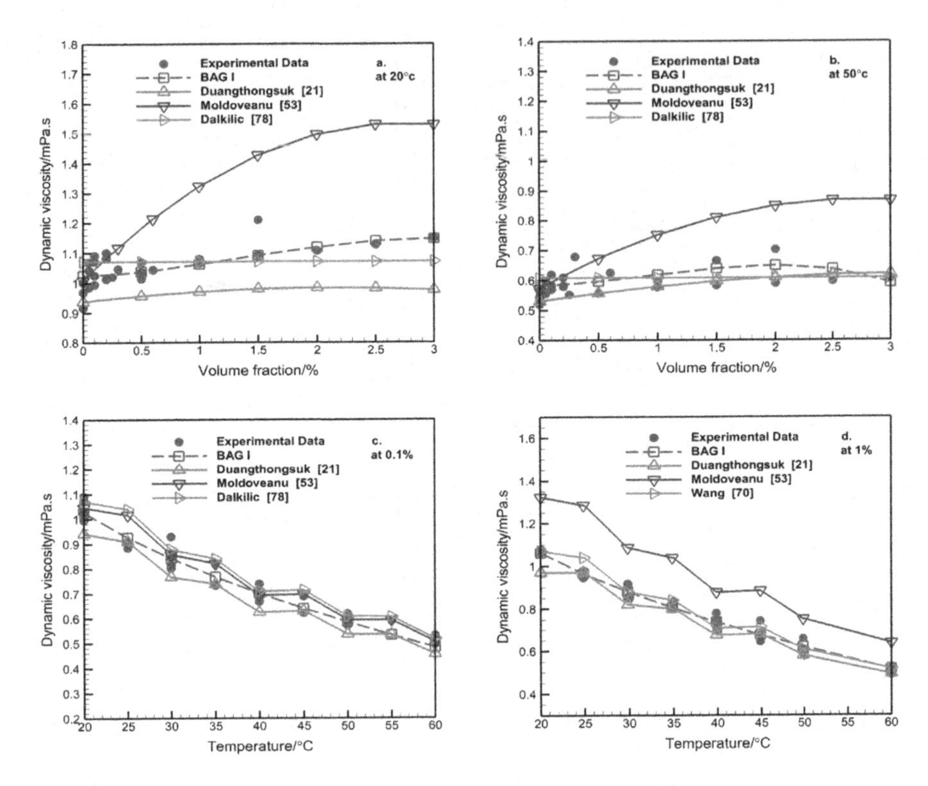

FIGURE 6.15 A comparison of the BAG I model with acceptable correlations appeared in Table 6.1.

been used. Also, the accuracy of the equations in estimating the experimental viscosity values is expressed by the R^2 index.

Table 6.8 shows the RMSE values obtained for the conventional correlations and acceptable correlations in Table 6.1 and the BAG I model on the experimental points for estimating the viscosity of the water-based nanofluid.

$$\text{RMSE} = \sqrt{\frac{\sum_{i=1}^{N}\left(\mu_{\text{pre}} - \mu_{\text{exp}}\right)^2}{N}} \tag{6.24}$$

$$R^2 = 1 - \frac{\sum_{i=1}^{N}\left(\mu_{\text{pre}} - \mu_{\text{exp}}\right)}{\sum_{i=1}^{N}\left(\mu_{\text{pre}} - \overline{\mu_{\text{exp}}}\right)} \tag{6.25}$$

Findings based on the RMSE value indicate that the BAG I model has a 35.82% lower performance error than the best correlation presented by the researchers to estimate the viscosity of water-based nanofluids.

In addition, the diagram of the results of the R^2 coefficient for the conventional and acceptable correlations of Table 6.1 and the BAG I model has been presented

TABLE 6.8
The RMSE Value of the BAG I Model Compared to Other Correlations for Water-Based Nanofluids

Equations	RMSE
Enestien [18]	**0.083422758**
Batchelor [20]	**0.083058883**
Wang [89]	**0.066209608**
Chen [90]	**0.062280061**
Ho [91]	**0.069636179**
Duangthongsuk and Wongwises [21]	**0.060700168**
Moldoveanu [53]	**0.173518889**
Dalkiliç [78]	**0. 081515186**
BAG I	**0.038959537**

in Figure 6.16. Besides the reasonable accuracy of other correlations in estimating nanofluid viscosity, the BAG I model has higher accuracy than other correlations in estimating nanofluid viscosity according to the trend of nanofluid viscosity changes.

In most studies on the viscosity of nanofluids, especially the conventional correlations, the correlations cannot predict the viscosity of nanofluids with the ethylene glycol-based nanofluid. One of the reasons for the weakness of these correlations is the high concentration of ethylene glycol-based nanofluid viscosity relative to water. On the other hand, usual correlations have been optimized for the viscosity of low concentration nanofluids. Therefore, conventional correlations do not respond proportionally to the nanofluid's viscosity with the ethylene glycol-based nanofluid.

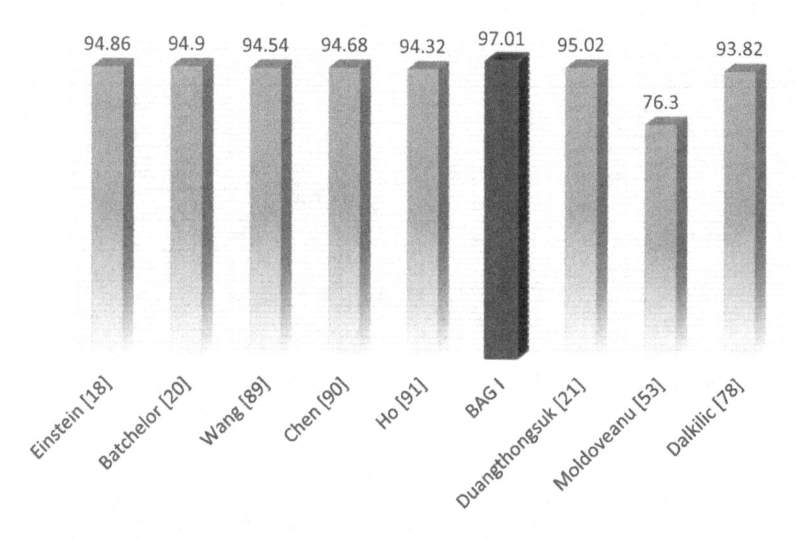

FIGURE 6.16 The value of R^2 in the BAG I model compared to other correlations for water-based nanofluids.

With these interpretations, the BAG II model for the viscosity of the nanofluids with ethylene glycol-based nanofluid has high accuracy for estimating viscosity.

Then, the results of estimating the viscosity of ethylene glycol-based nanofluid based on the BAG II model in row 1 of Table 6.7 have been evaluated with the conventional and selected correlations in Eqs. (6.17–6.23) according to the experimental data in Figure 6.13.

In Figure 6.17, parts (a) and (b), the viscosity of the nanofluid is respectively at 30°C and 50°C and the variable concentration. In parts (c) and (d), the results for the nanofluid viscosity have been plotted at volume fractions of 0.2% and 0.8% and variable temperature, respectively.

As expected, because the conventional correlations for the viscosity of nanofluids with the water-based nanofluid have been optimized, they cannot estimate the viscosity of nanofluids with ethylene glycol-based nanofluid. Considering the experimental data, the viscosity values predicted by the BAG II relation are much more accurate than the conventional relations.

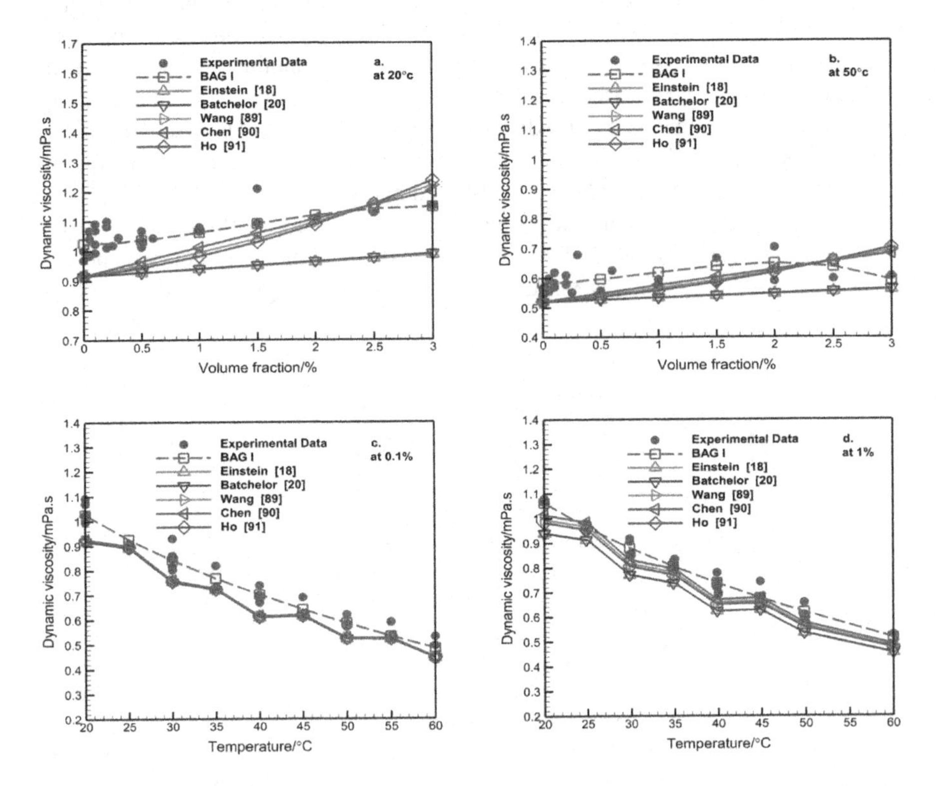

FIGURE 6.17 Comparison of the BAG II model with conventional correlations in Eqs. (6.17–6.23). (a) Dynamic viscosity by volume fraction at $T = 20°C$ and $0 < \varphi < 3\%$. (b) Dynamic viscosity by volume fraction at $T = 50°C$ and $0 < \varphi < 3\%$. (c) Dynamic viscosity by temperature °C at $\varphi = 0.1\%$ and $20°C < T < 60°C$. (d) Dynamic viscosity by temperature °C at $\varphi = 1\%$ and $20°C < T < 60°C$.

In another analysis, the BAG II model with the acceptable correlations in Table 6.1 for ethylene glycol-based nanofluids has been investigated based on the experimental data in Figure 6.13.

However, the accuracy range of the acceptable correlations in Table 6.1 is not the same as the range of experimental data in Figure 6.13, but to quantitatively present the estimation of the considered correlations relative to the BAG II model, which can be estimated in a wide range of temperature 20°C–60°C and volume fraction 0%–2%. Also, the results have been presented in Figure 6.18.

In Figure 6.18, , parts (a) and (b), at the volume fraction range of 1%–2% and temperatures of 30°C and 50°C, the BAG II model is significantly more accurately predicted than the other correlations.

Also, in Figure 6.18, parts (c) and (d), at volume fractions of 0.2% and 0.8% and in the temperature range of 20°C–60°C, the BAG II model has mainly provided better results than other correlations.

Table 6.9 shows the RMSE values obtained for the conventional correlations and acceptable correlations in Table 6.1 and the BAG II model on the experimental points for estimating the viscosity of the ethylene glycol-based nanofluid.

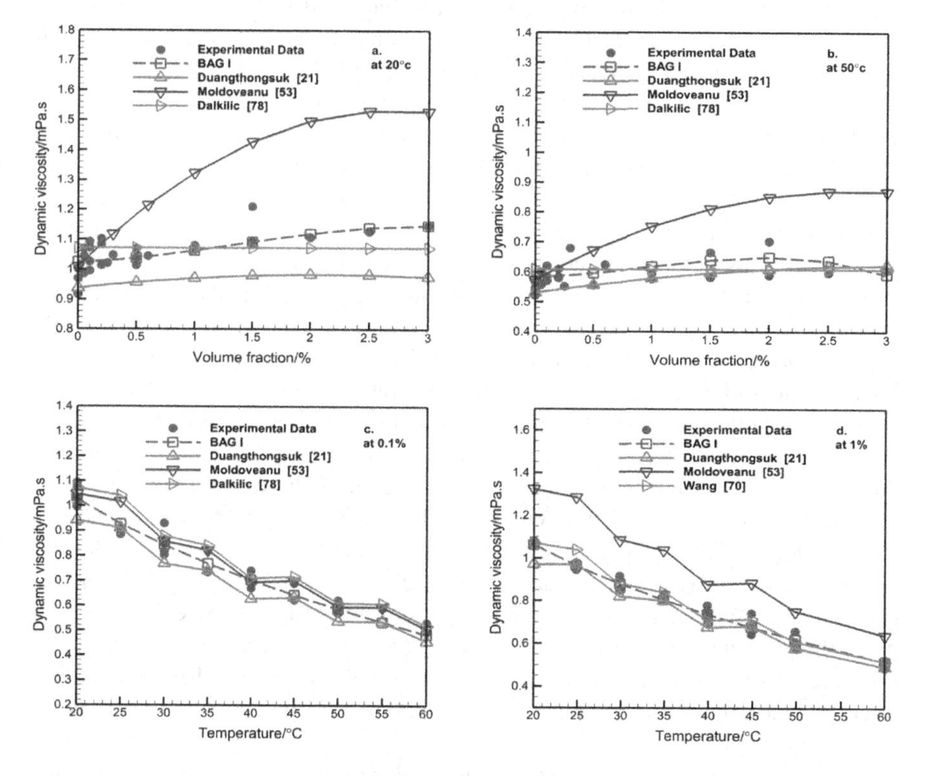

FIGURE 6.18 A comparison of the BAG II model with acceptable correlations appeared in Table 6.1. (a) Dynamic viscosity by volume fraction at $T=20$°C and $0<\varphi<3\%$. (b) Dynamic viscosity by volume fraction at $T=50$°C and $0<\varphi<3\%$. (c) Dynamic viscosity by temperature °C at $\varphi=0.1\%$ and 20°C $<T<$ 60°C. (d) Dynamic viscosity by temperature °C at $\varphi=1\%$ and 20°C $<T<$ 60°C.

TABLE 6.9
The RMSE Value of the BAG II Model Compared to Other Correlations for Ethylene Glycol-Based Nanofluid

Equations	RMSE
Enestien [18]	**5.26808813**
Batchelor [20]	**5.261787151**
Wang [89]	**5.101411188**
Chen [90]	**4.665411049**
Ho [91]	**4.860567618**
Saeedi [55]	**2.08799302**
Adio [73]	**4.863051208**
Li [79]	**2.63545833**
BAG II	**1.047228701**

Findings based on the RMSE value indicate that the BAG II model has a 49.84% lower performance error than the best correlation presented by the researchers to estimate the viscosity of ethylene glycol-based nanofluids.

To know the accuracy of the BAG II model for ethylene glycol in terms of R^2 coefficient, Figure 6.19 was plotted. The results show that the accuracy of the BAG II model is significantly higher than other correlations.

The base fluid viscosity parameter μ_{bf} is available in most of the correlations presented in Table 6.1 and the conventional correlations for calculating nanofluid viscosity. In this case, the viscosity of the base fluid plays the role of the variable temperature of the base fluid in addition to its role in the calculations so that the viscosity of the base fluid changes with the change of temperature. Therefore, the viscosity of the base fluid must also show the effect of temperature. Thus, by assuming the nanofluid type to be constant with temperature change, the base fluid viscosity in the nanofluid viscosity estimation correlation changes. On the other hand, in the nanofluid viscosity estimation correlations, the nanofluid temperature factor is not directly present in the above correlations. Therefore, these correlations alone are not able to estimate the nanofluid viscosity. Thus, when the nanofluid temperature is variable, the mentioned correlations increase the error probability in the calculations.

Also, while solving numerical problems of heat transfer due to temperature changes in the problem, it is sometimes impossible to change the base fluid's viscosity in the problem, which deviates the answer from the correct path. Therefore, the presence of a temperature variable in viscosity models is also felt here. Under such circumstances, the BAG model in Table 6.7 could be a turning point for other viscosity models in the future.

According to the above, another strength of BAG models is the ability to estimate the viscosity of the base fluid in the conventional temperature range used in heat transfer, which is less likely to provide a suitable response at zero concentration. To demonstrate the relationship ability of BAG, the values predicted by the BAG relation are examined with the experimental values of the viscosity of water-based nanofluid and ethylene glycol in Figure 6.20, parts (a) and (b), respectively.

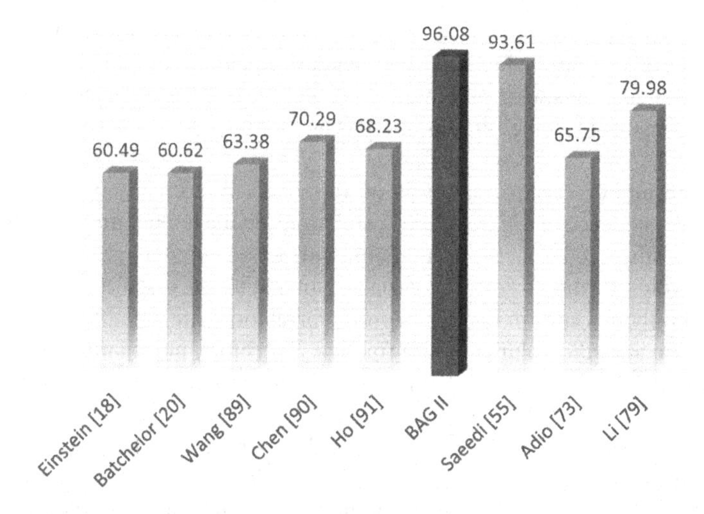

FIGURE 6.19 The value of R^2 in the BAG II model compared to other correlations for ethylene glycol-based nanofluid.

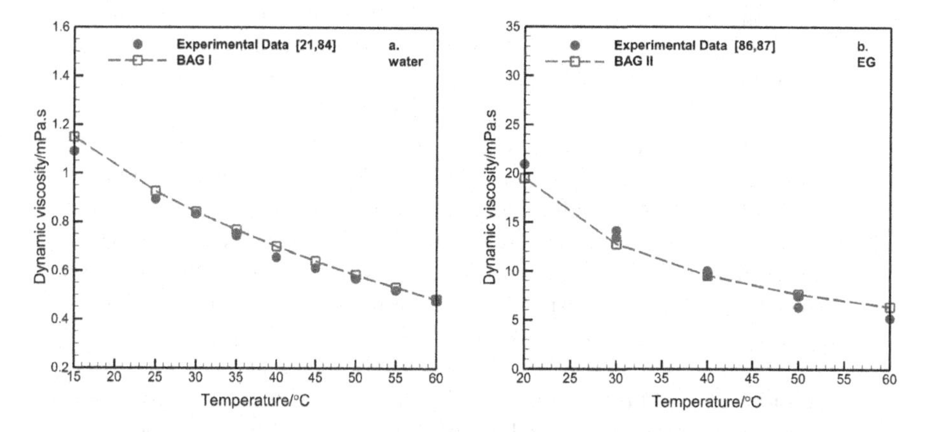

FIGURE 6.20 Estimation of water and ethylene glycol-based nanofluid viscosity using the BAG model. (a) Dynamic viscosity by temperature °C with water-based nanofluid at 15°C < T < 60°C. (b) Dynamic viscosity by temperature °C with ethylene glycol-based nanofluid at 20°C < T < 60°C.

6.6 CONCLUSION

The viscosity component plays a crucial role in heat transfer, especially convection heat transfer. The addition of nanoparticles to the base fluid is commonly considered to increase the viscosity rate. With the increase in the number of nanofluid viscosity models proposed, it has become necessary to review these models in the present study. We also evaluated the correlations in terms of physical compatibility with the viscosity of nanofluids and performed statistical tests of variance and sensitivity analysis on the viscosity models.

Finally, based on the weakness identified in previous models, through our statistical and correlation evaluations, two general equations for water- and ethylene glycol-based nanofluids were presented to predict the behavior of nanofluids. A summary of the results presented in this study is as follows;

- The results of variance analysis on the viscosity correlations showed the non-dependence of 42.2% of the correlations on the temperature component and another 27.3% on the volume (mass) fraction component.
- The volume (mass) fraction variables in nanofluid viscosity models are valid only in a certain range, and most correlations are not able to provide a solution at a zero volume (mass) fraction. Therefore, most nanofluid viscosity models do not cover the range of volume (mass) fractions used in heat transfer.
- In some of the viscosity models introduced by researchers, correlations sometimes have complex and long terms unrelated to viscosity physics. The presence of such terms in viscosity models only increases the likelihood of errors in calculations. However, many of these correlations can be corrected and modified with simple and short models.
- According to the study conducted physically and statistically on the viscosity models of nanofluids, only 53.6% of the correlations are statistically acceptable, and 73.2% of the correlations are physically reliable, and a total of 35.7% of the correlations have acceptable conditions.
- Sensitivity analysis revealed the significant contribution of temperature component in estimating nanofluid viscosity. It is while the effect of temperature in the form of nanofluid viscosity models is not directly considered. Instead, the effect of temperature is determined by the independent variable of base fluid viscosity μ_{bf} in the desired viscosity models. Therefore, the viscosity models cannot estimate the nanofluid's viscosity in proportion to the temperature variation trend, and this factor can cause problems.
- For modeling, the viscosity of nanofluids separated into water or ethylene glycol-based nanofluids, which is valid in a wide temperature and volume fraction range and function independent of the type of nanoparticles, BAG models were introduced.
- The BAG models presented for nanofluid viscosity for water and ethylene glycol-based nanofluid have 97.01% and 96.08% accuracy, respectively. Also, the RMSE value improved by 35.82% and 49.84% compared to the best correlation presented by the researchers for estimating the viscosity of water-based and ethylene glycol-based nanofluids, respectively.
- Most of the viscosity models have been optimized for nanofluids with water-based nanofluids. However, the BAG model has the ability to estimate the viscosity of nanofluid with ethylene glycol-based nanofluid with much higher accuracy than other correlations.
- Unlike other correlations, the results of BAG models showed that by changing the temperature of the nanofluid, BAG models maintain the ability to estimate the viscosity of the nanofluid accurately. Also, when the nanoparticle concentration is zero, the viscosity of the base fluid is well predicted.

Given that nanofluids with oil base such as the applied fluids in industry and few correlations have been provided for oil-based nanofluids. Therefore, developing correlations for oil-based nanofluids is a challenge and an open field of research.

In this chapter, whole content is presented from "Statistical study and a complete overview of nanofluid viscosity correlations: a new look" in "*Journal of Thermal Analysis and Calorimetry*" by A. Barkhordar, Ramin Ghasemiasl, and Taher Armaghani published in "Springer Nature" publications in the year 2022 with the volume and page number being 147.13 and 7099–7132, respectively [92].

REFERENCES

[1] S. Mondal, et al., A theoretical nanofluid analysis exhibiting hydromagnetics characteristics employing CVFEM, *J. Braz. Soc. Mec. Sci. Eng.* 42 (2020) 1–12.
[2] S.M. Seyyedi, et al., Second law analysis of magneto-natural convection in a nanofluid filled wavy-hexagonal porous enclosure, *Int. J. Numer. Method. Heat Fluid Flow* 30.11 (2020) 4811–4836.
[3] A.S. Dogonchi, et al., A modified Fourier approach for analysis of nanofluid heat generation within a semi-circular enclosure subjected to MFD viscosity, *Int. Commun. Heat Mass Transf.* 111 (2020) 104430.
[4] I. Tlili, et al., Analysis of a single-phase natural circulation loop with hybrid-nanofluid, *Int. Commun. Heat Mass Transf.* 112 (2020) 104498.
[5] A.S. Dogonchi, et al., Simulation of Fe_3O_4–H_2O nanoliquid in a triangular enclosure subjected to Cattaneo-Christov theory of heat conduction, *Int. J. Numer. Method. Heat Fluid Flow* 29.11 (2019) 4430–4444.
[6] S.M. Seyyedi, et al., Investigation of entropy generation in a square inclined cavity using control volume finite element method with aided quadratic Lagrange interpolation functions, *Int. Commun. Heat Mass Transf.* 110 (2020) 104398.
[7] Z. Abdelmalek, et al., Role of various configurations of a wavy circular heater on convective heat transfer within an enclosure filled with nanofluid, *Int. Commun. Heat Mass Transf.* 113 (2020) 104525.
[8] M.S. Sadeghi, et al., Analysis of thermal behavior of magnetic buoyancy-driven flow in ferrofluid-filled wavy enclosure furnished with two circular cylinders, *Int. Commun. Heat Mass Transf.* 120 (2021) 104951.
[9] M. Hashemi-Tilehnoee, et al., Magneto-fluid dynamic and second law analysis in a hot porous cavity filled by nanofluid and nano-encapsulated phase change material suspension with different layout of cooling channels, *J. Energy Storage* 31 (2020) 101720.
[10] A.S. Dogonchi, Z. Asghar, M. Waqas, CVFEM simulation for Fe_3O_4–H_2O nanofluid in an annulus between two triangular enclosures subjected to magnetic field and thermal radiation, *Int. Commun. Heat Mass Transf.* 112 (2020) 104449.
[11] A.S. Dogonchi, F. Selimefendigil, D.D. Ganji, Magneto-hydrodynamic natural convection of CuO-water nanofluid in complex shaped enclosure considering various nanoparticle shapes, *Int. J. Numer. Method Heat Fluid Flow* 29.5 (2019) 1663–1679.
[12] F. Selimefendigil, Natural convection in a trapezoidal cavity with an inner conductive object of different shapes and filled with nanofluids of different nanoparticle shapes, *Iran. J. Sci. Technol. Trans. Mech. Eng.* 42 (2018) 169–184.
[13] F. Selimefendigil, H.F. Öztop, Effects of a rotating tube bundle on the hydrothermal performance for forced convection in a vented cavity with Ag–MgO/water hybrid and CNT-water nanofluids, *J. Therm. Anal. Calorim.* 147.1 (2022) 939–956.
[14] A.J. Chamkha, et al., On the nanofluids applications in microchannels: A comprehensive review, *Powder Technol.* 332 (2018) 287–322.

[15] S. Izadi, et al., A comprehensive review on mixed convection of nanofluids in various shapes of enclosures, *Powder Technol.* 343 (2019) 880–907.
[16] M. Molana, A comprehensive review on the nanofluids application in the tubular heat exchangers, *Am. J. Heat Mass Transf* 3.5 (2016) 352–381.
[17] N.S Pandya, et al., Heat transfer enhancement with nanofluids in plate heat exchangers: A comprehensive review, *Eur. J. Mech.-B/Fluids* 81 (2020) 173–190.
[18] A. Einstein, *Eine neue bestimmung der moleküldimensionen*, Diss. ETH Zurich, 1905.
[19] H.C. Brinkman, The viscosity of concentrated suspensions and solutions, *J. Chem. Phys.* 20.4 (1952) 571.
[20] G.K. Batchelor, The effect of Brownian motion on the bulk stress in a suspension of spherical particles, *J. Fluid Mech.* 83.1 (1977) 97–117.
[21] W. Duangthongsuk, S. Wongwises, Measurement of temperature-dependent thermal conductivity and viscosity of TiO_2-water nanofluids, *Exp. Therm. Fluid Sci.* 33.4 (2009) 706–714.
[22] M.H. Esfe, S. Saedodin, An experimental investigation and new correlation of viscosity of ZnO-EG nanofluid at various temperatures and different solid volume fractions, *Exp. Therm. Fluid Sci.* 55 (2014) 1–5.
[23] M. Sharifpur, S.A. Adio, J.P. Meyer, Experimental investigation and model development for effective viscosity of Al_2O_3-glycerol nanofluids by using dimensional analysis and GMDH-NN methods, *Int. Commun. Heat Mass Transf.* 68 (2015) 208–219.
[24] S. Aberoumand, et al., Experimental study on the rheological behavior of silver-heat transfer oil nanofluid and suggesting two empirical based correlations for thermal conductivity and viscosity of oil based nanofluids, *Appl. Therm. Eng.* 101 (2016) 362–372.
[25] M. Akbari, et al., An experimental study on rheological behavior of ethylene glycol based nanofluid: Proposing a new correlation as a function of silica concentration and temperature, *J. Mol. Liq.* 233 (2017) 352–357.
[26] W. Li, C. Zou, Experimental investigation of stability and thermo-physical properties of functionalized β-CD-TiO_2-Ag nanofluids for antifreeze, *Powder Technol.* 340 (2018) 290–298.
[27] L. Yu, et al., Experimental investigation on rheological properties of water based nanofluids with low MWCNT concentrations, *Int. J. Heat Mass Transf.* 135 (2019) 175–185.
[28] S.-R. Yan, et al., Rheological behavior of hybrid MWCNTs-TiO_2/EG nanofluid: A comprehensive modeling and experimental study, *J. Mol. Liq.* 308 (2020) 113058.
[29] M. Molana, R. Ghasemiasl, T. Armaghani, A different look at the effect of temperature on the nanofluids thermal conductivity: Focus on the experimental-based models, *J. Therm. Anal. Calorim.* 147 (2022) 4553–4577. https://doi.org/10.1007/s10973-021-10836-w.
[30] R.A. Fisher, Statistical methods for research workers, in: Kotz, S., Johnson, N.L. (eds) *Breakthroughs in Statistics: Methodology and Distribution*, Springer, New York, 1992: pp. 66–70. https://doi.org/10.1007/978-1-4612-4380-9_6
[31] H. Scheffe, *The Analysis of Variance*, Vol. 72. John Wiley & Sons United States, 1999.
[32] A.S. Dalkılıç, et al., Experimental investigation on the viscosity characteristics of water based SiO_2-graphite hybrid nanofluids, *Int. Commun. Heat Mass Transf.* 97 (2018) 30–38.
[33] L. Li, et al., Stability, thermal performance and artificial neural network modeling of viscosity and thermal conductivity of Al_2O_3-ethylene glycol nanofluids, *Powder Technol.* 363 (2020) 360–368.
[34] I.M. Alarifi, et al., On the rheological properties of MWCNT-TiO_2/oil hybrid nanofluid: An experimental investigation on the effects of shear rate, temperature, and solid concentration of nanoparticles, *Powder Technol.* 355 (2019) 157–162.

[35] B. Ruhani, P. Barnoon, D. Toghraie, Statistical investigation for developing a new model for rheological behavior of silica-ethylene glycol/water hybrid Newtonian nanofluid using experimental data, *Physica A* 525 (2019) 616–627.
[36] A. Huminic, et al., Thermo-physical properties of water based lanthanum oxide nanofluid. An experimental study, *J. Mol. Liq.* 287 (2019) 111013.
[37] M.H. Esfe, S. Esfandeh, S. Niazi, An experimental investigation, sensitivity analysis and RSM analysis of MWCNT (10)-ZnO (90)/10W40 nanofluid viscosity, *J. Mol. Liq.* 288 (2019) 111020.
[38] M.H. Esfe, A.T.K. Abad, M. Fouladi, Effect of suspending optimized ratio of nano-additives MWCNT-Al_2O_3 on viscosity behavior of 5W50, *J. Mol. Liq.* 285 (2019) 572–585.
[39] M.H. Esfe, S. Esfandeh, The statistical investigation of multi-grade oil based nanofluids: Enriched by MWCNT and ZnO nanoparticles, *Physica A* 554 (2020) 122159.
[40] Z. Li, et al., Experimental study of temperature and mass fraction effects on thermal conductivity and dynamic viscosity of SiO_2-oleic acid/liquid paraffin nanofluid, *Int. Commun. Heat Mass Transf.* 110 (2020) 104436.
[41] R.R. Sahoo, V. Kumar, Development of a new correlation to determine the viscosity of ternary hybrid nanofluid, *Int. Commun. Heat Mass Transf.* 111 (2020) 104451.
[42] Z. Tian, et al., Prediction of rheological behavior of a new hybrid nanofluid consists of copper oxide and multi wall carbon nanotubes suspended in a mixture of water and ethylene glycol using curve-fitting on experimental data, *Physica A* 549 (2020) 124101.
[43] Z.X. Li, et al., Nanofluids as secondary fluid in the refrigeration system: Experimental data, regression, ANFIS, and NN modeling, *Int. J. Heat Mass Transf.* 144 (2019) 118635.
[44] P.G. Kumar, et al., Experimental study on thermal properties and electrical conductivity of stabilized H_2O-solar glycol mixture based multi-walled carbon nanotube nanofluids: developing a new correlation, *Heliyon* 5.8 (2019) E02385.
[45] A. Asadi, F. Pourfattah, Heat transfer performance of two oil-based nanofluids containing ZnO and MgO nanoparticles; a comparative experimental investigation, *Powder Technol.* 343 (2019) 296–308.
[46] A. Shahsavar, et al., A novel comprehensive experimental study concerned synthesizes and prepare liquid paraffin-Fe_3O_4 mixture to develop models for both thermal conductivity & viscosity: a new approach of GMDH type of neural network, *Int. J. Heat Mass Transf.* 131 (2019) 432–441.
[47] F. Li, et al., Effects of ultrasonic time, size of aggregates and temperature on the stability and viscosity of Cu-ethylene glycol (EG) nanofluids, *Int. J. Heat Mass Transf.* 129 (2019) 278–286.
[48] M.H. Esfe, et al., Viscosity and rheological properties of antifreeze based nanofluid containing hybrid nano-powders of MWCNTs and TiO_2 under different temperature conditions, *Powder Technol.* 342 (2019) 808–816.
[49] D.P. Soman, et al., Impact of viscosity of nanofluid and ionic liquid on heat transfer, *J. Mol. Liq.* 291 (2019) 111349.
[50] B. Ruhani, et al., Statistical investigation for developing a new model for rheological behavior of ZnO-Ag (50%–50%)/Water hybrid Newtonian nanofluid using experimental data, *Physica A* 525 (2019) 741–751.
[51] E.B. Elcioglu, et al., Experimental study and Taguchi Analysis on alumina-water nanofluid viscosity, *Appl. Therm. Eng.* 128 (2018) 973–981.
[52] S. Ghasemi, A. Karimipour, Experimental investigation of the effects of temperature and mass fraction on the dynamic viscosity of CuO-paraffin nanofluid, *Appl. Therm. Eng.* 128 (2018) 189–197.
[53] G.M. Moldoveanu, et al., Viscosity estimation of Al_2O_3, SiO_2 nanofluids and their hybrid: An experimental study, *J. Mol. Liq.* 253 (2018) 188–196.

[54] G.M. Moldoveanu, et al., Experimental study on viscosity of stabilized Al_2O_3, TiO_2 nanofluids and their hybrid, *Thermochim. Acta* 659 (2018) 203–212.
[55] A.H. Saeedi, M. Akbari, D. Toghraie, An experimental study on rheological behavior of a nanofluid containing oxide nanoparticle and proposing a new correlation, *Physica E* 99 (2018) 285–293.
[56] A. Karimipour, et al., A new correlation for estimating the thermal conductivity and dynamic viscosity of CuO/liquid paraffin nanofluid using neural network method, *Int. Commun. Heat Mass Transf.* 92 (2018) 90–99.
[57] A.A.A.A. Alrashed, et al., Effects on thermophysical properties of carbon based nanofluids: experimental data, modelling using regression, ANFIS and ANN, *Int. J. Heat Mass Transf.* 125 (2018) 920–932.
[58] H. Khodadadi, D. Toghraie, A. Karimipour, Effects of nanoparticles to present a statistical model for the viscosity of MgO-water nanofluid, *Powder Technol.* 342 (2019) 166–180.
[59] M.H. Esfe, A.A.A. Arani, An experimental determination and accurate prediction of dynamic viscosity of MWCNT (% 40)-SiO2 (% 60)/5W50 nano-lubricant, *J. Mol. Liq.* 259 (2018) 227–237.
[60] M.H.Esfe, et al., Optimization of MWCNTs (10%)-A12O3 (90%)/5W50 nanofluid viscosity using experimental data and artificial neural network, *Physica A* 512 (2018) 731–744.
[61] K.A. Hamid, et al., Experimental investigation of thermal conductivity and dynamic viscosity on nanoparticle mixture ratios of TiO_2-SiO_2 nanofluids, *Int. J. Heat Mass Transf.* 116 (2018) 1143–1152.
[62] M.F. Nabil, et al., An experimental study on the thermal conductivity and dynamic viscosity of TiO_2-SiO_2 nanofluids in water: ethylene glycol mixture, *Int. Commun. Heat Mass Transf.* 86 (2017) 181–189.
[63] G. Żyła, and J. Fal, Viscosity, thermal and electrical conductivity of silicon dioxide-ethylene glycol transparent nanofluids: An experimental studies, *Thermochim. Acta* 650 (2017) 106–113.
[64] M. Amani, et al., Experimental study on viscosity of spinel-type manganese ferrite nanofluid in attendance of magnetic field, *J. Magn. Magn. Mater.* 428 (2017) 457–463.
[65] O. Soltani, M. Akbari, Effects of temperature and particles concentration on the dynamic viscosity of MgO-MWCNT/ethylene glycol hybrid nanofluid: Experimental study, *Physica E* 84 (2016) 564–570.
[66] B. Ilhan, M. Kurt, H. Ertürk, Experimental investigation of heat transfer enhancement and viscosity change of hBN nanofluids, *Exp. Therm. Fluid Sci.* 77 (2016) 272–283.
[67] M.H. Esfe, et al., Designing an artificial neural network to predict dynamic viscosity of aqueous nanofluid of TiO_2 using experimental data, *Int. Commun. Heat Mass Transf.* 75 (2016) 192–196.
[68] D. Toghraie, S.M. Alempour, M. Afrand, Experimental determination of viscosity of water based magnetite nanofluid for application in heating and cooling systems, *J. Magn. Magn. Mater.* 417 (2016) 243–248.
[69] M.Kh. Abdolbaqi, et al., Experimental investigation and development of new correlation for thermal conductivity and viscosity of BioGlycol/water based SiO_2 nanofluids, *Int. Commun. Heat Mass Transf.* 77 (2016) 54–63.
[70] L.S. Sundar, et al., Thermal conductivity and viscosity of water based nanodiamond (ND) nanofluids: An experimental study, *Int. Commun. Heat Mass Transf.* 76 (2016) 245–255.
[71] R.M. Mostafizur, et al., Investigation on stability and viscosity of SiO_2-CH_3OH (methanol) nanofluids, *Int. Commun. Heat Mass Transf.* 72 (2016) 16–22.

[72] M. Asadi, A. Asadi, Dynamic viscosity of MWCNT/ZnO-engine oil hybrid nanofluid: an experimental investigation and new correlation in different temperatures and solid concentrations, *Int. Commun. Heat Mass Transf.* 76 (2016) 41–45.
[73] S.A. Adio, et al., Experimental investigation and model development for effective viscosity of MgO-ethylene glycol nanofluids by using dimensional analysis, FCM-ANFIS and GA-PNN techniques, *Int. Commun. Heat Mass Transf.* 72 (2016) 71–83.
[74] M.H. Esfe, et al., An experimental study on viscosity of alumina-engine oil: Effects of temperature and nanoparticles concentration, *Int. Commun. Heat Mass Transf.* 76 (2016) 202–208.
[75] M. Baratpour, et al., Effects of temperature and concentration on the viscosity of nanofluids made of single-wall carbon nanotubes in ethylene glycol, *Int. Commun. Heat Mass Transf.* 74 (2016) 108–113.
[76] M. Afrand, K.N. Najafabadi, M. Akbari, Effects of temperature and solid volume fraction on viscosity of SiO_2-MWCNTs/SAE40 hybrid nanofluid as a coolant and lubricant in heat engines, *Appl. Therm. Eng.* 102 (2016) 45–54.
[77] M.Kh. Abdolbaqi, et al., An experimental determination of thermal conductivity and viscosity of BioGlycol/water based TiO_2 nanofluids, *Int. Commun. Heat Mass Transf.* 77 (2016) 22–32.
[78] A.S. Dalkilic, et al., Prediction of graphite nanofluids' dynamic viscosity by means of artificial neural networks, *Int. Commun. Heat Mass Transf.* 73 (2016) 33–42.
[79] X. Li, et al., Stability and enhanced thermal conductivity of ethylene glycol-based SiC nanofluids, *Int. J. Heat Mass Transf.* 89 (2015) 613–619.
[80] S. Andrea, et al., *Global Sensitivity Analysis: The Primer*, Ist ed., John Wiley & Sons, The Atrium, Southern Gate, Chichester, England, 2008.
[81] R. Gangadevi, B.K. Vinayagam, Experimental determination of thermal conductivity and viscosity of different nanofluids and its effect on a hybrid solar collector, *J. Therm. Anal. Calorim.* 136 (2019) 199–209.
[82] A. Topuz, et al., Experimental investigation of optimum thermal performance and pressure drop of water-based $A12O_3$, TiO_2 and ZnO nanofluids flowing inside a circular microchannel, *J. Therm. Anal. Calorim.* 131 (2018) 2843–2863.
[83] H. Ghodsinezhad, M. Sharifpur, J.P. Meyer, Experimental investigation on cavity flow natural convection of Al_2O_3-water nanofluids, *Int. Commun. Heat Mass Transf.* 76 (2016) 316–324.
[84] L.S. Sundar, M.K. Singh, A.C.M. Sousa, Turbulent heat transfer and friction factor of nanodiamond-nickel hybrid nanofluids flow in a tube: An experimental study, *Int. J. Heat Mass Transf.* 117 (2018) 223–234.
[85] A.D. Zadeh, D. Toghraie, Experimental investigation for developing a new model for the dynamic viscosity of silver/ethylene glycol nanofluid at different temperatures and solid volume fractions, *J. Therm. Anal. Calorim.* 131.2 (2018) 1449–1461.
[86] L.S. Sundar, et al., Experimental investigation of the thermal transport properties of graphene oxide/Co_3O_4 hybrid nanofluids, *Int. Commun. Heat Mass Transf.* 84 (2017) 1–10.
[87] C. Selvam, D. Mohan Lal, S. Harish, Heat transport and pressure drop characteristics of ethylene Glycol-based Nano fluid containing silver nanoparticles, in: *IOP Conference Series: Materials Science and Engineering*, Vol. 402, No. 1. IOP Publishing, 2018.
[88] A.A. Nadooshan, H. Eshgarf, M. Afrand, Measuring the viscosity of Fe_3O_4-MWCNTs/EG hybrid nanofluid for evaluation of thermal efficiency: Newtonian and non-Newtonian behavior, *J. Mol. Liq.* 253 (2018) 169–177.
[89] X. Wang, X. Xu, S.U.S. Choi, Thermal conductivity of nanoparticle-fluid mixture, *J. Thermophys. Heat Transf.* 13.4 (1999) 474–480.
[90] H. Chen, Y. Ding, C. Tan, Rheological behaviour of nanofluids, *New J. Phys.* 9.10 (2007) 367.

[91] C.J. Ho, et al., Natural convection heat transfer of alumina-water nanofluid in vertical square enclosures: An experimental study, *Int. J. Therm. Sci.* 49.8 (2010) 1345–1353.
[92] A. Barkhordar, R. Ghasemiasl, T. Armaghani, Statistical study and a complete overview of nanofluid viscosity correlations: A new look, *J. Therm. Anal. Calorim.* 147.13 (2022) 7099–7132.

NOMENCLATURE

T: Temperature (°C or K)
φ: Concentration (%)
μ: Dynamic viscosity (mPa·s)
d: Diameter (nm)
$\dot{\gamma}$: Shear rate (s^{-1})
N: Number of Data
θ: Dimensionless temperature
a, b: Constant values
miobf (μbf): Dynamic viscosity of base fluid
phi (φ): Concentration (%)
Subscripts
bf: Base fluid
nf: Nanofluid
p: Particles
max: Maximum
min: Minimum
o: Reference value
exp: Experimental Data
pre: Predicted Data
w: Mass concentration

7 Other Thermal Properties of Nanofluids

7.1 ISOBARIC-SPECIFIC HEAT CAPACITY

One of the critical factors characterizing nanofluid's thermal properties is the isobaric-specific heat capacity c_p. It is important to note that most nanofluids research focuses on the k. In contrast, just around 5% of studies concentrate on $c_{p,nf}$ [1]. Since 2008, scientists have focused more on studying c_p. When nanoparticles and water are combined to create a colloidal suspension, the adequate isobaric-specific heat capacity is decreased since the nanoparticle-to-water c_p ratio is around 1:10.

7.1.1 THEORY

Pak and Cho [2] were the first to develop a theoretical formula based on the mixing rule for estimating the $c_{p,nf}$:

$$c_{p,nf} = \left(1 - \Phi_p\right) c_{p,bf} + \Phi_e c_{p,p} \tag{7.1}$$

$c_{p,nf}$, $c_{p,bf}$, and $c_{p,p}$ indicate the adequate isobaric-specific heat capacity of the nanofluids, base fluid, and nanoparticles, and Φ_p shows the volume fraction of the nanoparticles. This formula can only be used when diluted suspensions with nanoparticle densities are not significantly different from the first fluid. This formula would result in a significant inaccuracy for nanoparticles with a substantial volume fraction. Xuan and Roetzel [3] presented an enhanced c_p model that considers thermal equilibrium between the nanoparticles and the base fluid [4].

$$c_{p,nf} = \frac{\left(1 - \Phi_e\right) \rho_f c_{p,f} + \Phi_e \rho_p c_{p,p}}{\left(1 - \Phi_e\right) \rho_f + \Phi_e \rho_p} \tag{7.2}$$

where ρ_f and ρ_p represent the base fluid and nanoparticle densities, respectively. Many scholars have cited this equation, and experimental research has verified its dependability [5–8].

Using 81 sets of test data on water-based Al_2O_3, CuO, SiO_2, and TiO_2 nanofluids, Sekhar and Sharma [9] created a regression equation to estimate $c_{p,nf}$ at any concentration:

$$c_{p,nf} = 0.8429 \left(1 + \frac{T_{nf}}{50}\right)^{-0.3037} \left(1 + \frac{r_p}{50}\right)^{0.4167} \left(1 + \frac{\Phi_p}{100}\right)^{2.272} \tag{7.3}$$

DOI: 10.1201/9781032664118-7

where T_{nf} represents the temperature range of the nanofluids, which is between 20°C and 50°C, r_p indicates the diameter of the nanoparticles (15–50 nm). Φ_p denotes the volume fraction (between 0.01% and 4%). When the computed values are compared to the experimental data, the average absolute deviations (AAD) are 8%–10%, indicating that Eq. (7.3) is reliable. More tests on a wide variety of temperatures, particle sizes, and volume fractions are still required. Vajjha and Das [10] proposed a way to predict $c_{p,nf}$ using a correlation:

$$\frac{c_{p,nf}}{c_{p,bf}} = \frac{A\left(T_{nf}/T_0\right) + B\left(c_{p,p}/c_{p,bf}\right)}{C + \Phi_p} \tag{7.4}$$

$c_{p,nf}$ can be defined as the function of temperature, nanoparticle volume fraction, nanoparticle isobaric-specific heat capacity, and base fluid isobaric-specific heat capacity in this case. A, B, and C are experimentally obtained constants. The AAD of the experimental results for the SiO_2, Al_2O_3, and CuO nanofluids with the similar 2 vol % nanoparticle addition was 0.5%–1.9%, which agreed exceptionally well with Eq. (7.4). Then, a new correlation equation was reported by Cabaleiro et al. [11].

$$\frac{c_{p,bf}(T) - c_{p,nf}(T)}{c_{p,bf}(T)} = \frac{\Phi_p + A\dfrac{c_{p,p}(T)}{c_{p,bf}(T)}}{B + \Phi_p} \tag{7.5}$$

This agrees with the experimental c_p findings for five high concentrations of metal oxide nanofluids (AAD = 0.37%–0.46%).

7.1.2 Experimental Measurement Techniques

The differential scanning calorimetry (DSC) technique is the traditional experimental methodology for investigating the c_p of nanofluids. This approach includes measuring the amount of heat necessary to give to the sample and reference (both of which have a well-defined heat capacity) to raise their temperature. The heat capacity is determined based on those.

Zhou and Ni [12] utilized this approach to calculate the c_p of Al_2O_3/water nanofluids by examining the relationship between the heat flow rate and the rise in the sample temperature. At a nanoparticle concentration of 15%, c_p is reported to be around 2.75 J/gK, which is 1 J/gK lower than the nanofluid concentration of 10%. This outcome is consistent with Buongiorno's [13] paradigm. DSC was used by Shin and Banerjee [14] to determine the c_p of SiO_2/melted carbonate crystal nanofluids at high temperatures (>400°C). After adding SiO_2 nanoparticles with high specific surface energy, they saw a 25% rise in $c_{p,bf}$. They used a similar method to determine the c_p of Al_2O_3/alkali salt nanofluids [15]. The addition of Al_2O_3 nanoparticles increases $c_{p,bf}$ by 32%, which is thought to be due to the development of nanochain formations. Using the DSC technique, Kumaresan and Velraj [16] measured c_p in the temperature range of –50°C to 50°C using about 20 mg of CNT/water-EG nanofluids sample.

The reported c_p was 3.9 J/gK for 0.15 concentration at room temperature, showing that the present model still needed improvement. Ho and Pan [17] investigated how the c_p of Al_2O_3/molten Hitec salt nanofluids changed with various concentrations. At a concentration of 0.063 wt%, the maximal enhancing impact of c_p (19.9%) occurs. Raud et al. [18] conducted experiments using water-based metal oxide nanofluids to establish the validity of the DSC method for c_p measurements. It was found that when the concentration of nanoparticles increases, c_p drops.

The upgraded version of the classic DSC method is modulated temperature differential scanning calorimetry (MTDSC). It has better sensitivity and resolution, allowing for direct detection of c_p through modulated (mostly sinusoidal) heating signals [19]. Using this approach, Robertis et al. [20] measured the c_p of Cu/EG nanofluids and reported a c_p value of 2.3 J/gK. Cabaleiro et al. [11] measured c_p for nanofluids containing MgO/EG, ZnO/EG, and ZrO_2/EG at different concentrations of up to 15%. They discovered that the c_p-enhancing effects of ZnO and ZrO_2 on the base fluid are similar (30%) at the same settings. Żyła et al. [21] utilized MTDSC to study the c_p of nanofluids containing only a few nanometer-sized nanodiamonds. They used two different purity nanodiamonds (87% and 97%) to evaluate EG-based nanofluids with particle mass concentrations up to 10%. They concluded that the Xuan and Roetzel model (Eq. 7.2) could be used to forecast the c_p of such nanofluids.

Murshed [22] employed a transient double hot-wire approach to detect the c_p of metal oxide particles (TiO_2, $Al2O_3$) in deionized water, glycol, and other base fluids, in addition to the aforementioned techniques. This novel approach which needs essential equipment yet has excellent precision, with a measuring error of less than 2.77%. The results show that when the concentration of nanoparticles increases, the effective c_p drops considerably. Satti et al. [23] created a unique thermal property analyzer relying on the transient plane heat source technique called C-Therm TCi (Figure 7.1), to evaluate c_p for five metal oxide particles (Al_2O_3, ZnO, CuO, TiO_2, SiO_2)/propylene glycol nanofluids.

According to the experimental data for the five different types of nanofluids, it was concluded that the size of the nanoparticle does not influence c_p. However, the underlying needs to be further studied.

7.1.3 Experiments vs. Theory

Theoretical predictions and experimental evidence have been compared in multiple studies. Using DSC, Zhou and Ni [12] investigated the c_p of an Al_2O_3/water nanofluid. It was then compared to the theoretical formula (Eqs. 7.1 and 7.6). According to the results, it was discovered that when the volume fraction of the nanoparticles increased, $c_{p,nf}$ dropped. The experimental data were highly consistent with Eq. (7.2). However, they departed substantially from Eq. (7.1), whose predictions were more significant than the experimental data. The findings also validated the prediction of Eq. (7.2) prediction. Barbés et al. [4] measured the c_p of Al_2O_3/water and Al_2O_3/ethanol nanofluids as a function of volume fraction at 303.1 K and 330.4 K, respectively. The findings show that when the nanoparticle volume fraction increases, c_p decreases, consistent with the theoretical predictions of Eq. (7.2). The c_p of SiO_2/water, Al_2O_3/water, and CuO/water nanofluids with volume fractions ranging from 5% to 50% was investigated by O'Hanley et al. [24]. They demonstrated that Eq. (7.1) correctly predicted $c_{p,nf}$, but Eq. (7.2) did not. They also discovered that when the volume fraction of nanoparticles increases, the discrepancy between the predictions from Eqs. (7.1 and 7.2) grow more prominent. The c_p was shown to function temperature, nanoparticle size, and volume percent by Eq. (7.3). The theoretical values agreed with experimental results from other research [25,26], and the AAD was about 8%–10%, confirming the dependability of Eq. (7.3).

Vajjha and Das [10] examined the c_p of Al_2O_3, SiO_2, and ZnO nanofluids with a volume fraction of 2%. They got the correlation Eq. (7.4), in which the AAD of the experimental data is 0.5%–1.9%. Cabaleiro et al. [11] designed a novel model, Eq. (7.5), to link the c_p with temperature and nanoparticle volume fraction. MgO/EG, ZnO/EG, ZrO_2/EG nanofluids, and ZnO/EG-water and ZrO_2 EG-water, mixed nanofluids, had their c_p determined with mass fractions varying to about 15%. The findings show that the theoretical and experimental data agree well, and the ADD was just 0.37%–0.46%. After collecting $c_{p,nf}$ measurement data from multiple articles and comparing them, it is evident from the comparison findings that the c_p of traditional metal oxide-based [12,22,24] and non-metal oxide-based [22] nanofluids declines as the nanoparticle loading rises. This pattern differs significantly from that seen with CNTs/water [27,28] and graphene/water [29,30] nanofluids. There is no noticeable difference in c_p with increasing CNT volume fraction for any CNT-loaded nanofluids. However, graphene-loaded nanofluids exhibit a modest decreasing trend (see Figure 7.1).

This discrepancy is primarily because the c_p of conventional nanoparticles is about ten times lower than the c_p of the base liquid. At the same time, CNTs and graphene have greater c_p values than conventional nanoparticles. As a result, when the nanoparticle loading increases, the c_p of CNTs and graphene-loaded nanofluids does not drop significantly.

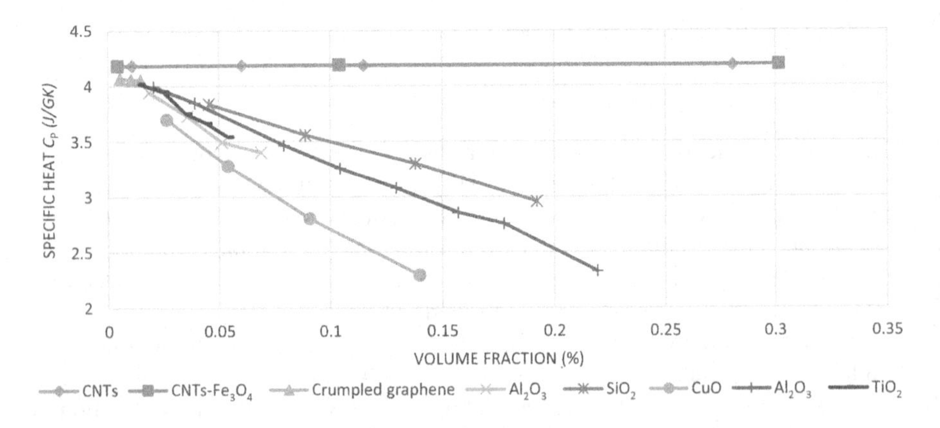

FIGURE 7.1 Schematic diagram of the c_p changes of suspensions in which various nanoparticles are dispersed in water with solid volume fraction as shown in different literature studies.

7.2 DENSITY

k_{nf} and μ_{nf} are the most investigated thermophysical characteristics of nanofluids in research. Researchers are interested in additional thermophysical characteristics such as ρ and c_p because of the broad application of nanofluids. We describe the current research work, including theoretical modeling, experimental measurement methodologies, and some meaningful conclusions, although the study of ρ is limited.

7.2.1 THEORY

One of the most fundamental physical characteristics of nanofluids is density, although theoretical studies on density are limited. The most widely used density model in this study is based on the physical principles that two substances must follow when they are combined [31,32]:

$$\rho_{nf} = \left(\frac{m}{V}\right)_{nf} = \frac{m_{bf} + m_p}{V_{bf} + V_p} = \frac{\rho_{bf} V_{bf} + \rho_p V_p}{V_{bf} + V_p} = \left(1 - \Phi_p\right)\rho_{bf} + \Phi_p \rho_p \tag{7.6}$$

$\Phi_p = \dfrac{V_p}{V_{bf} + V_p}$ shows the nanoparticle's volume fraction. At room temperature, this linear theoretical approach for density and volume fraction is in excellent agreement with experimental results [2,33]. Furthermore, Khanafer and Vafai [31] presented an empirical model based on experimental test data by Ho et al. [33] to describe the connection between density and temperature. The model is illustrated as follows:

$$\rho_{nf} = 1001.064 + 2738.6191\Phi_p - 0.2095T \tag{7.7}$$

The model's most significant relative error is 0.22% when Φ_p is between 0% and 4% and T is somewhere between 5°C and 40°C. Although the aforementioned model has been extensively utilized for calculating ρ [34–36], Sharifpur et al. [37] discovered that the anticipated ρ by the model was much higher than the experimental measurements. They linked this event to the fact that Eq. (7.6) overlooks the influence of the nanolayer on the particle surface, which creates a space between the nanoparticles and the base fluid. As a result, based on the aforementioned models and considering the nanolayer, the researchers created the following model based on experimental data:

$$\rho_{nf} = \frac{m_p + m_{bf}}{V_{bf} + V_p\left(r_p + d_{nl}\right)^3 / r_p^3} = \frac{\rho_{nf}}{\left(1 - \Phi_p\right) + \Phi_p\left(r_p + d_{nl}\right)^3 / r_p^3} \tag{7.8}$$

r_p is the average particle diameter, and d_{nl} shows the thickness of the nanolayer, and the nanolayer is considered to be completely void. Furthermore, based on the experimental results for nanofluids generated by spreading oxide particles (SiO_2, MgO, CuO) in the base fluid such as water or glycerol or EG, the researchers provided an approximate function connection between the thickness and the average nanoparticle diameter, which is shown below:

$$d_{nl} = -0.0002833 \times r_p^2 + 0.0475 \times r_p - 0.1417 \quad (7.9)$$

According to Eq. (7.9), the impact of the nanolayer thickness (equal to the gap) is strongly connected to the nanoparticle size. In comparison to the model provided by Eq. (7.9), the model in their study is more in agreement with the experimental data from Eq. (7.6). More study is needed to develop a density model with more precise forecasts and a comprehensive, adaptive range.

7.2.2 Experimental Measurement Techniques

During nanofluid flow and heat transfer operations, density directly impacts Re, friction coefficient, pressure loss, and Nu. We shall pay close attention to certain studies presenting measurements of the density of nanofluids for the measurement of ρ. In general, ρ measurements for nanofluids are split into two categories. The first is a direct measurement using a density meter, while the second is by computing the mass and volume of the nanofluid. The ρ of three distinct nanofluids was investigated by Vajjha et al. [38]. An Anton-Paar digital density meter was used, and the nanofluids were kept at a constant temperature using a circulating fluid temperature bath. A density meter is typically made up of U-shaped oscillating tubes and a system that includes an electrical excitation device, a frequency-counting device, and a display device. Typically, U-shaped tubes are made of glass because: first, it is transparent and let us see if there are any air bubbles inside the sample, which can change the outcome; second, it is unreactive and easy to clean; and third, it has a beneficial temperature deformation coefficient and sensitivity because of its low specific weight. The experimental ρ results for the Al_2O_3/EG-water and Sb_2O_5-SnO_2/EG-water nanofluids are in excellent accordance with the Pak and Cho equation (Eq. 7.6). The most significant difference between experimental and anticipated Eq. (7.6) values for ZnO/EG-water nanofluids is around 8%. Furthermore, Pastoriza-Gallego et al. [39] and Mariano et al. [40] utilized this density meter to test various nanofluids. The measured results were within the experimental range of error.

The thermophysical characteristics of nanofluids were measured by Pandey and Nema [41], in which the ρ is computed using the mass and volume of the nanofluid. A high-precision electronic balance was used by Kumaresan and Velraj [16] to weigh 25 and 50 mL of MWCNT/DI water/EG nanofluids. After repeating measurements three times, they used the average results of these three measurements to find the ρ. Pantzali et al. [42] computed ρ with an estimated accuracy of around 5% by weighing a known amount of nanofluid.

7.2.3 Experiment vs. Theory

Żyła et al. [43] experimentally studied the ρ of EG-based nanofluids containing different nitride nanoparticles, including AlN, Si_3N_4, and TiN. For ρ measurement, a

DMA 500 densimeter based on the U-type oscillation method was employed. It was concluded that the ρ progressively reduces as the sample temperature rises. However, it rises as the nanoparticle fraction in the nanofluid rises. Then, the findings from the measurements at RT were compared to theoretical expectations based on Eq. (7.6). The findings show that the experimental measurement and the theoretical model prediction are in agreement. The highest AAD of ρ measurements at RT for EG-based nanofluids containing different nitride nanoparticles was below 0.89%. Shoghl et al. [44] demonstrated the relationship between ρ and the previously mentioned concentration and temperature in another research. Using a DMA-35N portable density meter, they measured the ρ of nanofluids at various concentrations and temperatures. The experimental findings confirmed that Eq. (7.6) could adequately predict the ρ of nanofluids compared to the mixed model predictions. Ho et al. [33] investigated the volume fraction and temperature dependency of Al_2O_3/water nanofluid ρ. It was discovered that the decrease of ρ is not significant with rising temperature, which is because the ρ of Al_2O_3 nanoparticles is not sensitive to temperature shifts although the temperature is increased. It is worth mentioning that the experimental data correspond quite well with the mixing theory predictions (Eq. 7.6).

Furthermore, Michaelides [32] computed adjustments to the k data due to the different expansion coefficients of liquids and nanoparticles and differences in the ρ because of temperature changes. In other words, the previous discussion shows that we cannot overlook the impact of temperature and different factors on ρ. Most observations have shown that the ρ of nanofluids can be predicted quite precisely by classical mixed theory.

Lastly, it is obvious that nanoscience has made great progress in recent decades. Some research literature presents comprehensive reviews and analyses of recent advancements in nanofluid heat transfer phenomena across various geometries [45–48] and configurations. Drawing upon a wide range of scholarly contributions, including studies investigating natural convection, forced convection, and mixed convection heat transfer [49–53] processes, this review synthesizes findings from diverse research endeavors. Furthermore, investigations examine the implications of non-Newtonian behavior, hybrid nanofluids, and thermal radiation on heat transfer characteristics [54–57].

REFERENCES

1. R. Taylor, S. Coulombe, T. Otanicar, P. Phelan, A. Gunawan, W. Lv, G. Rosengarten, R. Prasher, H. Tyagi, Small particles, big impacts: A review of the diverse applications of nanofluids, *J. Appl. Phys.* 113 (2013) 011301. https://doi.org/10.1063/1.4754271.
2. B.C. Pak, Y.I. Cho, Hydrodynamic and heat transfer study of dispersed fluids with submicron metallic oxide particles, *Exp. Heat Transf.* 11 (1998) 151–170. https://doi.org/10.1080/08916159808946559.
3. Y. Xuan, W. Roetzel, Conceptions for heat transfer correlation of nanofluids, *Int. J. Heat Mass Transf.* 43 (2000) 3701–3707. https://doi.org/10.1016/S0017-9310(99)00369-5.
4. B. Barbés, R. Páramo, E. Blanco, M.J. Pastoriza-Gallego, M.M. Piñeiro, J.L. Legido, C. Casanova, Thermal conductivity and specific heat capacity measurements of Al2O3 nanofluids, *J. Therm. Anal. Calorim.* 111 (2013) 1615–1625. https://doi.org/10.1007/s10973-012-2534-9.

5. M. Chandrasekar, S. Suresh, A. Chandra Bose, Experimental investigations and theoretical determination of thermal conductivity and viscosity of Al_2O_3/water nanofluid, *Exp. Therm. Fluid Sci.* 34 (2010) 210–216. https://doi.org/10.103/j.expthermflusci.2009.10.022.
6. A.T. Utomo, H. Poth, P.T. Robbins, A.W. Pacek, Experimental and theoretical studies of thermal conductivity, viscosity and heat transfer coefficient of titania and alumina nanofluids, *Int. J. Heat Mass Transf.* 55 (2012) 7772–7781. https://doi.org/10.1016/j.ijheatmasstransfer.2012.08.003.
7. D. Shin, D. Banerjee, Enhancement of specific heat capacity of high-temperature silica-nanofluids synthesized in alkali chloride salt eutectics for solar thermal-energy storage applications, *Int. J. Heat Mass Transf.* 54 (2011) 1064–1070. https://doi.org/10.1016/j.ijheatmasstransfer.2010.11.017.
8. A.K. Starace, J.C. Gomez, J. Wang, S. Pradhan, G.C. Glatzmaier, Nanofluid heat capacities, *J. Appl. Phys.* 110 (2011) 093123. https://doi.org/10.1063/1.3672685.
9. Y.R. Sekhar, K.V. Sharma, Study of viscosity and specific heat capacity characteristics of water-based Al_2O_3 nanofluids at low particle concentrations, *J. Exp. Nanosci.* 10 (2015) 86–102. https://doi.org/10.1080/17458080.2013.796595.
10. R.S. Vajjha, D.K. Das, A review and analysis on influence of temperature and concentration of nanofluids on thermophysical properties, heat transfer and pumping power, *Int. J. Heat Mass Transf.* 55 (2012) 4063–4078. https://doi.org/10.1016/j.ijheatmasstransfer.2012.03.048.
11. D. Cabaleiro, C. Gracia-Fernández, J.L. Legido, L. Lugo, Specific heat of metal oxide nanofluids at high concentrations for heat transfer, *Int. J. Heat Mass Transf.* 88 (2015) 872–879. https://doi.org/10.1016/j.ijheatmasstransfer.2015.04.107.
12. S. Zhou, R. Ni, Measurement of the specific heat capacity of water-based Al_2O_3 nanofluid, *Appl. Phys. Lett.* 92 (2008) 093123. https://doi.org/10.1063/1.2890431.
13. J. Buongiorno, Convective transport in nanofluids, *J. Heat Transfer.* 128 (2006) 240–250. https://doi.org/10.1115/1.2150834.
14. D. Shin, D. Banerjee, Enhanced specific heat of silica nanofluid, *J. Heat Transfer.* 133 (2010) 024501. https://doi.org/10.1115/1.4002600.
15. D. Shin, D. Banerjee, Specific heat of nanofluids synthesized by dispersing alumina nanoparticles in alkali salt eutectic, *Int. J. Heat Mass Transf.* 74 (2014) 210–214. https://doi.org/10.1016/j.ijheatmasstransfer.2014.02.066.
16. V. Kumaresan, R. Velraj, Experimental investigation of the thermo-physical properties of water-ethylene glycol mixture based CNT nanofluids, *Thermochim. Acta.* 545 (2012) 180–186. https://doi.org/10.103/j.tca.2012.07.017.
17. M.X. Ho, C. Pan, Optimal concentration of alumina nanoparticles in molten hitec salt to maximize its specific heat capacity, *Int. J. Heat Mass Transf.* 70 (2014) 174–184. https://doi.org/10.1016/j.ijheatmasstransfer.2013.10.078.
18. R. Raud, B. Hosterman, A. Diana, T.A. Steinberg, G. Will, Experimental study of the interactivity, specific heat, and latent heat of fusion of water based nanofluids, *Appl. Therm. Eng.* 117 (2017) 164–168. https://doi.org/10.1016/j.applthermaleng.2017.02.033.
19. Z. Jiang, C.T. Imrie, J.M. Hutchinson, An introduction to temperature modulated differential scanning calorimetry (TMDSC): A relatively non-mathematical approach, *Thermochim. Acta.* 387 (2002) 75–93. https://doi.org/10.1016/S0040-6031(01)00829-2.
20. E. De Robertis, E.H.H. Cosme, R.S. Neves, A.Y. Kuznetsov, A.P.C. Campos, S.M. Landi, C.A. Achete, Application of the modulated temperature differential scanning calorimetry technique for the determination of the specific heat of copper nanofluids, *Appl. Therm. Eng.* 41 (2012) 10–17. https://doi.org/10.1016/j.applthermaleng.2012.01.003.

21. G. Żyła, J.P. Vallejo, J. Fal, L. Lugo, Nanodiamonds-ethylene glycol nanofluids: Experimental investigation of fundamental physical properties, *Int. J. Heat Mass Transf.* 121 (2018) 1201–1213. https://doi.org/10.1016/j.ijheatmasstransfer.2018.01.073.
22. S.M.S. Murshed, Determination of effective specific heat of nanofluids, *J. Exp. Nanosci.* 6 (2011) 539–546. https://doi.org/10.1080/17458080.2010.498838.
23. J.R. Satti, D.K. Das, D. Ray, Specific heat measurements of five different propylene glycol based nanofluids and development of a new correlation, *Int. J. Heat Mass Transf.* 94 (2016) 343–353. https://doi.org/10.1016/j.ijheatmasstransfer.2015.11.065.
24. H. O'Hanley, J. Buongiorno, T. McKrell, L.W. Hu, Measurement and model validation of nanofluid specific heat capacity with differential scanning calorimetry, *Adv. Mech. Eng.* 4 (2012) 181079. https://doi.org/10.1155/2012/181079.
25. S.W. Lee, S.D. Park, S. Kang, I.C. Bang, J.H. Kim, Investigation of viscosity and thermal conductivity of SiC nanofluids for heat transfer applications, *Int. J. Heat Mass Transf.* 54 (2011) 433–438. https://doi.org/10.1016/j.ijheatmasstransfer.2010.09.026.
26. S.P. Jang, J.H. Lee, K.S. Hwang, S.U.S. Choi, Particle concentration and tube size dependence of viscosities of Al_2O_3-water nanofluids flowing through micro- and minitubes, *Appl. Phys. Lett.* 91 (2007) 243112. https://doi.org/10.1063/1.2824393.
27. L.S. Sundar, M.K. Singh, A.C.M. Sousa, Enhanced heat transfer and friction factor of MWCNT-Fe_3O_4/water hybrid nanofluids, *Int. Commun. Heat Mass Transf.* 52 (2014) 73–83. https://doi.org/10.1016/j.icheatmasstransfer.2014.01.012.
28. S. Halelfadl, T. Maré, P. Estellé, Efficiency of carbon nanotubes water based nanofluids as coolants, *Exp. Therm. Fluid Sci.* 53 (2014) 104–110. https://doi.org/10.1016/j.expthermflusci.2013.11.010.
29. A. Amiri, G. Ahmadi, M. Shanbedi, M. Etemadi, M.N.M. Zubir, B.T. Chew, S.N. Kazi, Heat transfer enhancement of water-based highly crumpled few-layer graphene nanofluids, *RSC Adv.* 6 (2016) 105508–105527. https://doi.org/10.1039/c6ra22365f.
30. H. Yarmand, S. Gharehkhani, S.F.S. Shirazi, A. Amiri, M.S. Alehashem, M. Dahari, S.N. Kazi, Experimental investigation of thermo-physical properties, convective heat transfer and pressure drop of functionalized graphene nanoplatelets aqueous nanofluid in a square heated pipe, *Energy Convers. Manag.* 114 (2016) 38–49. https://doi.org/10.1016/j.enconman.2016.02.008.
31. K. Khanafer, K. Vafai, A critical synthesis of thermophysical characteristics of nanofluids, *Int. J. Heat Mass Transf.* 54 (2011) 4410–4428. https://doi.org/10.1016/j.ijheatmasstransfer.2011.04.048.
32. E.E. Michaelides, Variation of the expansion coefficient of nanofluids with temperature: A correction for conductivity data, *J. Nanotechnol. Eng. Med.* 3 (2013) 044502. https://doi.org/10.1115/1.4024100.
33. C.J. Ho, W.K. Liu, Y.S. Chang, C.C. Lin, Natural convection heat transfer of alumina-water nanofluid in vertical square enclosures: An experimental study, *Int. J. Therm. Sci.* 49 (2010) 1345–1353. https://doi.org/10.1016/j.ijthermalsci.2010.02.013.
34. S.Z. Heris, M.N. Esfahany, S.G. Etemad, Experimental investigation of convective heat transfer of Al_2O_3/water nanofluid in circular tube, *Int. J. Heat Fluid Flow.* 28 (2007) 203–210. https://doi.org/10.1016/j.ijheatfluidflow.2006.05.001.
35. J. Li, C. Kleinstreuer, Thermal performance of nanofluid flow in microchannels, *Int. J. Heat Fluid Flow.* 29 (2008) 1221–1232. https://doi.org/10.1016/j.ijheatfluidflow.2008.01.005.
36. B. Raei, F. Shahraki, M. Jamialahmadi, S.M. Peyghambarzadeh, Experimental study on the heat transfer and flow properties of γ-Al_2O_3/water nanofluid in a double-tube heat exchanger, *J. Therm. Anal. Calorim.* 127 (2017) 2561–2575. https://doi.org/10.1007/s10973-016-5868-x.

37. M. Sharifpur, S. Yousefi, J.P. Meyer, A new model for density of nanofluids including nanolayer, *Int. Commun. Heat Mass Transf.* 78 (2016) 168–174. https://doi.org/10.1016/j.icheatmasstransfer.2016.09.010.
38. R.S. Vajjha, D.K. Das, B.M. Mahagaonkar, Density measurement of different nanofluids and their comparison with theory, *Pet. Sci. Technol.* 27 (2009) 612–624. https://doi.org/10.1080/10916460701857714.
39. M.J. Pastoriza-Gallego, C. Casanova, J.L. Legido, M.M. Piñeiro, CuO in water nanofluid: Influence of particle size and polydispersity on volumetric behaviour and viscosity, *Fluid Phase Equilib.* 300 (2011) 188–196. https://doi.org/10.1016/j.fluid.2010.10.015.
40. A. Mariano, M.J. Pastoriza-Gallego, L. Lugo, A. Camacho, S. Canzonieri, M.M. Piñeiro, Thermal conductivity, rheological behaviour and density of non-Newtonian ethylene glycol-based SnO_2 nanofluids, *Fluid Phase Equilib.* 337 (2013) 119–124. https://doi.org/10.1016/j.fluid.2012.09.029.
41. S.D. Pandey, V.K. Nema, Experimental analysis of heat transfer and friction factor of nanofluid as a coolant in a corrugated plate heat exchanger, *Exp. Therm. Fluid Sci.* 38 (2012) 248–256. https://doi.org/10.1016/j.expthermflusci.2011.12.013.
42. M.N. Pantzali, A.G. Kanaris, K.D. Antoniadis, A.A. Mouza, S.V. Paras, Effect of nanofluids on the performance of a miniature plate heat exchanger with modulated surface, *Int. J. Heat Fluid Flow.* 30 (2009) 691–699. https://doi.org/10.1016/j.ijheatfluidflow.2009.02.005.
43. G. Żyła, J.P. Vallejo, L. Lugo, Isobaric heat capacity and density of ethylene glycol based nanofluids containing various nitride nanoparticle types: An experimental study, *J. Mol. Liquid* 261 (2018) 530–539. https://doi.org/10.1016/j.molliq.2018.04.012.
44. S.N. Shoghl, J. Jamali, M.K. Mostafa, Electrical conductivity, viscosity, and density of different nanofluids: An experimental study, *Exp. Therm. Fluid Sci.* 74 (2016) 339–346. https://doi.org/10.1016/j.expthermflusci.2016.01.004.
45. N. Alavi, T. Armaghani, E. Izadpanah, Natural convection heat transfer of a nanofluid in a baffle L-Shaped cavity, *J. Solid Fluid Mech.* 6.3 (2016) 311–321.
46. A.I. Alsabery, et al., Conjugate heat transfer of Al_2O_3-water nanofluid in a square cavity heated by a triangular thick wall using Buongiorno's two-phase model, *J. Therm. Anal. Calorim.* 135 (2019) 161–176.
47. T. Armaghani, et al., Effects of discrete heat source location on heat transfer and entropy generation of nanofluid in an open inclined L-shaped cavity, *Int. J. Numer. Methods Heat Fluid Flow* 29.4 (2019) 1363–1377.
48. M.S. Sadeghi, et al., On the natural convection of nanofluids in diverse shapes of enclosures: An exhaustive review, *J. Therm. Anal. Calorim.* (2020) 147 1–22. https://doi.org/10.1007/s10973-020-10222-y
49. R. Ghasemiasl, et al., Recent studies on the forced convection of nano-fluids in channels and tubes: A comprehensive review, *Exp. Tech.* 47.1 (2023) 47–81.
50. T. Armaghani, et al., Mixed convection and entropy generation of an ag-water nanofluid in an inclined L-shaped channel, *Energies* 12.6 (2019) 1150.
51. M.A. Ismael, T. Armaghani, A.J. Chamkha, Mixed convection and entropy generation in a lid-driven cavity filled with a hybrid nanofluid and heated by a triangular solid, *Heat Transfer Res.* 49.17 (2018) 1645–1665.
52. S. Izadi, et al., A comprehensive review on mixed convection of nanofluids in various shapes of enclosures, *Powder Technol.* 343 (2019) 880–907.
53. T. Armaghani, A. Kasaeipoor, U. Mohammadpoor, Entropy generation analysis of mixed convection with considering magnetohydrodynamic effects in an open C-shaped cavity, *Therm. Sci.* 23.6 Part A (2019) 3455–3465.

54. S. Hussain, et al., MHD mixed convection and entropy analysis of non-Newtonian hybrid nanofluid in a novel wavy elbow-shaped cavity with a quarter circle hot block and a rotating cylinder, *Exp. Tech.* 47.1 (2023) 17–36.
55. H.A. Nabwey, et al., A comprehensive review of non-Newtonian Nanofluid heat transfer, *Symmetry* 15.2 (2023) 362.
56. A.J. Chamkha, et al., MHD convection of an Al_2O_3-Cu/water hybrid nanofluid in an inclined porous cavity with internal heat generation/absorption, *Iran. J. Chem. Chem. Eng.* 41.3 (2022) 936–956.
57. A.M. Rashad, et al., Unsteady MHD hybrid nanofluid mixed convection heat transfer in a wavy porous cavity with thermal radiation, *J. Therm. Anal. Calorim.* (2024) 149 2425–2442. https://doi.org/10.1007/s10973-023-12690-4

8 Selecting a Thermophysical Model for Numerical Modeling of Nanofluids

8.1 INTRODUCTION

There is a need to increase heat transfer to improve thermal performance in many industries. As mentioned in previous chapters, one of the main candidates for this purpose is the use of nanofluids, which has attracted the attention of many researchers [1–9].

In this chapter, while briefly explaining the numerical modeling methods of heat and flow transfer of nanofluids (referred to as numerical modeling methods of nanofluids in this chapter), the effect of thermophysical models mentioned in previous chapters on the heat transfer of nanofluids will be discussed. The best viscosity and conductivity models for different nanofluids will be proposed.

8.2 AN OVERVIEW OF NUMERICAL MODELING METHODS OF NANOFLUIDS

There are various methods available for numerical modeling of nanofluids and their heat and flow transfer, which are presented in the following table (Figure 8.1).

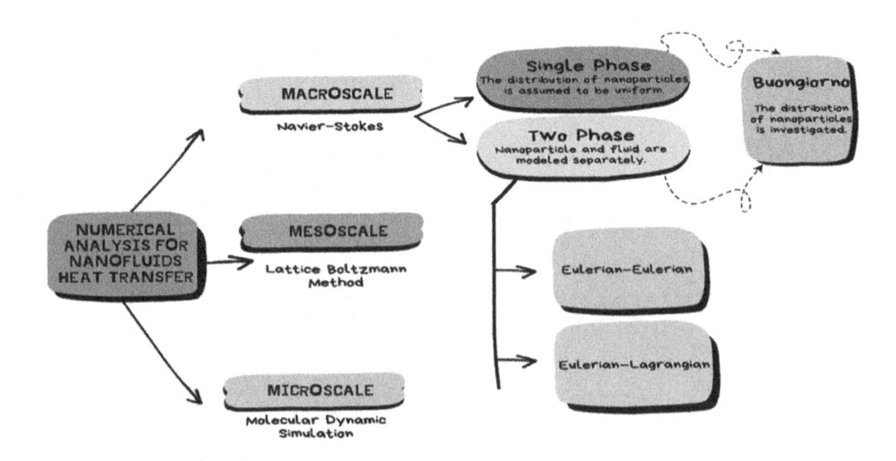

FIGURE 8.1 Numerical modeling methods of nanofluids.

DOI: 10.1201/9781032664118-8

In most cases, nanofluids are considered as a continuous fluid, and therefore modeling of nanofluids from a macroscopic perspective will yield reliable results [9–46], which will be briefly explained.

For the modeling and analysis of the heat transfer equations of nanofluids using the macroscopic approach, there are two perspectives: single phase and two phase.

8.2.1 Single-Phase Approach

In this method, the base fluid and nanoparticles are considered to be homogeneous, and it is assumed that both phases are in thermal and hydrodynamic equilibrium, and then the thermophysical properties of the nanofluid replace the base fluid properties in the Navier–Stokes and energy equations. As a result, the thermophysical properties of nanofluids play a decisive role in the obtained results.

In the single-phase approach, are neglected any interfacial force and relative velocity of nanoparticles to the base fluid, which simplifies the relevant equations and significantly reduces the calculation time.

Therefore, the mass conservation, momentum conservation, and energy conservation equations without considering external forces and viscous losses will be as follows.

$$\nabla \cdot (\rho \upsilon) = 0 \tag{8.1}$$

$$\nabla \cdot (\rho \upsilon \upsilon) = 0 \tag{8.2}$$

$$\nabla \cdot \left((\rho c_p) \upsilon T \right) = \nabla \cdot (k \nabla T) \tag{8.3}$$

Here, υ is the fluid velocity vector, p is the pressure, and T is the temperature. Also, thermal conductivity, dynamic viscosity, density, and specific heat capacity of the fluid are denoted by k, μ, ρ, and c_p, respectively.

8.2.2 Two-Phase Approach

In the two-phase approach, the base fluid and nanoparticles are considered as separate phases, each with their own specific velocity and temperature, in a way that allows relative motion between the solid particles and neighboring fluid molecules, a phenomenon referred to as "slip." As a result, the two-phase approach, due to the assumption of motion between nanoparticles and the base fluid, provides a better prediction in the study of nanofluids.

The two-phase approach has been interpreted in two distinct perspectives, leading to the development of various models. These two perspectives are known as the Eulerian–Eulerian and the Eulerian–Lagrangian methods. Figures 8.2 and 8.3 provide a schematic representation of these two methods.

For further study on the two-phase heat transfer modeling of nanofluids, Refs. [47–49] may be useful.

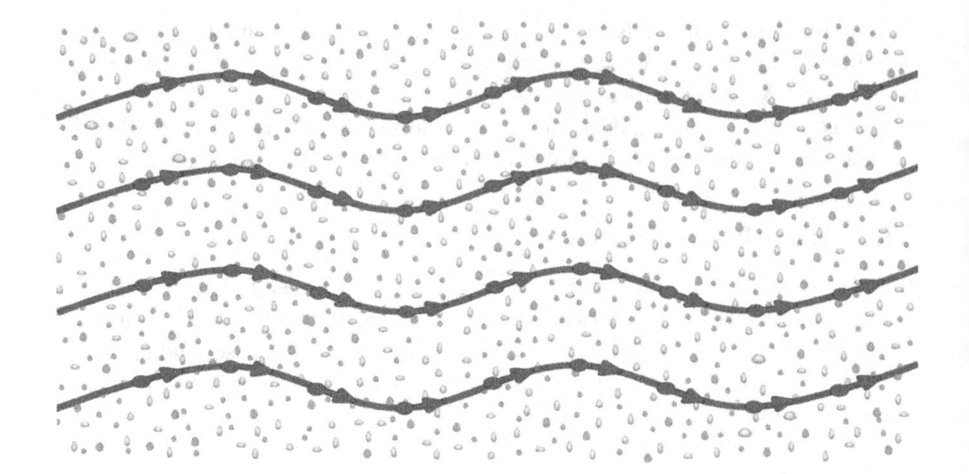

FIGURE 8.2 Dispersion and direction of motion of nanoparticles in the Eulerian–Eulerian two-phase approach.

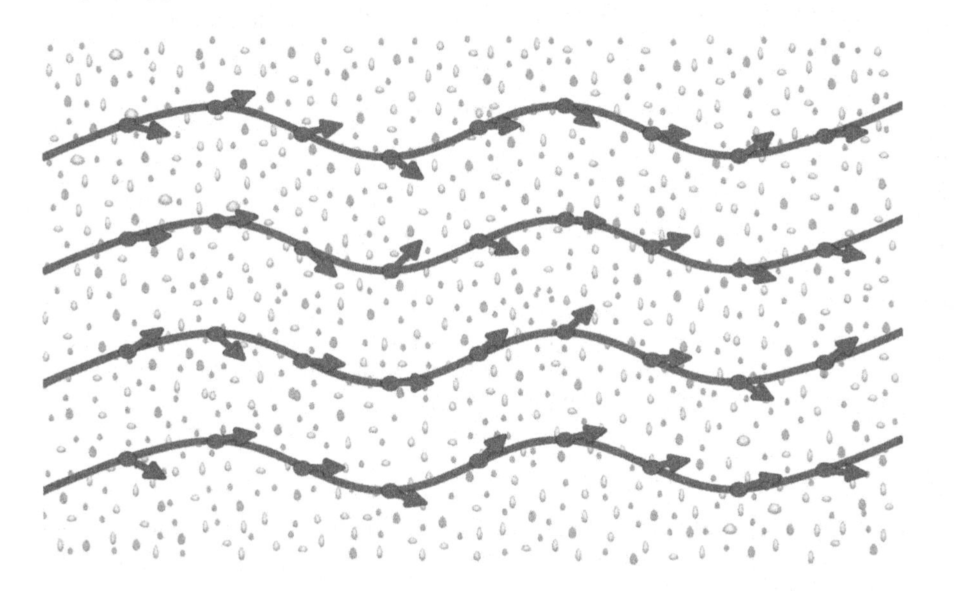

FIGURE 8.3 Dispersion and direction of motion of nanoparticles in the Eulerian–Lagrangian two-phase approach.

8.2.3 Boungiorno Semi-Two-Phase Model

Boungiorno [50] in theoretical exploration and analysis of nanofluids stated that the increase in heat transfer in nanofluids is not solely attributed to the increase in the heat conduction coefficient and this phenomenon cannot be predicted by conventional fluid equations.

By examining the physical phenomena that affect the heat transfer of nanofluids, Boungiorno identified seven potential factors. These factors, related to the slip mechanism of nanofluids, include inertia, Brownian diffusion, thermophoresis, diffusiophoresis, Magnus effect, fluid discharge, and gravitational settling, and he also believes that these mechanisms create a difference in velocity between the base fluid and nanoparticles.

Ultimately, Boungiorno observed that among these seven mechanisms, only Brownian diffusion and thermophoresis significantly affect the heat transfer of laminar flow nanofluids. Therefore, these two dominant and influential mechanisms are considered.

8.2.3.1 Brownian Diffusion

The random motion of nanoparticles in the base fluid is also one of the effective factors in increasing heat transfer in nanofluids. As the size of nanoparticles decreases, their Brownian motion increases, leading to an increase in heat transfer, and conversely, with an increase in nanoparticle size, Brownian motion decreases. The Brownian diffusion coefficient and mass flux resulting from this motion are obtained as follows:

$$D_{\mathrm{B}} = \frac{k_{\mathrm{B}} T_{\mathrm{nf}}}{3\pi \mu_{\mathrm{nf}} d_{\mathrm{p}}} \tag{8.4}$$

$$J_{\mathrm{p.B}} = -\rho_{\mathrm{p}} D_{\mathrm{B}} \nabla \varphi \tag{8.5}$$

In the discussed equations, the symbols k_{B} and d_{p} represent the Boltzmann constant and the average diameter of nanoparticles, respectively.

8.2.3.2 Thermophoresis

In a high-temperature environment, molecules have more kinetic energy, leading to frequent collisions with neighboring molecules. This interaction drives the molecules from the hotter environment toward the colder environment, thereby facilitating heat transfer. This process is known as thermophoresis. Thermophoresis and the mass flux resulting from this mechanism are equal to:

$$D_{\mathrm{T}} = \left(\frac{0.26 k_{\mathrm{bf}}}{2k_{\mathrm{bf}} + k_{\mathrm{p}}}\right)\left(\frac{\mu_{\mathrm{nf}}}{\rho_{\mathrm{nf}}}\right)\varphi \tag{8.6}$$

$$j_{\mathrm{p.T}} = -\rho_{\mathrm{p}} D_{\mathrm{T}} \frac{\nabla T}{T} \tag{8.7}$$

The Boungiorno model for nanofluid modeling is as follows:

$$\nabla \cdot v = 0 \tag{8.8}$$

$$\rho_{\text{nf}}\left[\frac{\partial \boldsymbol{v}}{\partial t}+\boldsymbol{v}\cdot\nabla\boldsymbol{v}\right]=-\nabla P-\nabla\cdot\tau \tag{8.9}$$

$$\tau=-\mu_{\text{nf}}\left[\nabla\boldsymbol{v}+(\nabla\boldsymbol{v})^{\text{t}}\right] \tag{8.10}$$

$$\left(\rho C_{\text{p}}\right)_{\text{np}}\left[\frac{\partial T}{\partial t}+\boldsymbol{v}\cdot\nabla T\right]=\nabla\cdot k_{\text{nf}}\nabla T-\rho_{\text{p}}c_{\text{p}}\left[D_{\text{B}}\nabla\varphi\cdot\nabla T+D_{\text{T}}\nabla T\cdot\frac{\nabla T}{T}\right] \tag{8.11}$$

$$\frac{\partial\varphi}{\partial t}+\boldsymbol{v}\cdot\nabla\varphi=\nabla\cdot\left[D_{\text{B}}\nabla\varphi+D_{\text{T}}\frac{\nabla T}{T}\right] \tag{8.12}$$

As can be seen from Eq. (8.12), this method attempts to model the distribution of nanoparticles in the base fluid. Therefore, the Boungiorno model is not only known as a semi-two-phase model but also known as a non-homogeneous single-phase model.

For further study on nanofluid modeling in various issues using the Boungiorno model, would be a useful reference to articles [51–57].

The use of nanofluids to increase heat transfer in free convection heat transfer and to some extent in combined heat transfer is challenging, and numerous studies indicate an increase in free convection heat transfer by adding nanoparticles to the base fluid, while in some cases, numerical results have reported a decrease in heat transfer.

In single-phase modeling, in many cases, the numerical results depend on the choice of thermophysical properties model for nanofluids, due to the nature of free convection heat transfer, where the nanofluid velocity is low and the effects of viscosity play a significant role in free convection heat transfer.

Furthermore, the percentage of dependency of numerical modeling results on thermophysical properties models for nanofluids has been statistically examined, and the dependency on thermophysical properties has been investigated for several experimental reports with the numerical modeling.

8.3 THE MOST WIDELY USED THERMOPHYSICAL EQUATIONS OF NANOFLUIDS

The main task of thermophysical models for nanofluids is to predict and measure the thermal and physical properties of nanofluids. They can also assist researchers in understanding the fundamental mechanisms of heat transfer and designing efficient heat transfer systems. Therefore, thermophysical models for nanofluids are widely used in heat transfer studies.

In the following section, an attempt will be made to compare the common thermophysical models available for numerical modeling of nanofluids with experimental results and recommend them to the readers.

In this section, 170 reputable scientific studies that simulated nanofluid heat transfer using computational fluid dynamics in single-phase conditions were randomly selected, evaluated, and classified. This analysis aimed to gain insight into the prevalent and commonly used models for the thermal conductivity and viscosity coefficients of nanofluids.

Tables 8.1 and 8.2 refer to the commonly used models for thermal conductivity coefficient and viscosity for single-phase nanofluid modeling. The most commonly utilized coupled models of dynamic viscosity and thermal conductivity for heat transfer modeling of nanofluids are presented in Table 8.3.

TABLE 8.1
Review of Thermal Conductivity Coefficient Models Used in Numerical Studies of Nanofluids

Thermal Conductivity Model	Percentage Used (%)	Equations
Maxwell	53	$k_{nf} = \frac{2k_{bf} + k_p + 2\varphi(k_p - k_{bf})}{2k_{bf} + k_p - \varphi(k_p - k_{bf})} k_{bf}$
Koo and Kleinstreuer	7.2	$k_{nf} = k_{static} + k_{brownian}$
Xue	5.9	$k_{nf} = \left[1 + 4.4\left(Re^{0.4} Pr^{0.66}\right)\left(\frac{T}{T_{fr}}\right)^{10}\left(\frac{k_p}{k_{bf}}\right)^{0.03}\right]$
Ho	5.2	$k_{nf} = \left(1 + 2.944\varphi + 19.672\varphi^2\right) k_{bf}$
Corcione	3.3	$k_{nf} = \left[1 + 4.4\left(Re^{0.4} Pr^{0.66}\right)\left(\frac{T}{T_{fr}}\right)^{10}\left(\frac{k_p}{k_{bf}}\right)^{0.03}\right] k_{bf}$
Others	25.4	

TABLE 8.2
Review of Viscosity Models Used in Numerical Studies of Nanofluids

Viscosity Model	Percentage Used (%)	Equations
Brinkman	61.4	$\mu_{nf} = \frac{1}{(1-\varphi)^{2.5}} \mu_{bf}$
Ho	5.9	$\mu_{nf} = \left(1 + 4.39\varphi + 222.4\varphi^2\right) \mu_{bf}$
Koo and Kleinstreuer	7.2	$\mu_{nf} = \mu_{static} + \mu_{brownian}$
Corcione	5.9	$\mu_{nf} = \left[\frac{1}{1 - 34.87\left(d_p / d_{bf}\right)^{-0.3} \varphi^{1.03}}\right] \mu_{bf}$
Others	19.6	

TABLE 8.3
Review of Coupled Thermophysical Models Used in Numerical Studies of Nanofluids

No.	Thermal Conductivity	Dynamic Viscosity	Thermophysical Property Equations	Percentage Used
Model 1	Maxwell [59]	Brinkman [60]	$k_{nf} = \frac{2k_{bf} + k_p + 2\varphi(k_p - k_{bf})}{2k_{bf} + k_p - \varphi(k_p - k_{bf})} k_{bf}$ $\mu_{nf} = \frac{1}{(1-\varphi)^{2.5}} \mu_{bf}$	43.4
Model 2 [61]	Koo & Kleinstreuer	Koo & Kleinstreuer	$k_{nf} = k_{static} + k_{brownian}$ $\mu_{nf} = \mu_{static} + \mu_{brownian}$	4.7
Model 3 [62]	Ho	Ho	$k_{nf} = (1 + 2.944\varphi + 19.672\varphi^2) k_{bf}$ $\mu_{nf} = (1 + 4.39\varphi + 222.4\varphi^2) \mu_{bf}$	5.2
Model 4 [63]	Corcione	Corcione	$k_{nf} = \left[1 + 4.4\left(\mathrm{Re}^{0.4}\mathrm{Pr}^{0.66}\right)\left(\frac{T}{T_{fr}}\right)^{10}\left(\frac{k_p}{k_{bf}}\right)^{0.03}\right] k_{bf}$ $\mu_{nf} = \left[\frac{1}{1 - 34.87(d_p / d_{bf})^{-0.3} \varphi^{1.03}}\right] \mu_{bf}$	2.7

(Continued)

TABLE 8.3 (*Continued*)
Review of Coupled Thermophysical Models Used in Numerical Studies of Nanofluids

No.	Thermal Conductivity	Dynamic Viscosity	Thermophysical Property Equations	Percentage Used
Model 5	Xue [64]	Brinkman [60]	$k_{nf} = \dfrac{1-\varphi+2\varphi\dfrac{k_p}{k_p-k_{bf}}\ln\dfrac{k_p+k_{bf}}{2k_{bf}}}{1-\varphi+2\varphi\dfrac{k_{bf}}{k_p-k_{bf}}\ln\dfrac{k_p+k_{bf}}{2k_{bf}}}k_{bf}$ $\mu_{nf} = \dfrac{1}{(1-\varphi)^{2.5}}\mu_{bf}$	2.2

Furthermore, the coupled relationships for the thermal conductivity coefficient and viscosity models used by researchers to estimate thermophysical properties in numerical studies in single-phase conditions are provided.

The most common coupled relationships used in the numerical modeling of nanofluids are the Maxwell model for thermal conductivity coefficient and the Brinkman model for viscosity.

8.4 EVALUATION OF THERMOPHYSICAL RELATIONSHIPS BASED ON NUMERICAL MODELING

The approach of this relevant section is aimed at introducing a suitable thermophysical model to achieve accurate results in studies and numerical modeling of nanofluids. To this end, common thermophysical models in the continuity, momentum, and energy equations will be applied to create a basis for evaluating and comparing theoretical and experimental results under various physical and thermal conditions.

In addition, studies have shown that temperature and concentration variables directly and independently affect the thermal conductivity coefficient and viscosity of nanofluids, however, among the published thermophysical models, some relationships cannot respond proportional to the conventional temperature and concentration ranges in heat transfer. Applying such relationships in heat transfer issues will lead to increased computational errors. Therefore, the use of an appropriate thermophysical model in heat transfer issues is very important.

Furthermore, an attempt is made to numerically model the experimental results of free convection heat transfer of nanofluids published in reputable journals and compare the numerical results based on different thermophysical relationships.

The OpenFOAM software has been used for numerical modeling, which operates based on the finite volume method (FVM) and solves the conservation equations for various problems.

For numerical modeling using single-phase and Boungiorno models in C++, a code has been written and added to the OpenFOAM software, making it possible to model based on the Boungiorno method [12,14,20,51]. For higher accuracy, have been considered the effects of temperature on thermophysical properties.

Subsequently, numerical modeling has been performed based on the most common thermophysical equations for titanium oxide–water, alumina–water, and silicon dioxide–water nanofluids, and compared with experimental results, and has been recommended as the best model.

8.4.1 TiO_2–Water Nanofluid

In the experimental study conducted by Sharifipour et al. [58] on titanium oxide–water nanofluid in the temperature range of 5°C–55°C and concentration of 0.5%–0.8%, the heat transfer rate in the cavity was obtained in proportion to changes in temperature and volume fraction.

TABLE 8.4
Results of Single-Phase Modeling of TiO_2 Nanofluid Based on the Pair of Most Common Relationships

Model	Temperature Range			
	$\Delta T=20$	$\Delta T=30$	$\Delta T=40$	$\Delta T=50$
Experimental Data	**300.773**	**304.074**	**308.007**	**324.378**
Maxwell + Brinkman	300.748	339.298	369.454	394.41
Koo & Kleinstreuer + Koo & Kleinstreuer	300.753	339.303	369.48	394.273
Ho + Ho	300.688	339.231	369.382	394.16
Corcione + Corcione	300.562	339.086	369.223	394.068
Xue + Brinkman	300.748	339.493	369.667	394.639
MAG+BAG	300.694	336.979	362.002	365.513

To evaluate the dependency of numerical modeling results on thermophysical models, the experimental study by Sharifipour et al. has been modeled.

As shown in Table 8.4, with an increase in temperature difference, the heat transfer rate in the experimental data also increases, but with an increase in temperature difference, the accuracy of the numerical computation system in determining the heat transfer rate decreases.

In Table 8.4, the numerical results of single-phase modeling based on the most common thermophysical relationships for nanofluids have been compared with experimental results. Additionally, for the evaluation of the proposed models in this book, the MAG and BAG models have also been included in the comparisons.

Considering the results obtained from the experimental study of titanium oxide nanofluid at a volume fraction of 0.05% and various temperature ranges, and as indicated in Table 8.4, the MAG+BAG model shows the lowest error compared to the experimental results.

Under similar conditions, applying the Boungiorno method to the numerical computation equations results in significantly improved heat transfer rates for nanofluids. The results of the MAG+BAG thermophysical model in Table 8.5 show a reduction in the average error coefficient.

8.4.2 Al_2O_3–Water Nanofluid

In another study conducted by Srinivas Rao and Srivastava [65] on aluminum oxide–water nanofluid, the heat transfer rate of the nanofluid in a cavity with temperature differences of 1, 1.8, and 2.3 K and volume fraction of 0.01%–0.04% was obtained experimentally.

Numerical calculations were performed for the aluminum oxide–water nanofluid in a single-phase manner in a cavity with temperature differences of 1, 1.8, and 2.3 K

TABLE 8.5
Numerical Results of TiO_2 Nanofluid Modeling Based on the Pair of Most Common Relationships Using the Boungiorno Method

Model	Temperature Range			
	$\Delta T = 20$	$\Delta T = 30$	$\Delta T = 40$	$\Delta T = 50$
Experimental Data	**300.773**	**304.074**	**308.007**	**324.378**
Maxwell + Brinkman	281.212	319.229	351.281	357.231
Koo & Kleinstreuer + Koo & Kleinstreuer	281.366	319.413	351.308	357.337
Ho + Ho	281.206	319.341	351.209	357.282
Corcione + Corcione	281.119	319.126	351.083	357.079
Xue + Brinkman	281.549	319.533	351.507	357.482
MAG+BAG	283.613	316.979	342.703	347.869

and a volume fraction of 0.04%, and by considering the commonly used thermophysical relationships mentioned, the heat transfer rate was obtained according to Table 8.6. Therefore, the MAG+BAG relationship provides higher accuracy in estimating the heat transfer rate for the aluminum oxide–water nanofluid in the cavity compared to the experimental data.

In modeling based on the Boungiorno method, the numerical results show a higher value than the experimental values. Therefore, Boungiorno's method may be overestimated in some cases.

Thus, applying these relationships increases the mean error coefficient. Accordingly, as shown in Tables 8.7, the MAG+BAG model exhibits the lowest average error for the aluminum oxide–water nanofluid.

TABLE 8.6
Results of Single-Phase Modeling of Al_2O_3 Nanofluid Based on the Pair of Most Common Relationships

Model	Temperature Range		
	$\Delta T = 1$	$\Delta T = 1.8$	$\Delta T = 2.3$
Experimental Data	**237.03**	**263.23**	**274.36**
Maxwell + Brinkman	215.85	216.118	216.489
Koo & Kleinstreuer + Koo & Kleinstreuer	215.861	216.123	216.533
Ho + Ho	215.353	215.728	215.907
Corcione + Corcione	214.989	215.255	215.61
Xue + Brinkman	216.873	217.127	217.401
MAG+BAG	218.838	219.52	220.253

TABLE 8.7
Numerical Results of Al_2O_3 Nanofluid Modeling Using the Boungiorno Method Based on the Pair of Most Common Relationships

Model	Temperature Range		
	$\Delta T = 1$	$\Delta T = 1.8$	$\Delta T = 2.3$
Experimental Data	**237.03**	**263.23**	**274.36**
Maxwell + Brinkman	262.832	291.815	306.374
Koo & Kleinstreuer + Koo & Kleinstreuer	262.884	291.889	306.401
Ho + Ho	262.719	291.709	306.337
Corcione + Corcione	262.609	291.61	306.218
Xue + Brinkman	262.898	291.903	306.491
MAG+BAG	264.119	288.72	303.107

8.4.3 SiO_2–Water Nanofluid

In the experimental study conducted by Torki et al. [66] on silicon dioxide and water nanofluid in a cavity, the heat transfer rate for nanofluids was obtained at concentrations below 1% and in different temperature ranges.

Now, considering a constant volume fraction of 0.05% for silicon dioxide nanofluid, have been evaluated all the most commonly used thermophysical models. According to the numerical results specified in Tables 8.8 and 8.9, the MAG+BAG thermophysical model in the single-phase state and the Boungiorno method have the best accuracy for predicting the heat transfer rate.

TABLE 8.8
Results of Single-Phase Modeling of SiO_2 Nanofluid Based on the Pair of Most Common Relations

Model	Temperature Range					
	$\Delta T = 2.3$	$\Delta T = 4.5$	$\Delta T = 5.2$	$\Delta T = 10.3$	$\Delta T = 15.5$	$\Delta T = 10.8$
Experimental Data	**154.661**	**175.847**	**183.051**	**207.627**	**236.864**	**246.61**
Maxwell + Brinkman	167.9097	188.7354	204.0129	237.7767	276.3726	283.9305
Koo & Kleinstreuer + Koo & Kleinstreuer	167.94645	188.74065	204.01815	237.804	276.5343	283.78665
Ho + Ho	167.83095	188.6724	203.94255	237.7011	276.3663	283.668
Corcione + Corcione	167.6976	188.5401	203.7903	237.53415	276.27495	283.5714
Xue + Brinkman	168.0462	188.7354	204.21765	238.00035	276.72645	284.17095
MAG+BAG	166.2087	186.4737	199.89795	232.9971	266.71365	276.26445

TABLE 8.9
Numerical Results of SiO_2 Nanofluid Modeling Using the Boungiorno Method Based on the Pair of Most Common Relations

Model	Temperature Range					
	$\Delta T = 2.3$	$\Delta T = 4.5$	$\Delta T = 5.2$	$\Delta T = 10.3$	$\Delta T = 15.5$	$\Delta T = 10.8$
Experimental Data	**154.661**	**175.847**	**183.051**	**207.627**	**236.864**	**246.61**
Maxwell + Brinkman	159.914	179.748	194.298	226.454	263.212	270.41
Koo & Kleinstreuer + Koo & Kleinstreuer	159.949	179.753	194.303	226.48	263.366	270.273
Ho + Ho	159.839	179.688	194.231	226.382	263.206	270.16
Corcione + Corcione	159.712	179.562	194.086	226.223	263.119	270.068
Xue + Brinkman	160.044	179.748	194.493	226.667	263.549	270.639
MAG+BAG	158.294	177.594	190.379	221.902	254.013	263.109

8.5 CONCLUSION

Since in this chapter, commonly used and proposed thermophysical models for nanofluids have been numerically compared with experimental results under various physical and thermal conditions, based on the obtained results, it is recommended that the readers consider the pair of thermal conductivity and viscosity relationships proportional to the studied nanofluids.

- In the single-phase modeling of TiO_2 nanofluid, the MAG+BAG relationship have the lowest error rates in numerical computations, respectively.
- In numerical modeling using the Boungiorno method, the error rate is significantly reduced, and the thermophysical pairs of MAG+BAG exhibit the lowest error for the TiO_2 + water nanofluid.
- The numerical calculations performed on the aluminum oxide–water nanofluid show that the MAG+BAG thermophysical pair have the highest accuracy in single-phase modeling.
- Contrary to expectations, in the numerical modeling of the aluminum oxide–water nanofluid, applying the Boungiorno method increases the error rate compared to the single-phase method.
- In the single-phase modeling of the silicon dioxide–water nanofluid, the MAG+BAG thermophysical model shows the highest accuracy in numerical computations, and a similar performance is observed with the Boungiorno method.

REFERENCES

1. M. Nazarahari, R. Ghasemi Asl, T. Armaghani, Experimental study of nanofluids natural convection heat transfer in various shape pores of porous media, *J. Therm. Anal. Calorim.* 149 (2024) 2331–2349. https://doi.org/10.1007/s10973-023-12808-8

2. H.A. Nabwey, et al., A comprehensive review of nanofluid heat transfer in porous media, *Nanomaterials* 13.5 (2023) 937.
3. R. Ghasemiasl, et al., Recent studies on the forced convection of nano-fluids in channels and tubes: A comprehensive review, *Exp. Tech.* 47.1 (2023) 47–81.
4. H.A. Nabwey, Hossam A., et al., A comprehensive review of non-newtonian nanofluid heat transfer, *Symmetry* 15.2 (2023) 362.
5. A. Barkhordar, R. Ghasemiasl, T. Armaghani, Statistical study and a complete overview of nanofluid viscosity correlations: A new look, *J. Therm. Anal. Calorim.* 147.13 (2022) 7099–7132.
6. M. Molana, R. Ghasemiasl, T. Armaghani, A different look at the effect of temperature on the nanofluids thermal conductivity: Focus on the experimental-based models, *J. Therm. Anal. Calorim.* 147 (2022) 4553–4577. https://doi.org/10.1007/s10973-021-10836-w
7. M.S. Sadegh, et al., On the natural convection of nanofluids in diverse shapes of enclosures: An exhaustive review, *J. Therm. Anal. Calorim.* 147 (2022) 1–22. https://doi.org/10.1007/s10973-020-10222-y
8. S. Izadi, et al., A comprehensive review on mixed convection of nanofluids in various shapes of enclosures, *Powder Technol.* 343 (2019) 880–907.
9. M. Nemati, S.D. Farahani, T. Armaghani, A LBM entropy calculation caused by hybrid nanofluid mixed convection under the effect of changing the kind of magnetic field and other active/passive methods, *J. Magn. Magn. Mater.* 566 (2023) 170277.
10. A.M. Rashad, et al., Unsteady MHD hybrid nanofluid mixed convection heat transfer in a wavy porous cavity with thermal radiation, *J. Therm. Anal. Calorim.* 149 (2024) 2425–2442. https://doi.org/10.1007/s10973-023-12690-4
11. T. Armaghani, et al., Hybrid nanofluid unsteady MHD natural convection in an inclined wavy porous enclosure with radiation effect, partial heater and heat generation/absorption, *Iran. J. Sci. Technol. Trans. Mech. Eng.* (2024) 1–18. https://doi.org/10.1007/s40997-023-00720-3
12. A. Shafiei, et al., Nanoparticles migration effects on enhancing cooling process of triangular electronic chips using novel E-shaped porous cavity, *Comput. Part. Mech.* 10.4 (2023) 793–808.
13. S. Hussain, S. Shoeibi, T. Armaghani, Impact of magnetic field and entropy generation of Casson fluid on double diffusive natural convection in staggered cavity, *Int. Commun. Heat Mass Transfer* 127 (2021) 105520.
14. R. Ghasemiasl, et al., The effects of hot blocks geometry and particle migration on heat transfer and entropy generation of a novel I-shaped porous enclosure, *Sustainability* 13.13 (2021) 7190.
15. T. Armaghani, et al., MHD mixed convection of localized heat source/sink in an Al_2O_3-Cu/water hybrid nanofluid in L-shaped cavity, *Alex. Eng. J.* 60.3 (2021) 2947–2962.
16. T. Armaghani, et al., Inclined magneto: Convection, internal heat, and entropy generation of nanofluid in an I-shaped cavity saturated with porous media, *J. Therm. Anal. Calorim.* 142.6 (2020) 2273–2285.
17. A.J. Chamkha, et al., Magnetohydrodynamic mixed convection and entropy analysis of nanofluid in gamma-shaped porous cavity, *J. Thermophys. Heat Transfer* 34.4 (2020) 836–847.
18. M. Molana, et al., Investigation of hydrothermal behavior of Fe_3O_4–H_2O nanofluid natural convection in a novel shape of porous cavity subjected to magnetic field dependent (MFD) viscosity, *J. Energy Storage* 30 (2020) 101395.
19. M.S. Sadeghi, et al., Analysis of hydrothermal characteristics of magnetic Al_2O_3–H_2O nanofluid within a novel wavy enclosure during natural convection process considering internal heat generation, Math. Methods Appl. Sci. (2020) 1–13. https://doi.org/10.1002/mma.6520

20. A. Baqaie Saryazdi, et al., Numerical study of forced convection flow and heat transfer of a nanofluid flowing inside a straight circular pipe filled with a saturated porous medium, *Eur. Phys. J. Plus* 131 (2016) 1–11.
21. S. Hussain, T. Armaghani, M. Jamal, Magnetoconvection and entropy analysis in T-shaped porous enclosure using finite element method, *J. Thermophys. Heat Transfer* 34.1 (2020) 203–214.
22. M.A. Mansour, et al., Entropy generation and nanofluid mixed convection in a C-shaped cavity with heat corner and inclined magnetic field, *Eur. Phys. J.Special Topics* 228.12 (2019) 2619–2645.
23. A.S. Dogonchi, et al., Natural convection analysis in a cavity with an inclined elliptical heater subject to shape factor of nanoparticles and magnetic field, *Arab. J. Sci. Eng.* 44 (2019) 7919–7931.
24. T. Armaghani, et al., Effects of discrete heat source location on heat transfer and entropy generation of nanofluid in an open inclined L-shaped cavity, *Int. J. Numer. Methods Heat Fluid Flow* 29.4 (2019) 1363–1377.
25. A. Abedini, S. Emadoddin, T. Armaghani, Numerical analysis of mixed convection of different nanofluids in concentric annulus, *Int. J. Numer. Methods Heat Fluid Flow* 29.4 (2019) 1506–1525.
26. A.M. Rashad, et al., MHD mixed convection and entropy generation of nanofluid in a lid-driven U-shaped cavity with internal heat and partial slip, *Phys. Fluids* 31.4 (2019) 042006. https://doi.org/10.1063/1.5079789
27. T. Armaghani, et al., Mixed convection and entropy generation of an ag-water nanofluid in an inclined L-shaped channel, *Energies* 12.6 (2019) 1150.
28. A. Abedini, A., T. Armaghani, A.J. Chamkha, MHD free convection heat transfer of a water-Fe_3O_4 nanofluid in a baffled C-shaped enclosure, *J. Therm. Anal. Calorim.* 135.1 (2019) 685–695.
29. T. Armaghani, A. Kasaeipoor, U. Mohammadpoor, Entropy generation analysis of mixed convection with considering magnetohydrodynamic effects in an open C-shaped cavity, *Therm. Sci.* 23.6 Part A (2019) 3455–3465.
30. T. Armaghani, et al., MHD natural convection and entropy analysis of a nanofluid inside T-shaped baffled enclosure, *Int. J. Numer. Methods Heat Fluid Flow* 28.12 (2018) 2916–2941.
31. T. Armaghani, et al., MHD mixed convection flow and heat transfer in an open C-shaped enclosure using water-copper oxide nanofluid, *Heat Mass Transfer* 54 (2018) 1791–1801.
32. A.J. Chamkha, et al., Effects of partial slip on entropy generation and MHD combined convection in a lid-driven porous enclosure saturated with a Cu-water nanofluid, *J. Therm. Anal. Calorim.* 132 (2018) 1291–1306.
33. A.M. Rashad, et al., Entropy generation and MHD natural convection of a nanofluid in an inclined square porous cavity: Effects of a heat sink and source size and location. *Chin. J. Phys.* 56.1 (2018) 193–211.
34. M.A. Ismael, T. Armaghani, A.J. Chamkha, Mixed convection and entropy generation in a lid-driven cavity filled with a hybrid nanofluid and heated by a triangular solid, *Heat Transfer Res.*49.17 (2018) 1645–1665.
35. A.J. Chamkha, et al., Effects of heat sink and source and entropy generation on MHD mixed convection of a Cu-water nanofluid in a lid-driven square porous enclosure with partial slip, *Phys. Fluids* 29.5 (2017) 052001. https://doi.org/10.1063/1.4981911
36. T. Armaghani, et al., Forced convection heat transfer of nanofluids in a channel filled with porous media under local thermal non-equilibrium condition with three new models for absorbed heat flux, *J. Nanofluids* 6.2 (2017) 362–367.
37. T. Armaghani, M.A. Ismael, A.J. Chamkha, Analysis of entropy generation and natural convection in an inclined partially porous layered cavity filled with a nanofluid, *Can. J. Phys.* 95.3 (2017) 238–252.

38. T. Armaghani, et al., Numerical investigation of water-alumina nanofluid natural convection heat transfer and entropy generation in a baffled L-shaped cavity, *J. Mol. Liq.* 223 (2016) 243–251.
39. A. Chamkha, et al., Entropy generation and natural convection of CuO-water nanofluid in C-shaped cavity under magnetic field, *Entropy* 18.2 (2016) 50.
40. M.A. Ismael, T. Armaghani, A.J. Chamkha, Conjugate heat transfer and entropy generation in a cavity filled with a nanofluid-saturated porous media and heated by a triangular solid, *J. Taiwan Inst. Chem. Eng.* 59 (2016) 138–151.
41. T. Armaghani, et al., Numerical investigation of flow and thermal pattern in unbounded flow using nanofluid-Case study: Laminar 2-D plane jet, *Therm. Sci.* 20.5 (2016) 1575–1584.
42. T. Armaghani, et al., Effects of particle migration on nanofluid forced convection heat transfer in a local thermal non-equilibrium porous channel, *J. Nanofluids* 3.1 (2014) 51–59.
43. T. Armaghani, et al., Numerical analysis of a nanofluid forced convection in a porous channel: A new heat flux model in LTNE condition, *J. Porous Media* 17.7 (2014) 637–646.
44. M. Nazari, et al., New models for heat flux splitting at the boundary of a porous medium: Three energy equations for nanofluid flow under local thermal nonequilibrium conditions, *Can. J. Phys.* 92.11 (2014) 1312–1319.
45. M.J. Maghrebi, M. Nazari, T. Armaghani, Forced convection heat transfer of nanofluids in a porous channel, *Transp. Porous Media* 93 (2012) 401–413.
46. T. Armaghani, M.J. Maghrebi, F. Talebi, Effects of nanoparticle volume fraction in hydrodynamic and thermal characteristics of forced plane jet, *Therm. Sci.* 16.2 (2012) 455–468.
47. M. Kalteh, et al., Eulerian–Eulerian two-phase numerical simulation of nanofluid laminar forced convection in a microchannel, *Int. J. Heat Fluid Flow* 32.1 (2011) 107–116.
48. W. Cai, et al., Eulerian–Lagrangian investigation of nanoparticle migration in the heat sink by considering different block shape effects, *Appl. Therm. Eng.* 199 (2021) 117593.
49. H. Esmaeili, et al., Turbulent combined forced and natural convection of nanofluid in a 3D rectangular channel using two-phase model approach, *J. Therm. Anal. Calorim.* 135 (2019) 3247–3257.
50. J. Buongiorno, Convective transport in nanofluids, 128.3 (2006) 240–250. https://doi.org/10.1115/1.2150834
51. T. Armaghani, et al., Studying alumina-water nanofluid two-phase heat transfer in a novel E-shaped porous cavity via introducing new thermal conductivity correlation, *Symmetry* 15.11 (2023) 2057.
52. A. Asadi, et al., A new thermal conductivity model and two-phase mixed convection of cuo-water nanofluids in a novel I-shaped porous cavity heated by oriented triangular hot block, *Nanomaterials* 10.11 (2020) 2219.
53. A. Asadi, et al., Two-phase study of nanofluids mixed convection and entropy generation in an I-shaped porous cavity with triangular hot block and different aspect ratios, *Math. Methods Appl. Sci.* (2020) 1–18. https://doi.org/10.1002/mma.7006
54. A.J. Chamkha, et al., Thermal and entropy analysis in L-shaped non-Darcian porous cavity saturated with nanofluids using Buongiorno model: Comparative study, *Math. Methods Appl. Sciences* (2020).
55. A.I. Alsabery, et al., Two-phase nanofluid model and magnetic field effects on mixed convection in a lid-driven cavity containing heated triangular wall, *Alex. Eng. J.* 59.1 (2020) 129–148.
56. A.I. Alsabery, et al., Conjugate heat transfer of Al_2O_3-water nanofluid in a square cavity heated by a triangular thick wall using Buongiorno's two-phase model, *J. Therm. Anal. Calorim.* 135 (2019) 161–176.

57. A.I. Alsabery, et al., Effects of two-phase nanofluid model on convection in a double lid-driven cavity in the presence of a magnetic field, *Int. J. Numer. Methods Heat Fluid Flow* 29.4 (2018) 1272–1299.
58. M. Sharifpur, et al., Optimum concentration of nanofluids for heat transfer enhancement under cavity flow natural convection with TiO_2-Water, *Int. Commun. Heat Mass Transfer* 98 (2018) 297–303.
59. J. C. Maxwell, A Treatise on Electricity and Magnetism: Vol. 1. Oxford: Clarendon Press (1873) 360-366.
60. H. C. Brinkman, The viscosity of concentrated suspensions and solutions, *The Journal of Chemical Physics* 20, no. 4 (1952) 571.
61. J. Koo and C. Kleinstreuer, A new thermal conductivity model for nanofluids, Journal of Nanoparticle research 6 (2004) 577–588.
62. C. J. Ho, et al., Natural convection heat transfer of alumina-water nanofluid in vertical square enclosures: An experimental study, *International Journal of Thermal Sciences* 49.8 (2010) 1345–1353.
63. M. Corcione, Empirical correlating equations for predicting the effective thermal conductivity and dynamic viscosity of nanofluids, Energy conversion and management 52.1 (2011) 789–793.
64. Z. Xue, Model for effective thermal conductivity of nanofluids, Physics letters A 307.5-6 (2003) 313–317.
65. S. Srinivas Rao, A. Srivastava, Interferometric study of natural convection in a differentially-heated cavity with Al_2O_3-water based dilute nanofluids, *Int. J. Heat Mass Transfer* 92 (2016) 1128–1142.
66. M. Torki, N. Etesami, Experimental investigation of natural convection heat transfer of SiO_2/water nanofluid inside inclined enclosure, *J. Therm. Anal. Calorim.* 139 (2020) 1565–1574.

Index

3ω method 8, 18, 19, 20, 36, 44, 50, 79, 80, 81
3ω-T 20, 21

AAO 39
AB-TDTR 23
AC calorimetry 45, 46
AC calorimetry technique 45
aggregating particles 70
Al_2O_3 74, 107, 218, 244, 251, 259, 261, 265, 306, 307
AMM 32
analytical model 14, 15
aspect ratio 37, 45, 65, 84, 92, 94, 96, 97, 100, 215, 216
Au 3, 26, 30, 38, 44, 47, 88, 89, 97

base fluid 2, 3, 65, 70, 82, 91, 105, 106, 123, 129, 130, 140, 141, 145, 148, 156, 164, 168, 170, 173, 183, 190, 207, 209, 230, 231, 235, 258, 276, 277, 278, 284, 297, 299, 300
$BaTiO_3$ 46
BDS 46
Bi_2S_3 20
Bi_2Te_3 39
Boltzmann 10, 17, 67, 71, 72, 74
Bose-Einstein 15
Brownian effect 66, 67, 122
Brownian motion 67, 68, 70, 81, 82, 85, 89, 90, 93, 96, 101, 106, 122, 123, 125, 205, 206, 207, 214, 215, 217, 270, 299
Brownian velocity 70, 89, 90
Bruggeman 66, 67, 74, 99, 122, 126
B_{type} 11, 12
BvKS 14, 15

CAPAD 13
carbon nanotubes 1, 2, 3, 14, 18, 22, 27, 40, 49, 100, 159, 231
CFRP 40
cluster 1, 2, 67, 68, 69, 71, 75, 91, 94, 98, 123
CNT 1, 18, 19, 21, 24, 26, 30, 40, 41, 42, 44, 49, 78, 79, 81, 84, 85, 88, 89, 90, 94, 97, 108, 125, 126, 133, 213, 216, 217, 222, 286, 288
CNT fibers 19, 26, 27, 33, 41, 44
coefficient 3, 13, 14, 20, 25, 70, 74, 77, 101, 130, 204, 206, 210, 217, 253, 272, 276, 290, 291, 298, 299, 301, 304, 305, 306
colloids 2, 5, 65, 92
COMSOL 21
concentration 9, 24, 67, 70, 72, 83, 92, 105, 107, 130, 144, 145, 147, 150, 156, 161, 163, 164, 166, 168, 170, 177, 179, 183, 204, 208, 210, 212, 216, 217, 219, 230, 258, 260, 266, 269, 270, 273, 274, 276, 284, 287, 291, 304
conductivity 1, 2, 3, 8, 9, 10, 13, 14, 16, 34, 37, 45, 47, 65, 86, 87, 88, 90, 100, 122, 183, 296, 297, 301, 302, 303, 304, 308
convection 37, 44, 71, 80, 82, 83, 84, 89, 90, 108, 112, 123, 129, 223, 230, 270, 277, 291, 300, 304
convective 3, 70
c_p 8, 16, 17, 20, 33, 34, 35, 37, 41, 42, 43, 44, 45, 46, 47, 48, 49, 50, 72, 77, 105, 285, 286, 287, 288, 289, 297, 300
CuO 46, 67, 70, 71, 72, 81, 83, 85, 87, 89, 90, 93, 94, 97, 107, 108, 124, 132, 134, 147, 148, 149, 163, 170, 180, 207, 210, 213, 216, 217, 218, 221, 235, 243, 245, 285, 286, 287, 288, 289
CVD 28
c_ω 10, 11, 15

Debye 13, 16, 17, 23, 32, 43, 44, 45, 47, 48, 49, 50
density 3, 15, 16, 20, 28, 37, 39, 40, 48, 71, 72, 81, 105, 106, 130, 205, 207, 208, 216, 289, 290, 291, 297
dimension 1, 2, 8, 27, 28, 31, 37, 38, 45, 70, 75, 76, 93, 101, 103, 130, 137, 140, 142, 145, 148, 151, 153, 156, 159, 161, 165, 168, 184, 185, 207, 208, 215, 231, 236, 255, 267, 268, 284
DMM 31, 32, 33
DSC 44, 46, 286, 287
Dulong-Petit law 42, 43, 50
dynamic viscosity 2, 3, 4, 67, 72, 77, 204, 217, 271, 274, 275, 277, 284, 297, 301, 302, 303

effective diameter 70
EG 71, 72, 80, 94, 97, 104, 105, 107, 124, 133, 134, 164, 207, 209, 210, 211, 213, 218, 221, 235, 236, 239, 240, 243, 244, 246, 247, 248, 250, 265, 266, 286, 290, 291
Einstein 42, 43, 48, 50, 204, 205, 206, 210, 214, 215, 230, 268, 270
electrical conductivity 3, 17, 19, 20, 22, 25, 65, 102
electrical resistance 38, 83
EMD 26, 31, 32, 105, 106, 221

empirical data 67
empirical functions 67
entropy 1, 47, 49
e–p 33
experiment 17, 24, 35, 39, 44, 46, 79, 84, 89, 96, 132, 186, 206, 208, 211, 220, 235, 267, 286, 287, 290, 300, 305

F–K model 77, 79

GNR 29
grain boundaries 8, 25, 47, 48
grain size 4, 9, 13, 19, 24, 30, 47, 48, 50
graphene 2, 8, 14, 16, 20, 21, 22, 24, 26, 27, 28, 29, 30, 31, 33, 34, 36, 40, 41, 42, 49, 72, 73, 88, 89, 96, 97, 107, 108, 124, 132, 142, 210, 217, 218, 222, 234, 288

Hamilton 65, 66, 68, 69, 74, 75, 81, 90, 91, 99, 102, 122, 124, 127, 130
harmonic oscillator 41, 42, 43
heat conduction 12, 13, 14, 17, 18, 27, 28, 33, 36, 39, 40, 75, 98, 99, 124, 127, 130, 298
heat flux 15, 17, 18, 39, 98, 106, 131
heat transfer 3, 23, 33, 68, 74, 79, 85, 98, 106, 123, 129, 132, 217, 220, 256, 270, 276, 290, 298, 300, 305
H-type 21
hybrid 26, 95, 156, 161, 163, 164, 183. 211, 217, 220, 221, 236, 237, 240, 255, 256, 258, 261, 265

infrared thermography 38, 39, 41
interfacial thermal resistance 71, 72, 78, 91, 92, 97, 99, 123
isobaric specific heat 10, 17, 41, 47

k_{acc} 10, 12, 16
k_{bf} 65, 67, 70, 72, 75, 81, 92, 95, 105, 301, 302
k_{bulk} 10, 11, 12, 15
k_{mem} 15, 16
k_{mm} 77, 125
k_{nano} 11, 12
Knudsen 11, 13
Koo and Kleinstreuer 67, 72, 101, 122, 127, 301
k_p 16, 65, 67, 72, 75, 76, 77, 95, 130, 131, 137, 145, 184
k_{RTA} 14
k_{static} 77, 125, 301, 302

laser flash Raman spectroscopy 35, 36
lattice 13, 23, 25, 32, 40, 45, 85, 97
LDOS 28
LVM 9

Maxwell 65, 66, 67, 69, 70, 74, 77, 95, 99, 122, 128, 131, 301, 302, 305, 306, 307, 308
MD 31, 104, 221
MFP 10, 15, 18, 28, 31, 40
microelectrothermal chips 45
molecular dynamics 8, 9, 26, 31
MoS_2 22, 36

nanocomposite 21, 36, 31, 39, 40, 41
nanocrystalline 13, 19, 25, 26, 30, 46, 47, 48, 49
nanofilm 8, 11, 15, 21, 24, 25, 44, 45
nanofluid 3, 65, 95, 129, 217, 285, 296, 300, 304, 305, 307
nanolayer 66, 68, 69, 75, 79, 92, 106, 122, 125, 205, 207, 223, 250, 289
nanolayer effect 66, 69, 122
nanomaterials 2, 3, 5, 9, 30, 33, 34, 41, 42, 43, 108
nanoparticle 3, 65, 73, 84, 90, 91, 92, 103, 213, 214
nanostructure 3, 4, 10, 20
nanotechnology 1, 5
nanowire 2, 17, 18, 20, 31, 37, 39, 45, 46, 47, 92
NC 47
NML 26
Nusselt 70, 75

OHETS 38

PAN 38
particle size 47, 67, 68, 70, 80, 81, 90, 94, 105, 123, 208, 214, 222, 290, 299
PEEK 39
Phonon 8, 9, 10, 13, 14, 15, 16, 17, 22, 23, 25, 26, 27, 28, 29, 30, 31, 33, 34, 37, 40, 41, 44, 46, 48, 49, 50, 96
photothermal resistance 38, 41
porosity 3, 9, 13, 39
porous 3, 4, 20, 31, 39, 40
Prandtl number 70, 75, 101, 102
PS 26

Raman 21, 33, 35, 36, 41, 47
r_{cl} 67
Reynolds number 67, 101, 102, 122
r_p 67, 70, 71, 75, 77, 207, 289, 290
RTA 14

SAM 26
Seebeck 20, 25
SiO_2 107, 218, 235, 244, 265, 307, 308
size effect 24, 25, 30, 46, 50, 70, 76, 90
solid state 2, 8
surfactant 65, 94, 213

TDTR 22, 23, 26
temperature 29, 40, 49, 71, 89, 212, 304
TET 33, 36, 37, 38, 41
thermal conductivity 9, 34, 65, 66, 74, 76, 80, 86, 98, 122, 129, 132, 186, 301

thermal diffusivity 16, 33, 36, 38, 42
thermophysical properties 9, 25, 65, 81, 130, 185, 297, 300, 304
thermodynamic property 46
thermoelectric 20, 25
thermophysical characteristics 36, 65, 289, 290
TiO_2 107, 218, 244, 265, 304, 305, 306
TTM 33
T-type 8, 17, 18, 20

UNCD 48

volume fraction 40, 65, 66, 70, 71, 75, 76, 77, 80, 90, 99, 122, 204, 207, 214, 217, 221, 231, 237, 248, 256, 266, 268, 271, 274, 285, 287, 288, 291, 304

ZnO 72, 84, 85, 87, 89, 91, 97, 107, 124, 211, 214, 220, 231, 237, 240, 287, 288, 290

ε 16, 201, 202
ζ 39, 40, 41, 76
Λbulk 10, 12
μ 68, 77, 122, 124, 125, 204, 205, 219, 235, 255, 265, 266, 269, 301, 302
Φp 65, 70, 76, 77, 95, 207, 220, 285, 286, 289

Printed in the United States
by Baker & Taylor Publisher Services